T0291061

ESG Innovation for Sustainable Manufacturing Technology

Other related titles:

You may also like

●   PBTE108 | Imran | The Role of 6G and Beyond on the Road to Net-Zero Carbon | 2023

We also publish a wide range of books on the following topics:
Computing and Networks
Control, Robotics and Sensors
Electrical Regulations
Electromagnetics and Radar
Energy Engineering
Healthcare Technologies
History and Management of Technology
IET Codes and Guidance
Materials, Circuits and Devices
Model Forms
Nanomaterials and Nanotechnologies
Optics, Photonics and Lasers
Production, Design and Manufacturing
Security
Telecommunications
Transportation

All books are available in print via https://shop.theiet.org or as eBooks via our Digital Library
https://digital-library.theiet.org.

IET MANUFACTURING SERIES 27

ESG Innovation for Sustainable Manufacturing Technology

Applications, designs and standards

Edited by
Wai Yie Leong

The Institution of Engineering and Technology

About the IET

This book is published by the Institution of Engineering and Technology (The IET).

We inspire, inform, and influence the global engineering community to engineer a better world. As a diverse home across engineering and technology, we share knowledge that helps make better sense of the world, accelerate innovation, and solve the global challenges that matter.

The IET is a not-for-profit organisation. The surplus we make from our books is used to support activities and products for the engineering community and promote the positive role of science, engineering, and technology in the world. This includes education resources and outreach, scholarships and awards, events and courses, publications, professional development and mentoring, and advocacy to governments.

To discover more about the IET please visit https://www.theiet.org/.

About IET books

The IET publishes books across many engineering and technology disciplines. Our authors and editors offer fresh perspectives from universities and industry. Within our subject areas, we have several book series steered by editorial boards made up of leading subject experts.

We peer review each book at the proposal stage to ensure the quality and relevance of our publications.

Get involved

If you are interested in becoming an author, editor, series advisor, or peer reviewer please visit https://www.theiet.org/publishing/publishing-with-iet-books/ or contact author_support@theiet.org.

Discovering our electronic content

All of our books are available online via the IET's Digital Library. Our Digital Library is the home of technical documents, eBooks, conference publications, real-life case studies, and journal articles. To find out more, please visit https://digital-library.theiet.org.

In collaboration with the United Nations and the International Publishers Association, the IET is a Signatory member of the SDG Publishers Compact. The Compact aims to accelerate progress to achieve the Sustainable Development Goals (SDGs) by 2030. Signatories aspire to develop sustainable practices and act as champions of the SDGs during the Decade of Action (2020–30), publishing books and journals that will help inform, develop, and inspire action in that direction.

In line with our sustainable goals, our UK printing partner has FSC accreditation, which is reducing our environmental impact on the planet. We use a print-on-demand model to further reduce our carbon footprint.

British Library Cataloguing in Publication Data

A catalogue record for this product is available from the British Library

ISBN 978-1-83724-011-1 (hardback)
ISBN 978-1-83724-012-8 (PDF)

Typeset in India by MPS Limited

Cover image: Eco-friendly building in the modern city; Fahroni/iStock via Getty Images

Contents

Preface

The complex interplay between environmental, social, and governance (ESG) factors and innovation in sustainable manufacturing technology presents a multifaceted landscape that demands meticulous consideration and strategic navigation. At its core, the pursuit of ESG innovation in sustainable manufacturing technology embodies a holistic approach to addressing pressing global challenges whilst fostering economic growth and societal well-being.

Initiating this transformative journey necessitates a deep understanding of the intricate dynamics inherent in sustainable manufacturing ecosystems. It entails leveraging cutting-edge technologies, such as advanced materials science, artificial intelligence, and digitalization, to optimize resource utilization, minimize environmental footprint, and enhance operational efficiency throughout the manufacturing value chain.

Key to this endeavor is the integration of ESG principles into the fabric of manufacturing processes, products, and systems. This entails embedding environmental considerations, such as energy efficiency, emissions reduction, and waste management, into design and production practices, whilst simultaneously addressing social dimensions, including labor rights, workplace safety, and community engagement. Moreover, robust governance mechanisms must be established to ensure compliance with regulatory requirements, promote ethical conduct, and uphold corporate accountability across the manufacturing ecosystem.

Navigating the complexities of ESG innovation in sustainable manufacturing technology demands a nuanced understanding of the systemic challenges and opportunities at play. It requires fostering collaboration and partnerships among diverse stakeholders, including industry players, government entities, research institutions, and civil society organizations, to co-create innovative solutions, share best practices, and drive collective action towards common goals.

Moreover, fostering a culture of continuous learning, experimentation, and adaptation is essential to unlock the full potential of ESG innovation in sustainable manufacturing technology. Embracing a mindset of agility and resilience enables organizations to navigate evolving market dynamics, technological disruptions, and stakeholder expectations whilst seizing emerging opportunities for value creation and competitive advantage.

In essence, ESG innovation in sustainable manufacturing technology represents a paradigm shift towards a more inclusive, equitable, and regenerative approach to industrial production. By embracing this transformative agenda, organizations can not only mitigate risks and enhance operational efficiency but also contribute to positive environmental and social impact, fostering a more sustainable future for generations to come.

Wai Yie Leong

About the editor

Wai Yie Leong is a senior professor at INTI International University, Malaysia. Wai Yie is currently the past chairperson of the IET Malaysia Local Network, council member of IET UK, and vice president of the Institution of Engineers Malaysia (IEM). She received her PhD in Electrical Engineering from The University of Queensland, Australia. She specialises in medical signal processing, industrial revolution 4.0 technology, 5G, and telecommunications.

Chapter 1

Challenges and impact of ESG and green sustainable technology

Nantha Genasan[1]

The United Nations introduced the Sustainable Development Goals (SDGs) in 2015, aiming to achieve global sustainability across various dimensions by the year 2030. These goals cover a wide range of issues, including poverty, inequality, climate change, environmental degradation, peace, and justice. Achieving these goals requires collaborative efforts from governments, businesses, individuals, and organizations worldwide. There is a growing recognition among various stakeholders that businesses can play a crucial role in addressing societal challenges outlined in the SDGs. In this context, environmental, social, and governance (ESG) factors have gained prominence in investment strategies and business management. ESG considerations evaluate a company's performance concerning its impact on the environment, society, and governance practices.

Integrating ESG factors into investment decisions and business operations is seen as beneficial not only for addressing social and environmental issues but also for potentially enhancing productivity and fostering long-term profitability. By evaluating and improving a company's sustainability practices, the ESG approach helps assess its contribution towards achieving the SDGs. Regarding financial models, while the traditional Fama-French three-factor model (focusing on market risk, size, and value) has been foundational in explaining average returns in finance, the integration of ESG factors builds upon this model. ESG factors provide additional dimensions that go beyond financial metrics, allowing investors and businesses to consider non-financial elements like environmental impact, social responsibility, and governance practices in their decision-making processes.

By incorporating ESG considerations into financial models, investors and organizations can better assess risk, identify opportunities for sustainable growth, and align their strategies with broader societal goals such as the SDGs. This integration signifies a shift towards more holistic evaluations of a company's performance, emphasizing its impact on society and the environment alongside financial metrics.

[1]Perdana University Graduate School of Medicine (PuGSOM), Perdana University, Kuala Lumpur, Malaysia

1.1   Overview of ESG and sustainability

The involvement of investors from all over the world in environmental, social, and governance (ESG) investments is growing. Traditional investing paradigms are rapidly being replaced by investments that place an emphasis on business responsibility, social responsibility, and environmental responsibility. At the moment, the field of digital transformation has a restricted capacity for analysing the aspects that contribute to sustainability. However, in addition to this, there are a number of problems with the modern method of collecting and analysing environmental data, which is insufficient and inconsistent. It is also possible for the variety and complexity of environmental controls to have a considerable influence on the quality, dependability, and impartiality of the data. An organization's ESG indicators can be evaluated using Industry 4.0, which has the potential to be a significant possibility for integrating SDGs with the digital transformation of modern technologies. Real-time energy management, waste management, automatic carbon footprint calculation, boosting human well-being, supply chain resiliency, real-time operational performance of manufacturing, and other capabilities are made possible by the technologies that are part of the Industrial 4.0 initiative.

The SDGs are becoming increasingly integrated with sustainability in the context of business and politics in today's world. These objectives were outlined by the United Nations in the chapter titled 'Transforming our World: The 2030 Agenda for Sustainable Development'. The 17 objectives that are depicted in Figure 1.1 make up the framework. When it comes to the SDGs, the concept of sustainability and a holistic approach to the development of society and individuals on a global scale are intricately connected. There is a significant distinction between the two frameworks, despite the fact that the SDGs are clearly connected to a variety of fundamental human rights. The SDGs involve a more comprehensive focus on what is known as the five P's, which are people, planet, prosperity, peace, and partnership [12].

Figure 1.1   The sustainable development goals [12]

A representation of the ESG framework, which includes stakeholders and standards, can be found in Figure 1.2. The ESG strategies and risk management will implement change management, which will allow for value creation monitoring. This will permit the organization to identify and evaluate ESG risks, and it will also assist in future proofing and scenario planning. On the other hand, the world of business and finance is increasingly taking an active role in the process of achieving the SDGs, despite the fact that development and environmental pre-servation can be considered as work for governments. This is accomplished in part by the use of force in the form of laws and regulations, but also in part through the use of more informal processes that are associated with the requirement to get a social licence to operate. These days, investors, business partners, and customers have a tendency to demand more from corporations in terms of their responsibilities than what is mandated by the law alone. This results in a situation in which sustainability is a combination of compliance (a 'hygiene factor') and proactive strategic business development. This is due to the fact that sustainable business models are increasingly being demonstrated to give enterprises with an advantage in the market.

For the first time in history, concerns of privilege, social justice, and the effects that enterprises have on society and the environment are more pertinent than they have ever been. The phrase corporate social responsibility (CSR) currently refers to a significantly wider range of concerns than it did in the early stages of the con-cept's development. Currently, environmental challenges that encompass a wide variety of issues, including but not limited to climate change, loss of biodiversity, pollution, and other related issues, are universally acknowledged as significant

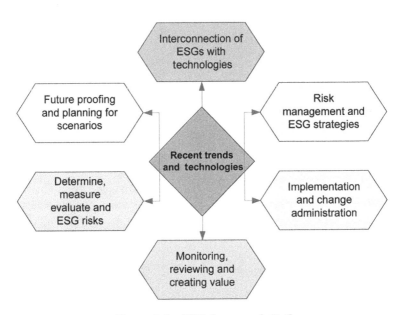

Figure 1.2 ESG framework [14]

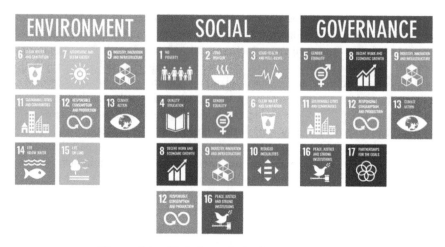

Figure 1.3   SDG through the lens of ESG [12]

challenges for humanity. Not only does this chapter highlight the impact of artificial intelligence, but it also draws attention to the unparalleled power and influence of the major technological companies, which prompts both examination and attention. GAFAM is an abbreviation that describes the big actors in the United States of America, which are Google, Amazon, Facebook, Apple, and Microsoft (MS). Big Tech is a term that describes the major technology corporations. Figure 1.3 shows the goals of sustainable development through the framework of ESG [12].

1.1.1   Terminology of ESG

Environment: The advancement of technology has resulted in the deterioration of a number of interrelated environmental issues, including soil pollution, water pollution, climate change, air pollution, loss of biodiversity, and over-exploitation of natural resources. A significant number of market participants, including institutional and private investors, have indicated a wish to take environmental sustainability into consideration when making investment decisions ever since the phrase 'sustainable development' was first used. *Social:* During the past few years, there has been a rise in global awareness of ESG risks, which include those that are associated with the environment, society, and governance. Additionally, nonfinancial investments are now taken into consideration when making decisions on investments. The focus is on the environment, and investments in bio and health are becoming more popular as a result of recent technology breakthroughs, while new investments and decisions are being made in the direction of the sequential termination or withdrawal of existing enterprises. Utilizing data from social networks and smart cities, it is possible to acquire real-time information regarding human behaviour in the social realm. These contribute to the enhancement of the ecosystem's intelligence and the creation of an intelligent space that can detect human actions or activities. As a result of the relevance of this, there has been an increase in the number of managers who are making the fight against climate change and sustainability a priority in their investment portfolios. Figure 1.4 shows the benchmarks involved in the ESG framework.

Corporate governance is defined as 'the rules and processes by which organizations are controlled or run'. *Governance* stands for 'corporate governance'. A wide range of philosophies can be traced back to the origins of corporate governance. There are several fundamental theories, and one of them is the agency hypothesis. Jensen and Meckling [7] provide the following description of an agency relationship: 'a contract whereby one or more persons engage another person to perform some service on their behalf and whereby the principal delegate some decision-making authority to the agent'. This definition highlights the nature of the connection between the primary and the agent.

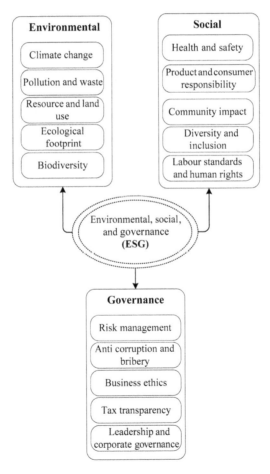

Figure 1.4   Benchmarks involved in ESG [14]

1.1.2   ESG investing

The activity of giving priority to the most beneficial ESG features or outcomes is known as ESG investment. Socially responsible investing, impact investing, and sustainable investing are some of the other names for this type of investment. ESG investing is a method of 'sustainably' investing, which means that investments are made while taking into consideration the economy, the environment, and the welfare of humans. This investing strategy is frequently referred to as 'sustainable' investing. ESG, which stands for 'corporate social responsibility', is a word that is frequently used to characterize the actions of corporations in this particular domain. Taking into account ESG aspects while making investment decisions is a commitment that the financial markets have made. According to a report published by the Governance and Accountability Institute in 2018, 86% of the companies that make up the S&P 500 submitted reports on their corporate responsibility or sustainability, which is an increase from just under 20% in 2011 [14].

The fact that this is the case suggests that corporations are interested in ESG and CSR. It is predicated on the rising concept that environmental and social issues are the ones that have a significant impact on the financial productivity of an organization. As a means of expediting the transition to ESG, institutional frameworks that encourage spending on government projects that improve social equality and natural capital are being developed. In light of the rise of stakeholder capitalism and responsible investment, a great number of companies are looking for ways to improve their ESG standards. To better understand how underlying social, economic, political, environmental, and technological conditions are changing and their potential effects on their company's operations and prospects, business leaders in this new era need to focus more intently on factors other than their company's short-term operations and financial results. This is necessary to gain a better understanding of how these conditions are changing. The term 'firm' refers to more than just an economic entity that generates profit. Because it is a part of the wider social structure, it satisfies the goals that humans and society have set for themselves. In addition to the return that an organization provides to its shareholders, performance evaluations should also take into account how well an organization achieves its environmental, social, and good governance objectives. Compensation for executives ought to take into account the duty they have towards stakeholders. There are five investment strategies [4] that are being crucially implemented in the ESG standard developments today as such:

- *Sustainable alpha investors*: Based on ESG data and sustainability leadership, these investors are looking to outperform the market. The objective of these investors, who are sometimes referred to as value-first investors or investors focused on the value driver model of portfolio construction is to identify businesses that are in a position to capitalize on what they consider to be large-scale market opportunities brought about by the emergence of sustainability issues and pressures. In particular, they invest in businesses that will be able to accomplish the reduction of risks, the reduction of costs, or the acceleration of growth by concentrating on the increasing necessity of sustainability. One example of such an opportunity would be investing money in businesses that are in a position to make progress towards the new energy economy and, as a result, contribute to the fight against climate change.

- *Smart beta investors*: The goal of these investors is to capitalize on market trends while maintaining a focus on social and environmental responsibility. ESG data are something they want to have so that they can reduce the amount of volatility in their portfolios and get more consistent results. They are looking for corporate indicators and information that will help them identify whether businesses are doing a consistent and complete evaluation of ESG risks, incorporating sustainability issues into management practices, and progressing risk mitigation measures with care. This group includes a significant number of institutional investors.

- *Activist investors*: By utilizing their position as shareholders, these investors want to influence corporate practices and, as a result, move society in the direction of a more sustainable future. Although they engage on a wide variety of issues, the overarching goal of their activism is to advance particular principles and outcomes in policy. When these investors are looking to affect the behaviour of corporations, they frequently desire greater corporate transparency and, as a result, more disclosure of ESG data, particularly in relation to what they view to be key concerns. Activist investors, for instance, frequently aim to get more and better data on the consequences and policies of firms regarding climate change. Some of these investors may threaten to divest from companies that have poor performance or weak disclosure practices.

- *Impact investors*: Influence investors place a higher priority on the potential for their investments in certain businesses, assets, or projects to have a positive influence on society or the environment, in addition to any financial rewards that may be generated. In point of fact, impact investors are characterized by their willingness to forego a certain degree of commercial success to pursue sustainability rewards. This group of investors is looking for impact measurements that can evaluate the extent to which the organization in which they have invested has been successful in addressing environmental and social issues.

- *Screening investors*: To evaluate the companies in their portfolios and include or reject shares or other assets based on a variety of qualities, these value-oriented investors are looking for ESG data and information. They frequently target entire sectors of the economy. Companies that are involved in the production or sale of tobacco, weapons, alcohol, ammunition, or fossil fuels are examples of well-known kinds of negative profiles. Renewable energy enterprises and accredited B-Corporations are examples of companies that could be considered positive screens. When it comes to ESG reports, screening investors have a relatively low level of interest. Instead, they concentrate on industry classifications and the primary financial streams of specific assets.

Every one of these categories of investors has a unique set of interests, concerns, questions, and expectations around sustainability, as well as data requirements. By way of illustration, investors who are looking to outperform the market by taking into account ESG issues focus their attention on the quality of ESG data as well as the methodological rigour of empirical studies that aim to identify a correlation between specific ESG scores and specific financial performance. Many of these investors believe that the complexity of the data and research that are now available provide potential for arbitrage. In light of this, they are content to sort through the jumbled material and to carry out their own research to come up with original findings. Their perspective is that Big Data, artificial intelligence, and machine learning are tools that have the potential to offer novel approaches to the collection or analysis of ESG issues, hence shedding light on potential sustainability-driven long- or short-term strategies. Comparability is handy for these investors; but, the quality of the information that lies behind the surface is of the utmost importance since it determines whether or not ESG measures offer a trustworthy foundation for their evaluations.

Activist investors and smart beta investors, to a certain extent, are looking for more decision-useful data on which to evaluate the possible exposure of an asset to ESG-based issues, which includes both potential downside risks and potential upside opportunities [10]. To better identify and address any exposure to environmental or social challenges and trends, these investors anticipate that corporations will broaden their management practices to better manage their operations. These investors, as a group, advocate for increased transparency in the processes by which companies evaluate risk, more robust risk assessment criteria that take into account the potential internalization of ESG externalities, and increased disclosure of the potential financial implications of these risks, particularly in regulated financial reports. When these investors are trying to construct investment strategies that overweight or underweight particular companies or industries based on ESG considerations over a broad spectrum of invest able assets, comparability is the most important principle for them to consider.

1.1.3   ESG for sustainability

When applied to sustainability, ESG techniques ought to be beneficial in all three aspects. We may make the assumption that a synergistic effect is manifested and effectively enriches our economic and social systems if every consequence that is offered by the ESG approaches is unquestionably accurate. There is a need to make sure that every effect is comprehended. Before anything else, ESG techniques make an effort to compute welfare losses in terms of numerical and numerical values. With this solution, it is anticipated that communities all around the world will be protected against catastrophic calamities. Second, ESG plans give rise to the possibility of competition and cooperation in both domestic and global societies. To put it another way, the ESG framework has the potential to steer communities located both locally and internationally towards sustainability. In the third place, ESG approaches are the spark that ignites social innovation.

The incorporation of the evaluations of some stakeholders into the articulations has the potential to boost the corporation's openness to both risks and innovations. ESG initiatives have the potential to greatly progress the fourth industrial revolution and sharing economies in the future. The ESG framework should be utilized during the development process to produce a wide range of investments that are geared towards meeting societal requirements. The disclosure of information regarding concerns pertaining to the environment, society, and corporate governance is what is known as sustainability reporting. Bringing attention to a company's ESG initiatives, increasing investor transparency, and encouraging other businesses to follow suit are the goals of this disclosure, as they are with other disclosures. Reporting is a strong tool that may be used to demonstrate that your ESG initiatives are genuine and not simply lip service or green-washing and that you are on the right path to achieve your goals.

One might get the impression that the idea of sustainability is not only easy to understand but also reasonably straightforward. The concept of sustainability, as well as the concept of sustainable development, can typically be traced back to the report titled 'Our Common Future', which was prepared in 1987 by a commission working under the leadership of Gro Harlem Brundtland of the United Nations [10]. The SDGs are becoming increasingly integrated with sustainability in the context of business and politics in today's world. In the chapter titled 'Transforming our World: The 2030 Agenda for Sustainable Development', the United Nations offered these objectives to the world today. The 17 objectives that are presented in Figure 1.3 make up the framework. When it comes to the SDGs, the concept of sustainability and a holistic approach to the development of society and individuals on a global scale are intricately connected. There is a significant distinction between the two frameworks, despite the fact that the SDGs are clearly connected to a variety of fundamental human rights. The SDGs involve a more comprehensive focus on what is known as the five P's, which are people, planet, prosperity, peace, and partnership.

For the first time in history, concerns of privilege, social justice, and the effects that enterprises have on society and the environment are more pertinent than they have ever been. The phrase CSR currently refers to a significantly wider range of concerns than it did in the early stages of the concept's development. Environmental concerns as they pertain to a variety of topics, including but not limited to climate change, loss of biodiversity, pollution, and other related issues, are now universally acknowledged as a significant challenge for humanity. GAFAM is an abbreviation that describes the big actors in the United States of America, which are Google, Amazon, Facebook, Apple, and MS. Big Tech is a term that describes the major technology corporations [12]. There are a variety of issues that are being discussed in relation to the operations of these firms. Some of the more prominent of these issues are their near-monopoly position and their utilization of surveillance and data to monitor and increasingly exercise control over individuals and society. In the second section, which is devoted to discussing ethical issues that are more closely associated with artificial intelligence systems, it is demonstrated how ESG-related risks need to be mapped to have a complete understanding of the impact that artificial intelligence (AI) will have.

1.2   The scopes of impacts of ESG

There are instances when the intricacies of ESG implications might be a barrier to conducting ESG analysis. When these barriers are not present, the analyses that are produced are frequently not very actionable. In an attempt to find a solution to this problem, a number of analytical techniques has been offered to the impacts that artificial intelligence has on sustainability. This strategy is partially influenced by the Greenhouse Gas Protocol (World Resources Institute, 2021), which differentiates emissions from Scopes 1, 2, and 3. In addition, the AI ESG protocol ties directly into the most popular and widely used methods that are utilized in the climate change section of all other frameworks and standards (World Resources Institute, 2021) [13]. These are some of the primary sources of risks and impacts in each scope, and they are addressed in greater detail below. Figure 1.5 illustrates some of these sources. The process of evaluating impact is somewhat distinct from the process of measuring merely emissions since it takes into account a wider variety of factors, despite the fact that the two processes are comparable and partially overlap. Specifically, the scopes of the AI ESG protocol are specified in the manner that is described below, and examples are provided in the subsequent sections [13].

Scope 1 addresses consequences that are directly tied to the fundamental activities and governance of a company. These impacts are confined to the social and governance impacts that occur within the firm as well as the environmental impacts that are associated with the computing infrastructure that the corporation directly controls (either on its own or through leasing). Scope 1 includes the information that was gathered by the entity. The challenges that are shown in Figure 1.5 include, for instance, those that pertain to cybersecurity, the influence of own machines and data, and the impact of own people. The upstream repercussions concerning the entity's supply chain are included in Scope 2, which incorporates all of these consequences. The procurement of development services, support, and algorithms are all included in Scope 2, as is the buying of cloud services and power.

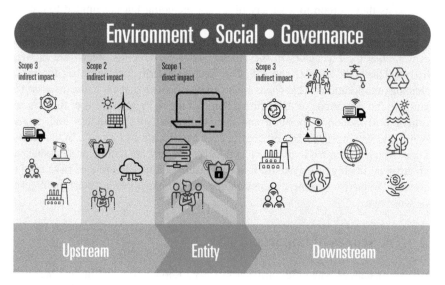

Figure 1.5   The effect and risk factors that are present in Scopes 1, 2, and 3 [13]

Scope 2 also includes the procurement of web hosting services. An essential component of Scope 2 is the collection of all the second-hand data that the corporation makes use of. This data serves as a supplement to the data that the entity itself has obtained, as described in Scope 1. Figure 1.5 illustrates this by utilizing various examples, such as the acquisition of cloud services, the participation of humans in the upstream supply chain, the implementation of upstream cybersecurity measures, and the procurement of energy.

Scope 3 encompasses the various upstream and downstream implications that are caused by the AI and data-based capabilities, assets, and activities of the firm. This covers, for instance, an algorithm that is used for artificial intelligence in the hiring process, and how this may involve the possibility of discrimination or the possibility of a reduction in the occurrence of bias in the hiring process. In addition to this, it encompasses the manner in which the operations of the business either encourage or discourage consumption. For instance, if the entity sells or distributes tools that, for instance, drive emissions upstream or downstream, this is included. Figure 1.5 illustrates this by illustrating, among other things, the datafication of human relationships, the growing use of the Internet of Things (IoT) in the business and private sectors, the increased targeting and surveillance of individuals and groups, the creation of value and innovation, transportation, the impact on water use, nature and biodiversity, and so on.

To summarize, the process of detailing the impacts in these three scopes encompasses all ESG-related impacts that are a result of artificial intelligence and data. This helps the entity and its stakeholders understand where in the value chain the impacts and risks occur, and consequently, what can and must be done to either minimize the negative impacts and risks or maximize and exploit the upside risk and positive impact. Although the three scopes are helpful in classifying impacts originating from a variety of sources both upstream and downstream, there is an additional important difference that needs to be made between the various types of impacts and hazards. In light of the fact that the protocol is founded on the ESG methodology, the primary categories that I will now discuss are those that pertain to ESG concerns.

1.2.1 Environmental impacts and risks

The environmental aspect is currently receiving the most attention in the energy, environment, and governance (ESG) world (ERM, 2022), and climate has been the obvious headline grabber recently. There is still a significant role for the Paris agreement in gaining an understanding of climate targets (UNFCCC, 2022). There has been a recent uptick in the number of businesses that are establishing NetZero targets and plans (ERM, 2022), and it is imperative that these strategies incorporate AI and data-related emissions. However, the integrity of natural systems is gaining more and more attention, not only because it is important in and of itself, but also because it is essential to the process of resolving issues that are related to climate change (ERM, 2022) [3].

The primary concern that pertains to artificial intelligence in this particular domain is the manner in which AI not only consumes energy but also generates emissions. Furthermore, AI has the potential to provide solutions that aid in the mitigation of climate change and the promotion of adaptation efforts. It is necessary to have an understanding of both sides of the issue to determine whether or not the use of artificial intelligence by an organization is sustainable. Artificial intelligence has the ability to concurrently have both good and negative impacts connected to emissions. When it comes to artificial intelligence, the direct emissions that are produced will typically be limited to Scopes 1 and 2, but Scope 3 is where an entity may demonstrate that it has a positive impact [11]. Additionally, there is a material basis for computing infrastructure. Because of this, it is necessary to take into account the environmental implications of the machinery that is utilized, as well as the materials that are used in it. This can be done either by the organization itself or by using data on or from cloud providers in the supply chain. Although emissions from the manufacturing of equipment are important, other factors, such as those pertaining to hazardous waste, rare minerals, and so on, are also important. It is now abundantly obvious that artificial intelligence is not only relevant with regard to the adaptation and mitigation of climate change, but it also has the potential to have impacts connected to, for instance, biodiversity, innovation, and the ability to make sense of data to address environmental difficulties, changes in land use, and variations in water use.

1.2.2   Social impacts and risks

A growing number of investors are concentrating their attention on the social aspects of the activities of an organization. There are a variety of factors that encourage this, and some examples of drivers include COVID-19 and the great resignation, the movement for black lives matter, and new regulations connected to contemporary slavery. All of this indicates that issues with (a) the satisfaction, engagement, and retention of employees, (b) issues concerning the supply chain and human rights, and (c) the larger impact connected to social justice and discrimination are crucial for investors to consider. It is common knowledge that all of these concerns are pertinent to the application of artificial intelligence and data. It is a fundamental component of mainstream digital or AI ethics to consider the broader consequences of artificial intelligence, as well as its problems, since discrimination and bias in such systems are increasingly concerning everyone. These kinds of problems are primarily classified as Scope 3 in the AI ESG procedure. Within the framework of the AI ESG protocol, concerns pertaining to the rights of data subjects are incorporated into the study of supply chain human rights issues in Scope 2, as well as the collection of one's own data in Scope 1.

As part of the social dimension of the ESG protocol for artificial intelligence, it is necessary to address not only the issue of discrimination and bias, but also the economic repercussions that are associated with issues like as poverty, inequality, and access to infrastructure, among other things. Artificial intelligence is a component of a larger sociotechnical system that may not be sustainable in the long run. This system may not be conducive to supporting all parts of the SDGs, such as 8, 9, and 10 [12]. Within the framework of this discussion, it is essential to inquire about who owns the data, who has access to the services, who benefits from the solutions that are developed, and so on. The issues that pertain to the utilization of artificial intelligence and data to foster growth and innovation are equally pertinent in this area; however, to be comprehensive, they need to be linked with an analysis of the societal repercussions.

Furthermore, it is imperative to examine matters pertaining to consumer activity and political institutions within the context of the social sphere. This phenomenon reflects the prevailing market trends that demand companies to assume responsibility and actively promote positive and sustainable conduct among their customers and partners. Additionally, it pertains to the correlation between their products, solutions, and systems with democracy and political institutions. This connection has gained significance, particularly in light of the influence exerted by social media on electoral processes.

1.2.3   Governance

An important advantage of utilizing the AI ESG protocol is its emphasis on governance-related matters, drawing from proven frameworks and tools from the finance and investor domain, which have a longstanding history of addressing such challenges. It is crucial to prioritize the governance of AI and data-intensive entities, considering their involvement in a sector that is still developing and experiencing significant expansion. These entities have challenges in establishing effective governance strategies. There is currently a significant increase in the number of frameworks designed for responsible, trustworthy, and ethical AI. Additionally, there are continuous discussions regarding the connections between ethics, politics, and regulation in relation to firms that utilize AI and data-driven solutions. However, it is important to acknowledge the growing governance approaches to AI that are worth mentioning. These approaches should be taken into account while utilizing the AI ESG protocol.

The protocol is impartial towards any particular approach and just mandates an entity to articulate and reveal its strategy for managing the governance of AI and the associated risks and opportunities. This strategy may draw upon several approaches to AI governance that are currently being formulated. It is crucial that AI governance is regarded as an integral component of an organization's current governance framework. Governance is also connected to stakeholders and emphasizes the necessity of employing multi-stakeholder approaches and collaboration to achieve effective AI governance. This aligns with established methods of promoting sustainability and ESG practices, such as the use of networks, forums, and guidance systems employed by the UN Global Compact [10]. Furthermore, the protocol can be integrated with other methods for auditing and ensuring the reliability of AI systems, in addition to addressing broader governance concerns.

Additionally, there have been suggestions and presentations regarding the use of impact assessments to evaluate and ensure the effectiveness of specific algorithms. Although there is a requirement for effect assessments, it is uncertain whether substituting topical assessments, such as 'environmental impact assessments', with technology-defined assessments, specifically algorithmic impact assessments, eliminates the necessity for auditing and assurance. In the realm of sustainability reporting and disclosure, it is quite probable that internal and external auditing processes, as well as limited or reasonable assurance work, will continue to serve their respective purposes. Within the AI ESG protocol, governance concerns encompass risk management, governance frameworks, auditing mechanisms, and the extent to which a corporation has formulated policies and goals pertaining to AI and data capabilities, assets, and operations.

Scope 1 primarily covers these issues; however, governance is also incorporated in the other scopes by means of indicators pertaining to conducting thorough investigations and evaluations of their suppliers and partners. Since the AI ESG protocol is not a comprehensive ESG reporting framework, it does not address specific governance issues like board composition. These matters will be addressed using a more general framework. The Task Force on Climate-Related Financial Disclosures (TCFD) framework offers guidelines for disclosing governance and risk management practices that might be effectively integrated into the reporting of governance issues in the AI ESG protocol. If the company fails to report on ESG factors using comprehensive frameworks, it may be necessary to incorporate specific indicators into the protocol. However, this will only apply to enterprises that heavily rely on AI and data, as the AI ESG protocol will primarily address the most significant challenges related to these areas.

1.3   Technology intervention for ESG

Conversations about ESG issues frequently revolve around subjects such as pollution management, biodiversity, health and safety, corporate ethics, and diversity in boardrooms in the majority of businesses. Companies that make representations about their ESG policies that are misleading or even wholly untrue are becoming increasingly unpopular among consumers who are concerned about ESG. The internet makes it easier to check facts, and social media platforms make it possible to denounce something in a short amount of time. Increasingly, investors will want concrete and verifiable evidence that ESG considerations are clearly included in the investment process of a manager. Furthermore, they will require evidence to support such assertions through stock selection, reporting, and portfolio structure.

1.3.1   IoT for ESG

An architecture for a smart ESG reporting platform that makes use of blockchain technology and the IoT is proposed in a study [16]. The purpose of this architecture is to enable corporate crowd-sensing for environmental data and to improve the transparency, security, and credit ability of the whole ESG reporting method. There is a growing amount of multidimensional environmental data being generated by weather forecasting models and high-resolution climate as well as global and regional sensor networks. This data is being produced in ever-increasing quantities. Therefore, in order for the information to be useful, it must be stored, maintained, and made available to a worldwide community of academics, politicians, and other individuals.

An architecture that is implemented for ESG that is based on the IoT is seen in Figure 1.6. Networking of things is the foundation of the IoT, which allows for intelligent items to communicate with one another through the exchange of data. First and foremost, the IoT is made up of five layers, which include the perception layer, the network layer, and the application layer. There is the possibility of data acquisition in the perception layer through the use of wireless sensors in conjunction with the computing unit. Oxygen, Catalytic gas sensors, particulate matter 2.5, particulate matter 10 (PM2.5), ozone (O3), carbon monoxide (CO), carbon dioxide (CO_2), sulphur dioxide (SO_2), nitrogen dioxide (NO_2), formaldehyde (HCHO), total volatile organic compounds (TVOCs), temperature, noise, humidity sensors, ultraviolet radiation index, and other factors are utilized to measure the carbon footprint and air quality (indoors and outdoors) of any premises and location. Real-time environmental data may be obtained with the assistance of these sensors, and the data can then be analysed to determine the impact it has on the sustainability parameter. It is possible to reach a higher level of life in terms of health for those who live in an environment of high quality. The IoT devices are integrated with sensors that measure airflow, pressure, oxygen, temperature, humidity, magnetic, thermometer, bar code, force, pulse, heart rate (HR), electromyography (EMG), electrocardiogram (ECG), electroencephalography (EEG), blood pressure (BP), pulse oximetry, photoplethysmography (PPG), glucose, gyroscope, and motion tracker to obtain real-time data for one of the social parameters in ESG [16].

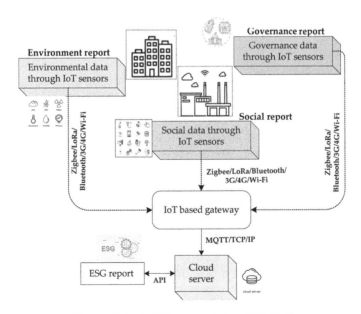

Figure 1.6   IoT framework for ESG [14]

Through the frequent alerts that they provide, these sensors contribute to the improvement of health monitoring and the maintenance of a person's health. Other types of sensors and identification technologies, such as vision cameras, motion sensors, radio frequency identification (RFID), barcodes, a global positioning system (GPS), ultrasonic sensors, proximity sensors, infrared (IR) sensors, pressure sensors, touch sensors, vibration sensors, and piezo sensors, have contributed to an increase in the monitoring of an organization's supply chain and operational functions through the utilization of real-time data (social reporting). The network layer is responsible for processing this sensor data and communicating it to the application layer where it is stored. Through the use of various communication protocols, the network layer is responsible for establishing the networking between the perception layer and the application layer. These protocols include Bluetooth low energy (BLE), Bluetooth 5, Zigbee, Long range (LoRa), wireless fidelity (Wi-Fi), Narrowband Internet of Things (NB-IoT), Sigfox, Advanced Message Queuing Protocol (AMQP), Data Distribution Service (DDS), Message Queue Telemetry Transport (MQTT), and others [8].

Concepts that are centred around the IoT, such as augmented reality, high-resolution video streaming, self-driving cars, smart environments, electronic health-care, and so on, are now present everywhere. High data rates, a huge bandwidth, enhanced capacity, reduced latency, and high throughput are all capabilities that are necessary for these applications. The IoT has revolutionized the world by allowing seamless connectivity between heterogeneous networks (HetNets), which is especially relevant in light of these new notions. One of the ultimate goals of the IoT is to bring about the implementation of plug-and-play technology, which will provide the end-user with ease of operation, remote access control, and the capacity to configure cellular networks of the fifth generation, often known as 5G, offer essential enabling technologies that are necessary for the widespread adoption of IoT technology. Carrier aggregation, multiple-input multiple-output (MIMO), massive-MIMO (M-MIMO), coordinated multipoint processing (CoMP), device-to-device (D2D) communications, centralized radio access network (CRAN), software-defined wireless sensor networking (SD-WSN), network function virtualization (NFV), and cognitive radios (CRs) are some of the technologies that fall under this category [17].

The IoT architecture comprises three components, which are as follows:

1. Hardware: It is made up of sensor nodes, as well as its embedded communication and interface circuitry.
2. Middleware is a collection of resources that are used for data storage, analysis, and management.
3. The presentation layer is comprised of effective visualization tools that are compatible with a variety of platforms for a variety of applications and present the data to the end-user in a form that is easily comprehensible.

The architecture of the IoT is influenced by a wide variety of parameters. As a result, the current research efforts have been directed towards the development of the most optimized architecture that addresses network challenges such as scalability, security, address ability, and efficient energy utilization. As for the future, there will be an increase in the number of devices that are connected to the network.

As a result, the design of the IoT needs to accommodate it. Scalability, energy consumption, and problem solving are all considered to be problems that must be overcome to successfully deploy the IoT. Through the development of a variety of multi-hop routing protocols that span a broader region and are capable of self-adaptation, research is implemented with the goal of resolving the scalability challenges. These can be broken down into three categories: Data-eccentric, location-based, and hierarchical are the three approaches.

In the current decade, the concept of a 'smart environment' has developed into a technology that is seeing rapid growth (Table 1.1). The idea encompasses a wide range of subject areas, including transport and logistics, healthcare, utilities, personal homes and workplaces, and a great deal more. In the field of transportation and logistics, innovations such as augmented maps, autonomous vehicles, mobile ticketing, and passenger counting have been successfully implemented during the course of this decade. There is also an ongoing process of improvement in these technologies that is currently being implemented. The IoT-enabled robot taxi is currently being developed as a potential application for the future. In a similar vein,

Table 1.1   IoT-enabled smart environment [16]

Applications	Communication enablers	Network types	Wireless local area network (WLAN) Standards	Modules
Smart cities	Wi-Fi, 3G, 4G, Satellite	MAN, WRANs	802.11	Architectures, protocols, and enabling technologies for urban IoT. Integrated information centre for the smart city
Smart homes	Wi-Fi	WLAN	802.11	Cloud-based home solution for detection of faulty location using software-defined networks (SDNs)
Smart grid	3G, 4G, Satellite	WLAN, WANs	802.11	Real-time monitoring system for powering transmission lines to avoid disasters. Smart-grid control
Smart buildings	Wi-Fi	WLAN	802.11	Access control for services inside a typical smart building
Smart transport	Wi-Fi, Satellite	WAN, WRANs, MANs	802.11	Smart ticketing, smart passenger counting
Smart health	Wi-Fi, 3G, 4G, Satellite	WLAN, WPANs, WANs	802.15.4	Remote healthcare
Smart industry	Wi-Fi, Satellite	WLAN, WPANs, WANs	802.11	Energy-efficient remote monitoring and optimized decision-making

the IoT has enabled numerous advancements in the field of healthcare, including innovations such as remote patient monitoring, smart biosensors, smart ambulances, wearable gadgets, and telemedicine. The use of smart metering and smart grid technologies has resulted in a significant improvement to the infrastructure involved in the provision of public utilities.

The IoT is a heterogeneous network that will connect approximately 7 billion devices by the year 2025. As part of this network, several wireless technologies and standards, including 2G, 3G, 4G, Bluetooth, and Wi-Fi, will be utilized. These technologies are used by the IoT networks that are currently in use, as listed in Table 1.2. Various initiatives are being taken all over the world for the purpose of adopting and standardizing the IoT that is enabled by 5G, as listed in Table 1.3. In a similar manner, International Mobile Telecommunications (IMTs) additionally initiated research and technology practices in 2013, and the standardization was carried out in 2016. 2015 was the year that the decision was made that the Third Generation Partnership Project (also known as 3GPP) would be responsible for forming a group known as the technical specification group (TSG), which would be in charge of building the 5G radio access network. Over the course of the same time period, the responsibility of designing and specifying 5G technology has been taken on by the International Telecommunication Union-Radio Communication (ITU-R) by the year 2020 [17].

A heterogeneous network, the IoT will connect approximately 7 billion devices by the year 2025 [8]. This will be accomplished by the utilization of several wireless technologies and protocols, including 2G, 3G, 4G, Bluetooth, and Wi-Fi. The process of standardization in the 5G IoT primarily comprises two different kinds of standards. In the first place, there are technological standards that are concerned with network technology, protocols, wireless communication, and data

Table 1.2   Wireless technology currently employed for IoT [16]

Type of technology	Wireless technology
Personal area network (PAN)	Bluetooth low energy, Zigbee
Wireless local area network (WLAN)	Wi-Fi
Cellular communication	3GPP, 4G (LTE)
Wide area network (WAN)	LP-WA

Table 1.3   5G initiatives in different countries [16]

Country	Initiative
United States	4G America
China	IMT-2050 (5G)
Japan	2020 and beyond
Korea	5G forum
Europe	5G Private Public Partnership (5GPPP)
United Kingdom	5G Innovation Centre (5GIC)

aggregation standards. In the second place, there is a regulatory standard that prioritizes the protection of data privacy and security. Several distinct 5G protocols are being developed for the IoT to accommodate the vast networking of devices. At the beginning of the first phase, the 3GPP Release 14 specifies M2M communication, NB-IoT, and a low-power wide area technology that is intended to make the vast deployment of IoT easier. Currently, the 3GPP standardization is working on upgrades to meet the ever-increasing demand for its services. As part of the challenges, these enhancements are outlined in Table 1.4.

In many respects, the traffic that is produced by IoT devices is distinct from the traffic that is produced by cellular systems. To begin, in contrast to the situation with broadband access, the majority of IoT traffic is in the uplink. Moreover, the communications that are transmitted over IoT networks are often of a tiny size and occur infrequently. In addition, IoT devices have limited resources in terms of both energy and computing. Because of the qualities that these IoT devices possess, their connectivity to 5G networks is distinct from that of traditional cellular devices. The identification of the appropriate configuration of system parameters for a particular IoT use case is a significant task. Table 1.5 shows the key research ideas as part of

Table 1.4   Enhancements proposed for 5G on cellular IoT to overcome the challenges [16]

Open standards	Enhancements proposed
Machine-type communications (MTC) or M2M communications	• Reduced complexity • Increased transport block size to improve the data rate • Ensure Voice over LTE • Ensure multi-cast transmission • Ensure enhanced coverage area • Enhance MTC related to reception and transmission in case of time difference measurements.
Narrowband Internet of Things (NB-IoT)	• Multi-cast support for downlink transmission • Mobility and enhanced service continuity • Lowering the maximum transmit power which enables reduced form factor of wearable devices • Increased transport block size with 1352 bits for downlink and 1800 bits for uplink to support high data rate, low latency, and reduced power consumption • Co-existence of NB-IoT User equipment with CDMA to lower the adjacent channel leakage ratio (ACLR) up to 49 dB • New band supports for NB-IoT for effective utilization • D2D communication is employed as an extension to NB-IoT to provide routing cellular links

Table 1.5 Trending researches on challenges of 5G-IoT [16]

Researches	Challenges	Solutions
Enabling massive IoT in 5G and beyond systems: PHY radio frame design considerations	Flexibility in the physical layer radio framework of 5G technology to satisfy the diverse requirements of IoT.	Suitable radio numerology is designed with a random-access channel to support massive connection density and is capable of dealing with channel impairment and imperfections in the transceiver.
A survey of client-controlled HetNets for 5G	Huge signalling overhead in network control schemes for network edge devices and leveraging Radio access technologies (RATs).	A review on client-controlled HetNets for 5G networks is provided in-depth with distributed and hybrid control approaches.
The next generation of IoT	Different companies such Tata communications, Dell IoT Services, and Sierra Wireless are presented, which are developing IoT technology all around the globe.	
Long-range IoT technologies: the dawn of LoRa™	Efficient low-power WAN (LPWAN) enabling technologies for IoT.	LoRa is presented as the latest and most promising technology.
Extracting and exploiting inherent sparsity for efficient IoT support in 5G: challenges and potential solutions	Scarcity in spectrum resources to provide support for IoT devices to enable 5G technology. Also, the radio access channel hash many limitations to handle IoT-enabled 5G devices.	Wide-spectrum management techniques, D2D communications, and the concept of edge cloud solutions are presented.
Energizing 5G	Keeping in mind the 5G-IoT scenarios, the efficient recharging of ubiquitous IoT devices is a tedious task.	Solutions for the wireless powering of IoT devices using near and far-field techniques are provided. Furthermore, a new networking model called a wireless power communication network is introduced that integrates wireless power transmission and communication.

A survey of 5G network: architecture and emerging technologies	To cope up with the needs of 5G enabled IoT such as better data rate, reduced latency, consistent quality of service, and huge spectrum resources.	5G cellular network architecture is presented with its enabling technologies such as M-MIMO and D2D communication. Other associated emerging technologies are also presented, such as ultra-dense networks, cognitive radios, millimetre-wave (mm-wave) solutions for 5G networks, and cloud technologies.
5G backhaul challenges and emerging research directions: a survey	The requirements for maintaining the high quality of service in a 5G paradigm present a bottleneck: backhaul. The requirements of backhaul must be addressed as it is responsible for connecting the highly dense and heavy traffic cells to the core.	A joint radio access and backhaul framework is presented which addresses the QoS issues efficiently. The framework called Backhaul as a Service (BHaaS) which is part of SDN with Radio access network (RAN) intelligence, Self-optimizing network, and caching capabilities has a complete vision of end-to-end network and also enables optimization.
A heuristic offloading method for deep learning edge services in 5G networks	The limited computing resources and battery consumption of mobile devices (MDs), mobile tasks are often offloaded to the remote infrastructure, like cloud platforms, which leads to the unavoidable offloading transmission delay.	Computation offloading is a key technique for deep learning edge services in 5G networks. To shorten the transmission delay of deep learning tasks, a heuristic offloading method is devised and shown.

the future improvisation of IoT technology and the solutions to the identified challenges. It is necessary for future intelligent systems to have IoT nodes that are able to function in a variety of operating situations, interact with the network and cloud, and maximize system intelligence while minimizing energy and information consumption. The incorporation of intelligence and self-adaptation into a self-organizing network is the method that will allow the network to perform at its highest possible level to the greatest extent possible. Increasing network performance while simultaneously lowering the amount of human participation is the primary objective of the service-oriented network, which aims to improve service quality and reduce the financial aspects related with network operations.

1.3.2   AI for ESG

John McCarthy was the first person to adopt the term 'artificial intelligence' in 1955, while Alan Turing was the first person to advocate for research on artificial intelligence in the 1950s [5]. A number of different fields, such as marketing, finance, engineering, healthcare, and education, have significantly benefited from the impact that artificial intelligence has had. In the same way that the human mind extracts, analyses, and reacts instinctively, AI produces objects that are able to do so with better accuracy, speed, and correctness. ESG analyses as resources for ethical auditing of artificial intelligence. It has become abundantly evident that AI-based systems need to be effectively managed to not only prevent, but at the very least limit, the potential hazards and damages that may arise as the utilization of AI expands across all industries and communities. The incorporation of AI into ESG evaluations may be justified given that it is seen as a possible material risk for a wide variety of businesses and industries. The implementation of standardized standards for the accountability of artificial intelligence is necessary. Additionally, it is essential to resolve asymmetric knowledge linkages and critical bottlenecks.

There is a growing need among portfolio managers and individual investors for efficient asset allocation models that take into account both the behavioural biases of investors and ESG considerations. Despite the fact that investors have a preference for ESG sustainability, loss aversion, and cognitive dissonance. The comprehensive artificial intelligence approach to the ESG datasets as presented in this study includes a total of five separate studies. The study also focuses on the application of natural language processing algorithms to the analysis of governance and social statistics, and it provides a straightforward method for predicting the ESG rankings of a certain company. There is a methodology that is adopted for the purpose of doing statistical analysis and applying techniques of machine learning to analyse the significance of ESG criteria for investment decisions and how they affect financial performance. Companies that had the highest ESG ratings beat their competitors in terms of return on value. The governance and auditing of artificial intelligence systems offer the potential to bridge the gap between the ethical principles that guide AI and the ethical use of AI systems.

However, to accomplish this, evaluation tools and metrics are required. By proactively analysing the risks that are inherent in their artificial intelligence systems, organizations have the opportunity to go above and beyond their legal duties. Compliance with the law is only one aspect of effective governance of artificial intelligence. In addition, this highlights three main areas of finance scholarship that are roughly equivalent for both types of research. These categories include sentiment inference, forecasting and planning, valuation, portfolio design, investor behaviour, financial fraud, and distress. Utilizing co-occurrence and confluence studies, the author also identifies trends and future study areas for machine learning and artificial intelligence in the field of finance research. Artificial intelligence capabilities have shown to be beneficial to ESG investing, which formerly relied on self-disclosed, annualised firm information that was prone to inherent data errors and biases. Increasing amounts of pressure are being placed on investment managers to incorporate ESG considerations into their portfolios. Artificial intelligence has the potential to provide a solution to this problem by serving as the catalyst for large-scale sustainable investing through the use of analysis tools that filter significant data.

Figure 1.7 depicts the use of artificial intelligence to ESG, which involves the combination of data pertaining to governance, social issues, and the environment to produce a dataset. The design illustrates the flow of data being generated by the IoT. Various sensors, communication protocols, and actuators are utilized to collect data from the devices that make up the IoT. Web application programming interface is used to transport these data to the database. A pre-processing stage is performed on the data after it has been received from IoT devices. During this stage, the data is filtered and cleaned. The pre-processing of data makes it possible to obtain standardized data, which is structured data that can be easily applied to the programme and in which it is possible to construct a data set [6]. The artificial intelligence and machine learning model will be chosen and trained with previous data using a variety of learning algorithms to determine which method is the most appropriate for the data set once it has been produced. Following the completion of the training process, the data set is put through a test using a particular learning

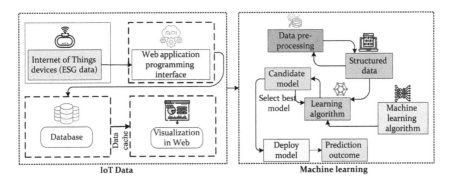

Figure 1.7   ML implementation for ESG [14]

algorithm to discover previously concealed patterns and forecast potential opportunities for the future. These artificial intelligence frameworks serve as a catalyst for people to analyse the necessary data on the ESG parameters. This allows investors to have a better understanding of the organizations in which they have invested. Artificial intelligence algorithms are used in sentiment analysis and natural language processing to apply and analyse context, texture, and tone.

These algorithms also detect patterns at which senior executives adopt strategies to implement ESG concepts in their organizations. The same strategies can be utilized by different organizations to successfully apply ESG standards. A list of the research that have used artificial intelligence in ESG reporting is included in Table 1.6. The majority of the studies have investigated and used AI with the purpose of evaluating the influence that ESG has on the performance of the company. In addition to this, the subsequent three models, which include feed-forward neural networks, gradient-boosted trees (XG Boost), and categorical gradient-boosted trees (CatBoost), are used to the core data to provide predictions regarding the ESG rating. A limited number of research have utilized AI for the purpose of risk assessment and valuation.

Table 1.6   Research into the application of AI for environmental, social, and governance [6]

Objective	Purpose of AI	AI model
An unsupervised network is put into operation for the purpose of detecting anomalies and indexing material with a high likelihood of validity	Data authenticity	Artificial neural network
An AI ESG protocol has been created with the purpose of evaluating and exposing the influence on sustainability	Valuation and risk assessments	NA
A structure that allows for the evaluation and disclosure of environmental, social, and governance	Sustainability-related impacts of AI	NA
Providing in-depth insight into the influence that ESG has on the performance of a company through the use of machine learning	Firm's operational	Machine learning techniques
ESG ratings can be predicted with the help of a heterogeneous ensemble model while employing fundamental data	ESG rating	Feed-forward neural networks, gradient-boosted trees (XG Boost), and categorical gradient-boosted trees (CatBoost)
Conducting an analysis of the structural balance sheet to determine how it affects the ESG score	Traded stocks	Random forest algorithm

There are still a great deal of obstacles to overcome, despite the fact that deep learning has achieved remarkable success during the third wave of artificial intelligence. In contrast to the knowledge-based AI paradigm that was developed during the first rise, the deep learning approaches that are currently in use lack interpretability. It has been demonstrated that deep learning techniques can achieve recognition accuracy that is comparable to or even beyond that of people in a variety of applications; however, these techniques demand a significantly greater amount of training data, power consumption, and computer resources than humans do. The accuracy findings are very excellent from a statistical standpoint, but they are frequently unreliable when viewed on an individual basis. Furthermore, the majority of the deep learning models that are currently available lack the ability to generate reasoning and explanations, which leaves them susceptible to catastrophic failures or attacks because they lack the ability to anticipate and, as a result, prevent them. Research that is both basic and applied is required to overcome the problems that have been outlined above and to attain ultimate success in the field of general artificial intelligence. Until fresh conceptual, algorithmic, and hardware breakthroughs are made, the next new wave of artificial intelligence will not come into existence.

In the field of artificial intelligence research [5], one potential breakthrough could be the development of interpretable deep learning models that can be taught and used without the need for a black box. In the event that this endeavour is successful, new artificial intelligence algorithms and methods will be developed. These algorithms and methods will be able to overcome the current limitation of AI systems, which is that they are unable to explain the actions, decisions, and outcomes to human users. However, they promise to perceive, learn, decide, and act on their own. The new kind of artificial intelligence systems will, ideally, make it possible for users to comprehend and, consequently, have faith in the outputs of the AI system, as well as to anticipate and forecast the behaviour of the systems in the future. To achieve this objective, it is necessary to effectively combine neural networks and symbolic systems, allowing the AI systems to autonomously develop models that can provide explanations for the functioning of the world. Specifically, individuals will independently uncover the fundamental factors or rational principles that influence their ability to anticipate and make decisions, which may be comprehended by human users through symbolic and natural language expressions.

Early research in this field utilizes a combined neural-symbolic representation known as tensor-product neural memory cells. These cells may be converted back to symbolic form without any loss of information after undergoing considerable learning in the neural-tensor domain. Continual research is currently being conducted in this field. If the AI system constructed using tensor-product representations is successful, it will acquire the ability to comprehend extensive natural language documents. Then, the system will possess the capability to not only respond to inquiries in a logical manner but also to genuinely comprehend the content it reads to such a degree that it can effectively communicate this comprehension to human users by explaining the process it follows to arrive at the answer. The processes can be presented as logical reasoning articulated in a comprehensible manner using natural language for human users of machine reading comprehension AI systems.

1.3.3   Big Data for ESG

The fundamental architecture of Big Data that can be utilized for ESG reporting is depicted in Figure 1.8. In addition to data visualization, the architecture includes data sources, data collecting, data storage and analytic, and data visualization. Data sources for ESG reporting often come primarily from remote sensing (geospatial) data, end devices of the IoT, operational systems, and social media. The quantity of information that is produced by various sources is incredible. The discovery of previously unknown patterns, correlations, trends, and preferences of customers through the identification of hidden patterns. Air quality, weather data, forest data, water resource data, waste management data, and other environmental data are all included in environmental data. Data collected through remote sensing constitute a significant portion of Big Data, and the volume of this data is rapidly growing – at least by 20% each year. In the course of this investigation, the researcher investigates both the advantages and the challenges that geospatial Big Data has presented to us. Geographic Big Data analytics has a number of advantages, some of which include the reduction of time and fuel consumption, the expansion of revenue, the improvement of urban planning, and the improvement of healthcare. After that, they present brand-new technologies that are currently in the process of being developed for the purpose of sharing all of the geospatial Big Data that has been accumulated and monitoring human movements utilizing mobile devices [6].

In addition to health, diversity, freedom of expression, and human rights, social data also includes other topics. Information pertaining to data openness, taxation, and policies of the government are all included in the governance data. The IoT gateway, the web application programming interface, and Node.js (which is a JavaScript framework that enables the server to function) are the means by which data gathering is received from the data source. When it comes to Big Data, the most significant challenge that arises with the flow of the data ingestion process is ensuring that the data is stored in the appropriate area according to its utilization.

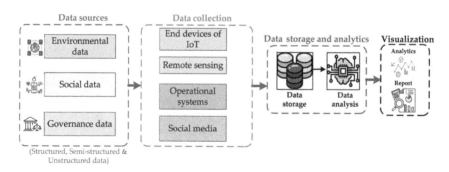

Figure 1.8   Big Data architecture [14]

We have been successful in storing our data in relational databases; however, these databases do not have the consistency in data input that is necessary for being relational. On the other hand, the polygot persistence approach is able to divide the data over numerous databases by utilizing the combined power of the different databases. When it comes to Big Data, the data analytics layer is extremely important since it allows for the utilization of various analytical models, which can not only be used to recognize existing trends but also to forecast the future. Data mining, predictive analytics, and deep learning are some of the techniques that are utilized. An application of data mining is the identification of links and patterns through the use of anomalies and data clusters. Utilizing historical data, predictive analytics is able to make accurate forecasts and identify potential possibilities in the future. AI and machine learning are utilized in deep learning, which involves layering algorithms to recognize patterns in complicated and abstract data. When it comes to visualization, the results are presented in the form of reports and analytics.

It is possible to make the case that Big Data is the driving force behind the current surge in artificial intelligence, and it is also indisputable that the field of cognitive computing will be incomplete if it does not make use of the advantages that Big Data analytics provides. As a result of the advent of the Big Data era, other types of data that were not previously utilized in research have been included, such as data from social media. On the other hand, AI gives meaning to Big Data through cognitive computing [12]. This is due to the fact that the analysis of Big Data by humans can be exceedingly time-consuming, and as a result, the usage of AI techniques offers assistance in making sense of Big Data. AI, on the other hand, is just one of the many applications that may be found for Big Data. However, there is still a significant need to investigate and comprehend the synergy that exists between artificial intelligence and Big Data. It is necessary to conduct additional studies to determine the specific benefits that may be acquired through the integration of these technologies and to gain an understanding of how artificial intelligence can be further improved with the growing availability of Big Data, which is characterized by its volume, diversity, and velocity.

1.3.4 Blockchain for ESG

Through the implementation of its distributed ledger technology, blockchain has brought about a full transformation in the banking sector. Blockchain technology has the potential to be utilized in a variety of applications, including asset management and peer-to-peer networks for the open exchange of data. For the purpose of international climate funding, these recent advancements in financial technology will prove to be especially beneficial. As of right now, there is a lack of a standardized reporting system, which would enable the information to be scalable and comparative. The ESG reporting system has been established by reporting methods in such a way that it is visible, discontinuous, inconsistent, and difficult to compare data disclosures. This is because the market offers a vast variety of ESG grading models. Blockchain has the potential to streamline the processing and packaging of

ESG reports by building data collection procedures that are nimble, transparent, and automated.

The blockchain is used to execute ESG reports and evaluations of a company's sustainability, as shown in Figure 1.9, which is an illustration of an architecture that is used to implement the block chain. A wireless communication protocol is used to collect ESG data from the various IoT devices and then transmit it to the block chain gateway. An IoT edge gateway that is based on blockchain technology makes it possible to bypass authentication concerns in IoT data [10]. The data from IoT devices is transmitted to the block chain network through this gateway, which functions as a node. This ensures that both privacy and transparency are maintained. It is possible to automate the process of putting a contract into effect using the concept of a smart contract. This allows all parties involved to be certain of the outcomes immediately, without the need for a middleman or the possibility of a time failure. A further application of smart contracts is the evaluation of the continuity between the raw data and the final ESG report. In the end, the measurements of sustainability that were gathered through the smart contract are utilized for the purpose of evaluating and analysing the behaviour of businesses.

Various research that focused on adopting blockchain technology for ESG purposes are shown in Table 1.7. Blockchain technology is utilized in the bulk of the research with the purpose of boosting transparency, security, trustworthiness, and data authentication. The vast majority of research has suggested using tokens and coins as means for evaluating ESG reports. The IoT and AI are being utilized in conjunction with blockchain technology to ensure the authenticity of data. This is accomplished by verifying and cleaning the data. Additionally, it has been discovered from the studies that blockchain technology has been mostly utilized for the purpose of energy trading, with the intention of transparently monitoring carbon emissions. A limited number of studies have carried out a pilot study to investigate the significance of blockchain technology for the collecting of data automatically [9].

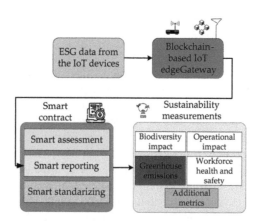

Figure 1.9   Blockchain for ESG reporting [14]

Table 1.7 Studies of blockchain technology implementation for ESG [10]

Objective (challenges)	Purpose of blockchain	Sector
A proposed framework based on blockchains and a token-based mechanism to assist in the evaluation of enterprises' environmental, social, and governance (ESG) sustainability.	Transparency, data authentication, and consistency	Supply chain
Utilizing the Internet of Things (IoT) and blockchain technology to combat green-washing in businesses and to achieve intelligent and trustworthy environmental, social, and governance.	Security, transparency, and creditability	Apparel industry
ESG disclosures from businesses along the whole value chain are being cross-validated with the use of a Life Cycle Assessment system that is based on blockchain technology.	Life cycle assessment	Tesla's electric vehicles
Assimilation of environmental, social, and governance (ESG) data with financial data and reporting in real time through the use of blockchain technology, as well as automation of assurance service through the utilization of smart contracts.	Credibility, transparency, and traceability	Peer-2-Peer Energy Trading
To determine the significance of distributed ledger technology, a pilot research is carried out.	Agile, transparent, and automated data collection	Asset management firms
A blockchain investment known as a carbon coin is a cryptocurrency that tokenizes the privilege of energy producers to generate carbon.	On-chain assets	Carbon trading
To automate the collection of environmental data and generate reporting that is relatively credible, a framework for the environmental smart reporting system is now being built. This framework is based on technologies such as blockchain and the IoT.	The authenticity of the data	NA

To begin, the framework that is based on blockchain technology will lead to major improvements in the quality, consistency, efficiency, and transparency of ESG reporting. In particular, by utilizing the blockchain platform and other related technologies, publicly traded companies will be able to provide data that is more reliable, ESG professional service providers will be able to issue the ESG report in a more effective manner, and external stakeholders, such as investors, non-governmental organizations, and associations, will be able to refer to the ESG reporting in a manner that is more transparent. In the second place, the use of early project deliverables will greatly enhance the competitiveness of the ESG reporting industry in general. It is anticipated that the organizations who collaborate will be able to decrease their overall direct expenses and increase their competitiveness by increasing their efficiency and enhancing their relationships with their customers.

Thirdly, the method that is provided by blockchain technology not only considerably expands the capabilities of the ESG reporting industry, but it also offers an interoperable platform that can integrate and coordinate the usage of other fragmented information systems in the various phases and occasions involved with the ESG business. Consequently, it enhances the decision-making mechanism and streamlines the reporting process, which results in improved management of the report preparation, report generation, and report publication processes.

1.4   Discussion and recommendations

When determining the level of sustainability that an organization possesses, ESG has become an important factor to consider. A company's process flow is used to collect ESG data, which is then used to evaluate sustainability measurements. The incorporation of technology that is part of Industry 4.0 is extremely helpful in terms of producing the finest ESG reports. A framework that combines all of the technologies that enable Industry 4.0 to acquire ESG data is depicted in Figure 1.10. This framework also provides an intelligent platform for the evaluation of ESG reports and metrics through its presentation. The following is a list of the few

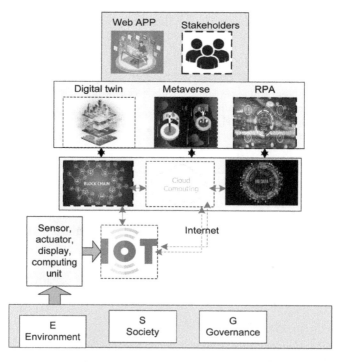

Figure 1.10   Emerging ESG technology phase [14]

restrictions and suggestions that can be adopted as a part of the ongoing study work for ESG projects.

• Previous studies have shown that there is a lack of consistency in ESG practices, and the standards of multiple dimensions are measured by a variety of database metrics for an organization. This is because the vast majority of evaluation methods are dependent on the grading of experts, which means that there is some degree of subjectivity involved. At the same time, the database is not applicable and cannot be implemented because of the dissimilarities in the national systems and the backgrounds of the various industries. An ESG evaluation framework that is both consistent and developed serves as the foundation for the research of the sustainable development of corporations. An effective ESG evaluation system that maintains uniformity in its evaluation due to its intelligent algorithms and data analysis can be empowered to be implemented through the integration of AI and Big Data technology [14].

• It has been proved through the use of hash cryptography algorithms that blockchain technology is capable of delivering security, immutability, and transparency. An additional demonstration of blockchain technology's transparency has been made. In a general sense, the data that is generated by the respective devices during the processing of supply chain tracking energy trading, monitoring of greenhouse gas emissions, financial transactions, and other activities that are comparable need to be protected by blockchain to obtain standard data concerning the environment of the particular nation under consideration. As a result of the distributed ledger and other consensus mechanisms that are incorporated into the blockchain, the parties involved are able to visualize the data and standards that were followed during the process of data collection for the purpose of evaluating ESG issues. A blockchain-enabled architecture was developed for the goal of obtaining general ESG data in the energy sector, according to a study that was conducted not too long ago. For the purpose of tokenizing the right of energy producers to emit carbon, this concept calls for the assignment of a carbon coin. The study also came to the conclusion that blockchain-based carbon markets can be established in a more comprehensive manner by utilizing ESG data and on-chain assets; however, this comes at the expense of a lower degree of performance. The research was conducted to determine whether or not this is possible. Another study, in addition to this one, made use of the IoT and blockchain technologies to improve the processes of ESG reporting in terms of security, credibility, and transparency. The purpose of this study was to facilitate corporate crowdsourcing for environmental data [15].

• Owing to the fact that the three factors of environment, society, and governance are simply diverse from one another owing to the nature of their relationship, there are a lot of issues that are related with ESG data. Data pertaining to the environment are more quantitative in nature and exhibit a greater degree of standardization than other types of data. On the other hand, it is difficult to anticipate the occurrence of these natural disasters and pandemics. Because they lay a larger emphasis on the social sciences, the statistics that pertain to social and governance are qualitative and not standardized. This is because of the context in which they are collected. Each country has its own unique social and political structure, which sets it apart from the others worldwide. The implementation of the same kind of approach for the collection of data pertaining to ESG merely presents an issue in terms of accuracy. The amount of noise that happens during the process of obtaining the unstructured data that is found in public firms, such as regulatory filings, government research, and industry publications, can be reduced via the utilization of two approaches: deep learning and natural language processing. Both of these methods can be used to minimize the amount of noise that occurs. The fact that this is the case will make it possible to successfully resolve these challenges. A clear method for predicting the ESG rankings of a specialist company was the objective of this research project, which aimed to investigate governance and social datasets through the application of natural language processing algorithms [2].

• At the moment, the IoT is generating a significant quantity of data, the vast majority of which is being stored on a server that is situated in the cloud. It is a truth that the storage of enormous amounts of data leads to an increase in the amount of carbon that is burned on a yearly basis, which in turn has an effect on the climate. On the other side, the development of the digital twin has made it feasible to make use of data from the IoT to improve the administration of an organization while simultaneously lowering the amount of carbon emissions that are produced. In addition, the digital twins are responsible for determining which data must be collected, stored, and exploited by the business to generate ESG data. Not only that, but it also identifies the data that is missing from the IoT and provides suggestions for the sensors that are appropriate for deployment in a manner that is environmentally friendly. Because it necessitates the employment of highly qualified personnel, such as data scientists and software engineers, the implementation of a digital twin in conjunction with the IoT necessitates a substantial financial commitment [16].

1.5   Conclusions

In order for the organization to effectively meet the SDGs, the significance of sustainability and ESG issues is growing in a wide spectrum. Challenges have arisen in the process of evaluating the ESG report due to the absence of high-accuracy ESG data. However, the technologies that are part of Industry 4.0 have the potential to overcome the difficulties that are associated with ESG data and reports. It is well known that the AI framework serves as a driving force for humans to evaluate the necessary data in ESG metrics so that investors can understand which organizations they have decided to invest in. In any case, AI can provide a solution by acting as a catalyst for scaled sustainable investing via analysis technologies that filter important data; data mining is used to explore relationships and patterns in data by using anomalies and data clusters. The use of historical data to make predictions and discover new opportunities is an example of predictive analytic. Deep learning is a technique that employs artificial intelligence and machine learning to layer algorithms to recognize patterns in complicated and abstract data. As of right now, there is a lack of a standardized reporting system, which would enable the information to be scalable and comparative.

Because there is such a wide range of ESG rating models available on the market, reporting practices have designed the ESG reporting system in such a way that it is visible, discontinuous, inconsistent, and difficult to compare data disclosures. Blockchain has the potential to streamline the processing and packaging of ESG reports by building data collection procedures that are nimble, transparent, and automated. Since there is a lack of consistency in the overall ESG, as well as the standards of various dimensions measured by various databases, blockchain enables the organization to keep track of sustainable efforts carried out by the organization with the intention of being responsible to the sustainability, and the implementation of the same kind of approach for ESG data collection merely causes a challenge in terms of accuracy. Various technologies can be examined related to Industry 4.0, such as robotics, digital twins, and edge computing, for ESG purposes in the future.

References

[1]   Ali, O., Jaradat, A., Kulakli, A., and Abuhalimeh, A. (2021). A comparative study: Blockchain technology utilization benefits, challenges and functionalities. *IEEE Access*, 9, 12730–12749. https://doi.org/10.1109/access.2021.3050241.

[2]   Bora, I., Duan, H. K., Vasarhelyi, M. A., Zhang, C. (Abigail), and Dai, J. (2021). The transformation of government accountability and reporting. *Journal of Emerging Technologies in Accounting*, 18(2), 1–21. https://doi.org/10.2308/jeta-10780.

[3]   Bose, S. (2020). Evolution of ESG reporting frameworks. In *Values at Work* (pp. 13–33). Cham: Palgrave Macmillan. https://doi.org/10.1007/978-3-030-55613-6_2.

[4]   Cort, T., and Esty, D. (2020). ESG standards: Looming challenges and pathways forward. *Organization & Environment*, 33(4), 491–510. https://doi.org/10.1177/1086026620945342.

[5]   Deng, L. (2018). Artificial Intelligence in the rising wave of deep learning: The historical path and future outlook [perspectives]. *IEEE Signal Processing Magazine*, 35(1), 180–177. https://doi.org/10.1109/msp.2017.2762725.

[6]   Duan, Y., Edwards, J. S., and Dwivedi, Y. K. (2019). Artificial intelligence for decision making in the era of big data – evolution, challenges and research agenda. *International Journal of Information Management*, 48, 63–71. https://doi.org/10.1016/j.ijinfomgt.2019.01.021.

[7]　Jensen, M. C., and Meckling, W. H. (2019). Theory of the firm: Managerial behavior, agency costs and ownership structure. In *Corporate Governance* (pp. 77–132). Aldershot, UK: Gower.

[8]　Landaluce, H., Arjona, L., Perallos, A., Falcone, F., Angulo, I., and Muralter, F. (2020). A review of IOT sensing applications and challenges using RFID and wireless sensor networks. *Sensors*, 20(9), 2495. https://doi.org/10.3390/s20092495.

[9]　Liu, X., Wu, H., Wu, W., Fu, Y., and Huang, G. Q. (2020). Blockchain-enabled ESG reporting framework for sustainable supply chain. *Sustainable Design and Manufacturing* 2020, 403–413. https://doi.org/10.1007/978-981-15-8131-1_36.

[10]　Li, T.-T., Wang, K., Sueyoshi, T., and Wang, D. D. (2021). ESG: Research progress and future prospects. *Sustainability*, 13(21), 11663. https://doi.org/10.3390/su132111663.

[11]　Nitlarp, T., and Kiattisin, S. (2022). The impact factors of Industry 4.0 on ESG in the energy sector. *Sustainability*, 14(15), 9198. https://doi.org/10.3390/su14159198.

[12]　Sætra, H. S. (2021). A framework for evaluating and disclosing the ESG related impacts of AI with the SDGs. *Sustainability*, 13(15), 8503. https://doi.org/10.3390/su13158503.

[13]　Sætra, H. S. (2022). The AI ESG protocol: Evaluating and disclosing the environment, social, and governance implications of artificial intelligence capabilities, assets, and activities. *Sustainable Development*, 31(2), 1027–1037. https://doi.org/10.1002/sd.2438.

[14]　Saxena, A., Singh, R., Gehlot, A., *et al.* (2022). Technologies empowered environmental, social, and governance (ESG): An industry 4.0 landscape. *Sustainability*, 15(1), 309. https://doi.org/10.3390/su15010309.

[15]　Schulz, K., and Feist, M. (2021). Leveraging blockchain technology for innovative climate finance under the green climate fund. *Earth System Governance*, 7, 100084. https://doi.org/10.1016/j.esg.2020.100084.

[16]　Shafique, K., Khawaja, B. A., Sabir, F., Qazi, S., and Mustaqim, M. (2020). Internet of things (IOT) for next-generation smart systems: A review of current challenges, future trends and prospects for emerging 5G-IOT scenarios. *IEEE Access*, 8, 23022–23040. https://doi.org/10.1109/access.2020.2970118.

[17]　Zanella, A., Bui, N., Castellani, A., Vangelista, L., and Zorzi, M. (2014). Internet of things for smart cities. *IEEE Internet of Things Journal*, 1(1), 22–32. https://doi.org/10.1109/jiot.2014.2306328.

Chapter 2

Leading the ESG and green sustainable technology transformation: a technology roadmap and standards

Wai Yie Leong[1]

In recent years, environmental, social, and governance (ESG) considerations have become increasingly important for businesses, investors, and policymakers alike. As society recognizes the urgent need to address environmental challenges such as climate change, sustainable technology solutions have emerged as critical enablers for achieving ESG goals. This chapter presents a comprehensive technology roadmap and standards framework aimed at guiding organizations in leading the ESG and green sustainable technology transformation. By leveraging innovative technologies and adhering to standardized practices, businesses can enhance their ESG performance while contributing to a more sustainable future.

2.1 Introduction

A comprehensive technology roadmap and standards framework for companies seeking to lead the transformation toward ESG integration and the adoption of green sustainable technology will be highlighted [1,2]. By providing a structured approach, this roadmap aims to guide organizations through the implementation of sustainable practices, leveraging technology to achieve ESG goals. Drawing on existing best practices and emerging technologies, the proposed standards aim to foster innovation, transparency, and accountability in the pursuit of a more sustainable and responsible future.

The integration of ESG principles and green sustainable technology is critical for organizations aiming to lead in the era of sustainable business. This section outlines a technology roadmap and standards framework to guide companies in their transformative journey.

[1]Faculty of Engineering and Quantity Surveying, INTI International University, Malaysia

2.2   Technology roadmap

A technology roadmap for ESG and green sustainable technology outlines a strategic plan for organizations to integrate ESG principles into their technological initiatives, with a focus on sustainability. It provides a structured approach to identify, prioritize, and implement technology-driven solutions that contribute to ESG objectives while addressing environmental and social challenges, Figure 2.1.

Key components of a technology roadmap for ESG and green sustainable technology typically include the following:

Current landscape assessment: Evaluating the organization's current technological capabilities and ESG performance, identifying strengths, weaknesses, opportunities, and threats (SWOT analysis), and understanding market trends and regulatory requirements.

Key focus areas: Identifying specific areas where technology can make a significant impact on ESG goals, such as energy efficiency, renewable energy, waste management, sustainable transportation, and supply chain sustainability.

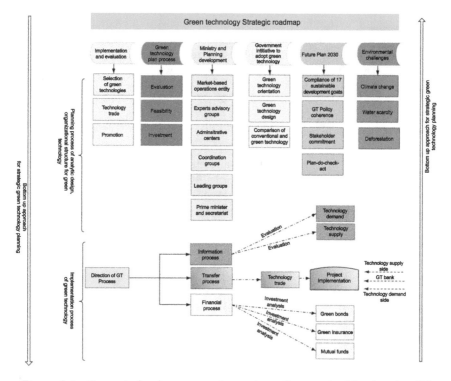

Figure 2.1   Green technology strategic roadmap for sustainable planning [1]

Short-term goals (0–2 years): Setting immediate, achievable objectives to kickstart the organization's ESG transformation, often focusing on quick wins and low-hanging fruit to build momentum and demonstrate commitment.

Medium-term goals (2–5 years): Establishing more ambitious targets and initiatives that require investment in technology adoption and innovation, aiming to drive substantial improvements in ESG performance over a longer timeframe.

Long-term goals (5+ years): Defining transformational goals and strategies aimed at achieving long-term sustainability and resilience, often involving radical shifts in technology, business models, and organizational culture.

Integration of ESG considerations: Ensuring that ESG considerations are integrated into the organization's technology development and deployment processes, from product design and supply chain management to operations and customer engagement.

Stakeholder engagement and collaboration: Engaging with stakeholders including employees, customers, investors, suppliers, and communities to gain support and alignment for ESG initiatives, and collaborating with industry peers, NGOs, and government agencies to share best practices, address common challenges, and advocate for supportive policies.

Monitoring and reporting mechanisms: Establishing robust monitoring, measurement, and reporting mechanisms to track progress against ESG goals, identify areas for improvement, and communicate outcomes to internal and external stakeholders transparently.

Continuous improvement: Emphasizing the importance of continuous learning and improvement, fostering a culture of innovation and adaptation to evolving technological, environmental, and social trends.

A technology roadmap for ESG and green sustainable technology serves as a strategic guide for organizations to align their technological investments and initiatives with their broader ESG objectives, driving positive environmental and social impact while ensuring long-term business sustainability and competitiveness.

2.3   Standards framework for ESG and sustainable technology

The standards framework for ESG and sustainable technology provides a set of guidelines, criteria, and best practices to help organizations integrate ESG considerations into their technological initiatives and operations. It aims to establish standardized approaches for measuring, reporting, and improving ESG performance, while promoting transparency, accountability, and comparability across industries and sectors, Figure 2.2.

Key components of a standards framework for ESG and sustainable technology typically include the following:

Definition of ESG metrics: Defining a comprehensive set of ESG metrics and indicators relevant to the organization's industry, operations, and stakeholder

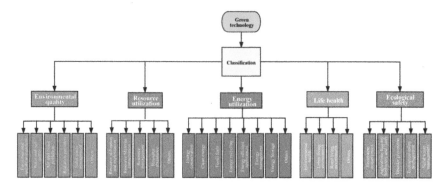

Figure 2.2   Classification of green technology [1]

interests. These metrics may include environmental factors such as carbon emissions, energy consumption, water usage, and waste generation; social factors such as labor practices, human rights, community engagement, and diversity and inclusion; and governance factors such as board diversity, executive compensation, ethics and compliance, and risk management.

Reporting guidelines: Establishing guidelines and templates for ESG reporting, including the disclosure of relevant data, methodologies, assumptions, and qualitative narratives. This may involve aligning with internationally recognized reporting frameworks such as the Global Reporting Initiative (GRI), the Sustainability Accounting Standards Board (SASB), and the Task Force on Climate-related Financial Disclosures (TCFD), as well as industry-specific standards and guidelines.

Stage 1: Assessing current state
- Conduct ESG risk assessment: Evaluate current environmental, social, and governance practices to identify areas for improvement and set a baseline for ESG performance.
- Technology audit: Assess the existing technology infrastructure, identifying opportunities for green sustainable technology adoption and integration.

Stage 2: Goal setting
- Establish clear ESG goals: Define specific and measurable ESG goals aligned with the organization's overall strategy.
- Technology alignment: Identify and prioritize green sustainable technologies that align with the established ESG goals.

Stage 3: Implementation
- Invest in renewable energy: Explore and invest in renewable energy sources such as solar and wind power to reduce the organization's carbon footprint.

- Adopt sustainable manufacturing technologies: Implement advanced manufacturing technologies, including 3D printing and eco-friendly materials, to minimize waste and enhance resource efficiency.
- Deploy smart supply chain technologies: Utilize Internet of Things (IoT) and blockchain to create transparent, efficient, and sustainable supply chains.

Stage 4: Monitoring and reporting
- Implement ESG data management systems: Deploy systems for collecting, analyzing, and reporting ESG performance data in a transparent and standardized manner.
- Continuous improvement: Use data analytics and feedback mechanisms to continuously improve ESG and sustainable technology practices.

Certification and verification processes: Developing processes for certifying and verifying ESG performance, including independent audits, third-party certifications, and verification of data accuracy and integrity. This helps build trust and credibility with stakeholders by providing assurance that the organization's ESG claims are accurate, reliable, and transparent (Figure 2.3).

Figure 2.3   The different aspects of the environmental, social, and governance (ESG) framework/approach/standard

Technology standards and best practices: Identifying technology standards and best practices that support ESG objectives, such as energy efficiency standards for buildings and appliances, renewable energy certification schemes, waste management guidelines, sustainable procurement criteria, and product lifecycle assessments. This may involve collaborating with industry consortia, standardization bodies, and regulatory agencies to develop and promote consensus-based standards that drive continuous improvement in ESG performance.

Standards framework

1. Environmental standards
   - ISO 14001: Environmental management systems—Requirements with guidance for use.
   - Greenhouse gas protocol: A widely used standard for accounting and reporting greenhouse gas emissions [3].

2. Social standards
   - ISO 26000: Guidance on social responsibility.
   - SA8000: Social Accountability International's standard for decent working conditions [4].

3. Governance standards
   - ISO 37001: Anti-bribery management systems.
   - GRI standards: GRI standards for comprehensive sustainability reporting [5].

Supply chain requirements: Addressing ESG considerations throughout the supply chain, from raw material sourcing and manufacturing to distribution and end-of-life disposal. This includes assessing supplier ESG performance, promoting responsible sourcing practices, and fostering transparency and accountability across the supply chain ecosystem.

Regulatory compliance: Ensuring compliance with relevant laws, regulations, and voluntary standards related to ESG and sustainable technology, including environmental regulations, labor laws, anti-corruption statutes, and industry-specific codes of conduct. This may involve conducting regular audits, risk assessments, and compliance reviews to identify and mitigate legal and regulatory risks.

Continuous improvement and stakeholder engagement: Establishing mechanisms for continuous improvement and stakeholder engagement, including feedback loops, stakeholder consultations, and multi-stakeholder initiatives. This helps ensure that the standards framework remains relevant, responsive, and adaptive to evolving ESG trends, stakeholder expectations, and industry developments (Figure 2.4).

The standards framework for ESG and sustainable technology serves as a roadmap for organizations to integrate ESG considerations into their technological strategies, operations, and decision-making processes, driving positive environmental and social impact while enhancing long-term business value and resilience (Figure 2.5).

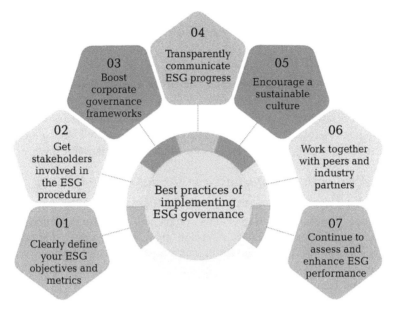

Figure 2.4   ESG governance best practices and examples

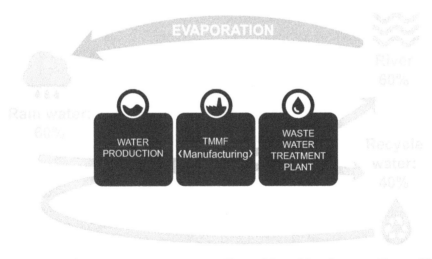

Figure 2.5   The water treatment process at Toyota Motor Manufacturing, France [6]

2.4   Case studies and best practices of successful ESG and sustainable technology initiatives across different industries

Many organizations across different industries are leveraging ESG principles and sustainable technology solutions to drive innovation, create value, and contribute to a more sustainable community. Here are case studies and best practices of successful ESG and sustainable technology initiatives across various industries:

2.4.1   Renewable energy integration in data centers

Case study: Google's data center renewable energy integration

Best practice: Google has achieved 100% renewable energy sourcing for its global data center operations. They have invested in wind and solar farms, entered into long-term power purchase agreements (PPAs), and implemented innovative energy storage solutions to ensure the continuous availability of renewable energy.

2.4.2   Zero-waste manufacturing in the automotive industry

Case study: Toyota's zero-waste manufacturing

Best practice: Toyota has implemented a comprehensive waste reduction strategy across its manufacturing facilities, focusing on recycling, reuse, and waste-to-energy conversion [6]. By optimizing production processes and engaging employees in waste reduction initiatives, Toyota has achieved significant cost savings and environmental benefits.

2.4.3   Sustainable supply chain management in the consumer goods industry

Case study: Unilever's sustainable sourcing program

Best practice: Unilever has implemented a sustainable sourcing program to ensure responsible sourcing of agricultural raw materials such as palm oil, soy, and cocoa. They work closely with suppliers to improve farming practices, protect biodiversity, and promote social development in farming communities while enhancing supply chain transparency and traceability.

2.4.4   Circular economy initiatives in the fashion industry

Case study: H&M's garment collection program

Best practice: H&M has launched a garment collection program to collect and recycle used clothing from customers [7]. They partner with recycling companies to turn old garments into new textiles, reducing the environmental impact of textile production and waste disposal while fostering a circular economy model in the fashion industry (Figure 2.6).

2.4.5   Smart building technologies in the real estate sector

Case study: The Edge, Amsterdam

Best practice: The Edge, a sustainable office building in Amsterdam, incorporates smart building technologies such as energy-efficient lighting, sensors, and automation systems to optimize energy usage, indoor air quality, and occupant comfort. It has achieved Leadership in Energy and Environmental Design (LEED) platinum certification and serves as a model for sustainable building design and operation.

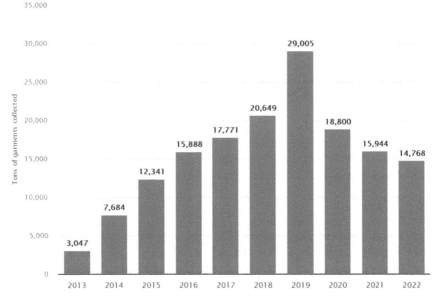

Figure 2.6   H&M's reuse and recycling collection worldwide 2013–22 [7]

2.4.6   Electric vehicle adoption in the transportation industry

Case study: Tesla's electric vehicle revolution

Best practice: Tesla has pioneered the mass adoption of electric vehicles (EVs) by developing high-performance EVs with long-range battery technology and building a global network of supercharger stations for fast charging. Their innovative business model and product design have accelerated the transition to sustainable transportation and reduced greenhouse gas emissions [8].

2.4.7   Community renewable energy projects in the energy sector

Case study: Ørsted's offshore wind farms

Best practice: Ørsted, a leading renewable energy company, engages local communities in the development of offshore wind farms through stakeholder consultations, community benefit agreements, and shared ownership models [9]. By involving communities in the planning and implementation process, Ørsted builds trust, creates economic opportunities, and fosters social acceptance of renewable energy projects.

These case studies demonstrate how organizations across different industries are leveraging ESG principles and sustainable technology solutions to drive innovation, create value, and contribute to a more sustainable future. By adopting best practices and learning from successful initiatives, businesses can enhance their ESG performance, mitigate risks, and seize opportunities in a rapidly evolving global economy.

2.5 Challenges and opportunities encountered in implementing ESG and sustainable technology initiatives

Implementing ESG and sustainable technology initiatives presents both challenges and opportunities for organizations. Here are some of the key challenges and opportunities encountered in this process:

2.5.1 Challenges

Cost and investment: Implementing ESG and sustainable technology initiatives often requires significant upfront investment, which can be a barrier for some organizations, especially smaller businesses. The initial cost of technology adoption, infrastructure upgrades, and operational changes may outweigh the immediate financial benefits, requiring long-term strategic planning and budget allocation [10–12].

Complexity and integration: ESG and sustainable technology initiatives involve multiple stakeholders, technologies, and processes, making them inherently complex to implement and manage. Integrating sustainability considerations into existing business practices, supply chains, and value chains requires coordination across departments, functions, and external partners, posing organizational and logistical challenges.

Regulatory compliance and standards: Meeting regulatory requirements and industry standards related to ESG and sustainable technology can be challenging, especially in jurisdictions with complex or evolving regulatory frameworks. Keeping up with changing regulations, reporting requirements, and certification criteria requires ongoing monitoring, interpretation, and compliance efforts, which can strain resources and expertise [13–15].

Supply chain complexity: Managing ESG risks and opportunities throughout the supply chain, from raw material sourcing to product disposal, is a complex and multifaceted challenge. Ensuring ethical sourcing, responsible manufacturing, and sustainable procurement practices across a global supply chain requires transparency, collaboration, and due diligence, particularly in industries with extensive and diverse supplier networks.

Data availability and quality: Collecting, analyzing, and reporting relevant ESG data is a fundamental aspect of sustainable business practices, but it can be challenging due to data availability, consistency, and reliability issues. Organizations may encounter difficulties in accessing accurate and timely data from internal and external sources, particularly for non-financial metrics such as environmental performance and social impact [16–18].

2.5.2   Opportunities

Competitive advantage and brand reputation: Embracing ESG and sustainable technology initiatives can provide organizations with a competitive advantage by enhancing their brand reputation, attracting environmentally and socially conscious consumers, investors, and employees, and differentiating themselves in the market. Companies that demonstrate leadership in sustainability can build trust, loyalty, and long-term value with stakeholders.

Innovation and differentiation: ESG and sustainable technology initiatives drive innovation and differentiation by fostering creativity, collaboration, and experimentation in product development, process optimization, and business model innovation. By investing in sustainable technologies, companies can create new revenue streams, enter new markets, and adapt to changing consumer preferences and market trends.

Risk mitigation and resilience: Addressing ESG risks such as climate change, resource scarcity, and social inequality can help organizations mitigate operational, reputational, and regulatory risks, enhance business resilience, and future-proof their operations. Proactively managing ESG risks can reduce vulnerability to external shocks and disruptions, ensuring long-term viability and stability.

Cost savings and efficiency: ESG and sustainable technology initiatives can deliver cost savings and efficiency improvements by optimizing resource usage, reducing waste, and increasing productivity. Energy efficiency measures, waste reduction strategies, and sustainable supply chain practices can lower operating costs, improve resource productivity, and enhance profitability over time.

Access to capital and investment opportunities: Embracing ESG and sustainable technology initiatives can attract capital and investment opportunities from socially responsible investors, impact investors, and green finance institutions seeking to support environmentally and socially responsible businesses. Companies that demonstrate strong ESG performance and commitment to sustainability are more likely to access capital at favorable terms and secure long-term investment partnerships.

While implementing ESG and sustainable technology initiatives presents challenges, it also offers significant opportunities for organizations to create value, drive innovation, and contribute to positive environmental and social impact. By overcoming barriers and seizing opportunities, businesses can enhance their competitiveness, resilience, and long-term sustainability in a rapidly changing global landscape.

2.6   Standard, policy, and governance on ESG and green sustainable technology transformation

Establishing standards, policies, and governance frameworks is essential for guiding ESG and green sustainable technology transformation initiatives effectively. Here is an outline of the components:

2.6.1   ESG standards and reporting frameworks

Identify internationally recognized ESG reporting frameworks such as the GRI [5], SASB, and TCFD. Develop internal ESG standards and reporting guidelines tailored to the organization's specific industry, operations, and stakeholder expectations. Ensure transparency, consistency, and comparability in ESG data collection, measurement, and reporting practices to facilitate informed decision-making and accountability.

2.6.2   Sustainable technology standards and certification

Adopt industry standards and best practices for sustainable technology design, development, and deployment, such as energy efficiency standards, eco-labeling schemes, and green building certifications. Seek certification from recognized bodies such as LEED, Energy Star, and Cradle to Cradle for products, buildings, and operations that meet sustainability criteria. Establish internal benchmarks and performance targets for sustainable technology adoption and implementation, aligned with organizational ESG goals and objectives (Figure 2.7).

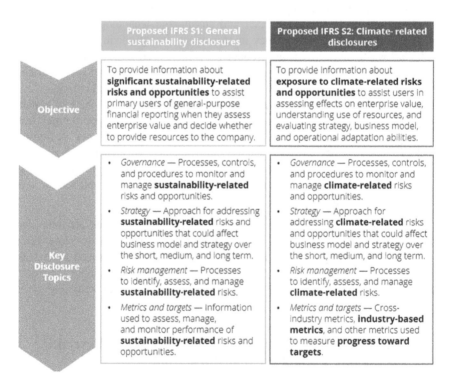

Figure 2.7   The International Sustainability Standards Board (ISSB) released two exposure drafts (EDs) of proposed standards on general sustainability and climate reporting [19]

2.6.3   ESG policy framework

Develop an ESG policy framework that articulates the organization's commitment to environmental stewardship, social responsibility, and corporate governance principles. Define clear goals, objectives, and performance targets for ESG initiatives, reflecting the organization's values, priorities, and stakeholder expectations. Integrate ESG considerations into strategic planning, risk management, and decision-making processes across all levels of the organization.

2.6.4   Governance structure and oversight

Establish a dedicated governance structure and oversight mechanisms for ESG and sustainable technology transformation initiatives, with clear roles, responsibilities, and accountability. Appoint a senior executive or board committee responsible for overseeing ESG strategy, performance, and reporting, ensuring high-level leadership and commitment to sustainability. Integrate ESG oversight into existing governance frameworks, such as corporate governance committees, risk management committees, and sustainability committees (Table 2.1).

2.6.5   Stakeholder engagement and dialogue

Engage with stakeholders including investors, customers, employees, suppliers, regulators, and communities to gather input, feedback, and support for ESG and sustainable technology initiatives. Foster open and transparent dialogue with stakeholders on ESG-related issues, challenges, and opportunities, seeking to build trust, alignment, and collaboration toward shared goals. Incorporate stakeholder perspectives into ESG strategy development, performance measurement, and reporting processes to ensure relevance and responsiveness to diverse interests and concerns.

2.6.6   Regulatory compliance and advocacy

Monitor and comply with relevant laws, regulations, and industry standards related to ESG and sustainable technology, staying informed about emerging trends and regulatory developments. Advocate for supportive policies, incentives, and regulatory frameworks that promote ESG integration, innovation, and investment, engaging with policymakers, industry associations, and advocacy groups to influence positive change.

2.6.7   Continuous improvement and innovation

Foster a culture of continuous improvement and innovation in ESG and sustainable technology practices, encouraging experimentation, learning, and adaptation to evolving trends and challenges. Invest in research and development (R&D) initiatives to drive technological innovation, product/service differentiation, and competitive advantage in the marketplace. Establish mechanisms for monitoring, evaluating, and reporting on the effectiveness of ESG and sustainable technology initiatives, seeking opportunities for optimization, replication, and scalability.

By implementing robust standards, policies, and governance frameworks for ESG and green sustainable technology transformation, organizations can enhance their environmental and social performance, mitigate risks, and unlock new opportunities for value creation and long-term sustainability (Figure 2.8).

Table 2.1   Compares key provisions of the proposed SEC rule, the proposed IFRS S1 and IFRS S2 standards, and the proposed ESRS [8]

| | | Regulation *(Proposed)* | Standards *(EDs)* | |
		SEC	**IFRS**	**ESRS**
Guiding Requirements	Type	Authoritative regulatory disclosure requirements proposed by the SEC	Standards to be jurisdictionally regulated, developed by ISSB	Standards regulated by CSRD, developed by EFRAG
	Scope	**Climate-related disclosures** (industry-agnostic)	**Sustainability (S1) and climate-related (S2) disclosures** (industry-specific)	**Sustainability disclosures** (industry-specific)
Guiding Requirements	Materiality	Investor-focused: Consistent with U.S. Supreme Court's definition of the term as "a substantial likelihood that a reasonable shareholder would consider [an omitted fact] important in deciding how to vote" or "that the disclosure of the omitted fact would have been viewed by the reasonable investor as having significantly altered the 'total mix' of information made available"	Investor-focused: Omitting, misstating, or obscuring information could be reasonably expected to influence decisions that primary users make on an entity's enterprise value	Double materiality: • Impact of climate issues on business strategy • Impact of business strategy on people and environment
	Assurance	• Limited assurance, followed by reasonable assurance (Scope 1 and Scope 2 GHG emissions) • Financial statement audit/ICFR (financial impact metrics)	Subject to jurisdictional requirements	Limited assurance, followed by reasonable assurance (reported sustainability information)
Climate-Related Disclosures	Governance	• Board oversight of and expertise in climate-related risks • Process by which management is informed about and monitors climate-related risks	• Reflection of risks and opportunities within the board mandates and policies • Climate-related remuneration policies	• Roles of governance and management regarding sustainability • Sustainability matters addressed during reporting period • Integration of sustainability into incentive schemes
	Strategy *(Climate Risks, Opportunities)*	• How climate-related risks had or are likely to have material impact on the business or financial results • Effect of risks on strategy, business model, and outlook • Scenario analysis details, if performed	• Direct and indirect responses to climate risk • Changes to financial position and resourcing impacts • Assets aligned with risks and opportunities • Resiliency of strategy, informed by scenario analysis	• Identified sustainability risks and opportunities • Description of due diligence process, adverse value chain impacts, and remediation actions • Resilience to different climate scenarios
	Risk Management	• Processes to identify, assess, and manage climate-related risks • Significance and materiality of climate-related risks • Consider actual or potential regulations for transition risks	• Processes to identify, assess, and manage climate-related risks • Extent of integration into overall risk management processes	• Processes to identify, assess, and manage climate-related risks • Climate change mitigation and adaptation risk management plans and policies

Table 2.1 (Continued)

Climate-Related Metrics	Financial	• Financial impact and expenditure metrics for climate-related events and transition activities • Impact on financial estimates and assumptions	Impact of climate-related risks and opportunities on financial position, performance, and cash flows for reporting period, and anticipated impacts over short, medium, and long term	Financial impact of physical and transition risks and reference to affected financial statement line items and anticipated impacts over short, medium, and long term
	GHG Emissions	• Scope 1 and Scope 2 GHG emissions • Scope 3 GHG emissions, if material, or Scope 3 target • Gross emissions disaggregated by each GHG • GHG emissions intensity	Scope 1, Scope 2, and Scope 3 GHG emissions required	• Scope 1, Scope 2, and Scope 3 GHG emissions required, and any material change to methodology • Granular qualitative and quantitative GHG disclosures
	Targets and Other Metrics	• Climate-related targets or goals, interim targets, and supporting qualitative disclosures • Carbon offsets or renewable energy credits (RECs)	• Metrics used to assess climate risks and opportunities • Targets, performance, and approach to target setting • Industry-based metrics	• Target aligned with Paris Agreement (1.5°C), net zero target (if applicable), and methods and frameworks used • If net zero target, methodologies and frameworks applied • Targets and target-setting process, industry-based metrics, offsets, renewable energy certificates, energy metrics, and internal carbon pricing

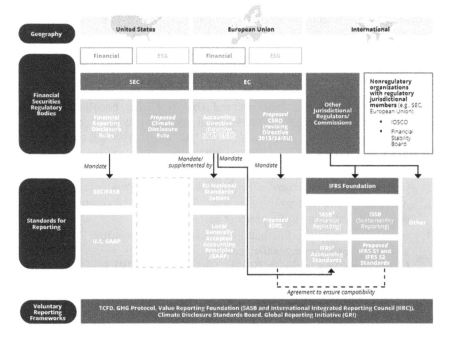

Figure 2.8 Emerging regulations and standards on sustainability and climate disclosure and organizations involved [19]

2.7   Conclusions

The chapter presents a comprehensive strategy for organizations to navigate the ESG landscape and drive sustainable technology transformation effectively. It begins by emphasizing the growing importance of ESG considerations in modern business practices and the critical role of technology in addressing environmental and social challenges.

The technology roadmap outlines a phased approach for organizations to integrate ESG and sustainable technology initiatives into their operations. It identifies key focus areas such as energy efficiency, renewable energy, waste management, and sustainable transportation, providing short-term, medium-term, and long-term goals for implementation.

In parallel, the standards framework establishes guidelines, criteria, and best practices for organizations to adhere to as they embark on their ESG journey. It includes standards for ESG reporting, sustainable technology certification, policy development, governance structure, stakeholder engagement, regulatory compliance, and continuous improvement.

The roadmap and standards emphasize the importance of collaboration, transparency, and accountability in driving positive environmental and social change through technology innovation. By aligning with international best practices and engaging with stakeholders, organizations can enhance their ESG performance, mitigate risks, and seize opportunities for long-term sustainability and competitiveness.

Overall, the chapter provides a roadmap for organizations to lead the ESG and green sustainable technology transformation, guiding them toward a future where business success is intertwined with environmental stewardship and social responsibility.

References

[1]   Ikram, M., Sroufe, R., Awan, U., and Abid, N. (2021). Enabling progress in developing economies: A novel hybrid decision-making model for green technology planning. *Sustainability*, 14, 258. doi:10.3390/su14010258.

[2]   International Organization for Standardization (ISO). (Various years). ISO Standards. https://www.iso.org/standards.html.

[3]   Greenhouse Gas Protocol. *The Greenhouse Gas Protocol: A Corporate Accounting and Reporting Standard*, 2001. https://ghgprotocol.org/sites/default/files/standards/ghg-protocol-revised.pdf.

[4]   Social Accountability International (SAI). SA8000 Standard, 2014. https://sa-intl.org/resources/sa8000-standard/.

[5]   Global Reporting Initiative (GRI). (Various years). GRI Standards. https://www.globalreporting.org/standards/.

[6]   Toyota. *Water Usage*. https://www.toyota.ie/discover-toyota/environment/better-earth/reuse.

[7] Smith, P. *Amount of clothing reused and recycled by H&M worldwide from 2013 to 2022, Statista 2024*, Sep 14, 2023.

[8] Elkind, P. Inside Elon Musks $1.4 Billion Score. *Fortune Magazine*. http://fortune.com/inside-elon-musks-billion-dollar-gigafactory/, Accessed APR 2024.

[9] Orsted. *Our Offshore Wind Farms*. https://orsted.com/, Accessed APR 2024.

[10] Leong, W. Y., Leong, Y. Z., and Leong, W. S. Green Policies and Programmes in ASEAN. In *The 9th Scientific Conference on "Applying new Technology in Green Buildings, The ATiGB 2024"*, 2024, Danang City, Vietnam.

[11] Leong, W. Y., Leong, Y. Z., and Leong, W. S. Green building initiatives in ASEAN countries. In *2023 Asia Meeting on Environment and Electrical Engineering (EEE-AM)*, Hanoi, Vietnam, 2023 (pp. 01–06). doi: 10.1109/EEE-AM58328.2023.10395261.

[12] Leong, W. Y. Investigating and enhancing energy management policy and strategies in ASEAN. In *2023 Asia Meeting on Environment and Electrical Engineering (EEE-AM)*, Hanoi, Vietnam, 2023 (pp. 01–06). doi: 10.1109/EEE-AM58328.2023.10395060.

[13] Leong, W. Y., Leong, Y. Z., and Leong W. S. Accelerating the Energy Transition via 3D Printing. In *2023 Asia Meeting on Environment and Electrical Engineering (EEE-AM)*, Hanoi, Vietnam, 2023 (pp. 01–05). doi: 10.1109/EEE-AM58328.2023.10395773.

[14] Leong, W. Y., and Kumar, R. 5G intelligent transportation systems for smart cities. In *Convergence of IoT, Blockchain, and Computational Intelligence in Smart Cities* 2023 (pp. 1–25). CRC Press.

[15] Leong, W. Y., Heng, L. S., and Leong, W. Y. Malaysia renewable energy policy and its impact on regional countries. In *Proc. 12th International Conference on Renewable Power Generation (RPG 2023)* 2023 Oct 14 (Vol. 2023, pp. 7–13). IET.

[16] Leong, W. Y., Leong, Y. Z., and Leong, W. S. Blockchain Technology in Next Generation Energy Management System. In *The 7th International Conference on Green Technology and Sustainable Development (GTSD)*, Ho Chi Minh City, 2024.

[17] Leong, W. Y., Leong, Y. Z., and Leong, W. S. Green Communication Systems: Towards Sustainable Networking. In *The 5th International Conference on Information Science, Parallel and Distributed Systems (ISPDS 2024)*, 2024.

[18] Leong, W. Y., Leong, Y. Z., and Leong, W. S. ESG Imperatives for Healthcare Sustainability. In *IEEE 6th Eurasia Conference on Biomedical Engineering, Healthcare and Sustainability (IEEE ECBIOS 2024)*, Taiwan, 2024.

[19] Miani, G., Zhang, Y., Rutenbar, M. H., and Sullivan, K., Deloitte & Touche LLP, #DeloitteESGNow — The Disclosure Heat Is On: The Move Toward International Standardization of Sustainability and Climate Reporting, May 26, 2022.

Chapter 3

Challenges and impact of ESG and green sustainable technology

Alex Looi Tink Huey[1,2]

Our global community faces a myriad of pressing issues, including climate change, the imperative to shift from a linear to a circular economy, rising inequality, and the delicate balance between economic prosperity and societal and environmental well-being [1].

How real is global warming and climate change? Fossil fuels emerge as the primary culprit behind global climate change, responsible for more than 75% of worldwide greenhouse gas emissions and nearly 90% of all carbon dioxide emissions [2]. These emissions form a layer around the Earth, trapping the sun's heat and subsequently fueling global warming and climate change. Currently, our planet is experiencing warming at an unprecedented rate compared to any period documented in history. The gradual rise in temperatures is altering weather patterns and unsettling the traditional equilibrium of nature. Such changes pose numerous risks to human beings and all other life forms on Earth. The United Nations Climate Change Conference COP28 in Dubai, United Arab Emirates, November 2023, reported that "the world is not on track to limiting temperature rise to 1.5 °C by the end of this century" [3]. The 1.5 °C is widely recognized globally as the tipping point or climate threshold which is capable of causing worsening and irreversible effects of climate change.

The linear economy focuses on profitability, irrespective of the product life cycle, and is characterized by a high volume of manufacturing, whereas the circular economy focuses on sustainability which is built on three principles or 3R: reduce, reuse, and recycle [4]. The linear economy was the dominant economic model since the 20th century. Nonetheless, it has adverse effects on the environment and the climate, further contributing to the loss of biodiversity. As we continue to extract natural resources from the earth at an unsustainable rate, surpassing the Earth's capacity for replenishment, we exacerbate the already alarming levels of global waste generation (the global waste generation is expected to increase by 70% by 2050) [4]. In a circular economy, the emphasis lies on minimizing waste,

[1]Malim Consulting Engineers Sdn Bhd, Selangor, Malaysia
[2]IEC Young Professional, International Electrotechnical Commission (IEC), Busan, Korea

Figure 3.1   Safeguarding the environment for our future generations

conserving natural resources, and mitigating pollution, thus, reducing climate change. The initial phase involves curtailing the over-extraction of natural resources by utilizing secondary resources obtained from recycled existing products.

The escalating inequality and associated disparities and concerns have fueled social dissatisfaction, serving as a significant catalyst for the heightened political polarization and surge in populist nationalism observable today [5]. Income and wealth inequality bring about adverse economic, social, and political repercussions.

Sustainable development is a concept aimed at meeting the pressing requirement for economic advancement, all while safeguarding the environment for the benefit of our future generations and fostering social equity (Figure 3.1).

The late Kofi Annan, the former Secretary-General of the United Nations (UN), garnered global attention with his assertion that the UN's objectives in peacekeeping and human rights cannot be attained without the involvement of the business community. He emphasized that the business sector stands to gain from a peaceful world and well-functioning global markets, forging a "creative partnership" that would benefit both parties [6].

3.1   The three pillars of ESG

The environmental pillar of ESG, denoted by "E," encompasses various factors such as energy consumption (brown energy or green energy), waste generation, resource utilization, and impacts on living beings attributed to the business's operations (i.e., greenhouse gasses emissions; air, water, and ground pollution; deforestation; biodiversity disclosure). Of significant importance, it includes considerations regarding carbon emissions and their contribution to climate change. Every business engages in energy usage and resource consumption, consequently exerting an influence on, and being influenced by, the environment [7].

The social pillar, denoted by "S," pertains to the connections the business cultivates and the reputation it builds with individuals and institutions within the communities where it operates. This encompasses aspects such as labor relations, diversity, and inclusion (i.e., providing access to products and services to underprivileged social communities; managing employee development; labor practices). Every company operates within the context of a diverse and interconnected society such as reporting on product liabilities regarding the safety and quality of their products or reporting on their supply chain labor and health and safety standards [1].

The governance pillar, denoted by "G," refers to the internal framework of practices, controls, and procedures implemented by the business to oversee its operations, facilitate sound decision-making, adhere to legal requirements (i.e., anti-competitive practices; anti-corruption), and address the concerns of external stakeholders. Governance is essential for every business, as it serves as the foundation for its operations and compliance within the legal framework.

Rather than insisting on the moral obligation of businesses to consider non-financial factors such as climate change and social issues, ESG has presented a "business case" for this approach, framing it as a risk management tool that can also unveil new business opportunities. Businesses have historically utilized natural resources and human capital to fuel growth, yet this has led to adverse effects on the environment and society, which will in turn threaten the existence of businesses if left unaddressed. The UN has empowered banks and institutional investors to take the lead, recognizing their influence over fund allocation and their ability to sway companies through shareholding. Additionally, the UN has demonstrated that prioritizing ESG issues benefits a company's financial performance, distinguishing it from traditional corporate social responsibility (CSR) and faith-based investing, which are often driven solely by ethical considerations. The development and reporting frameworks and measurable metrics in ESG are also contributing factors to why ESG has become more popular than concepts like CSR.

McKinsey & Company reported the value of sustainable investment in major financial markets worldwide reached $30 trillion, marking a 68% increase since 2014 and a tenfold surge since 2004 [7]. The surge in sustainable investment has been propelled by increased societal, governmental, and consumer attention on the broader impact of corporate activities. Additionally, investors and executives recognize that a robust ESG framework can safeguard a business's long-term success and viability (Figure 3.2). The substantial influx of investment underscores that ESG initiatives are not merely passing trends or feel-good exercise.

Figure 3.2   Strong ESG proposition can safeguard a business's long-term success

3.2   Challenges and impact

The global business market transition from a low-interest-rate, relatively stable business environment to one characterized by the upheaval caused by the COVID-19 pandemic, historically high inflation, intensifying geopolitical conflicts, escalating climate-related incidents, and the emergence of generative artificial intelligence (AI) [8]. Since the outbreak of the COVID-19 pandemic in 2019, governments and legislative bodies have increasingly prioritized policies centered around "eco-friendliness," "social value," and "ethical behavior." ESG has emerged as a crucial factor sought after by consumers, businesses, and civil societies in products and services, leading to continuously expanding demand for ESG management [9].

The World Economic Forum (WEF) categorizes ESG risks for businesses into three main categories: physical risk, reputational risk, and transition risk [10]. Physical risks stemming from climate change have the potential to disrupt businesses and communities, leading to both human and financial losses. Events such as floods, drought, and pandemics can cause significant disruptions to the global value chain and world economy, as evidenced by the challenges faced during the COVID-19 pandemic. As a component of their risk management strategy, companies must integrate mitigation and adaptation plans to manage environmental risks effectively.

Governance is the foundation of ESG, necessitating precise attention to prevent it from being overlooked, as its neglect has historically led to some of the most significant corporate scandals covered by the media facing reputational risks. In our increasingly interconnected world, the spotlight on ESG and the importance of human rights throughout the value chain is brighter than ever, with both the law and media playing crucial roles. This heightened attention compels businesses to more effectively monitor their ESG factors. Failure to do so may result in reputation damage, increased compliance costs, potential revenue loss, and challenges in attracting top talents. Furthermore, non-compliance with ESG policies, legislation, and regulations could hinder a company's access to secure funding and its ability to attract potential investors and customers [11].

Businesses face transition risks associated with shifting to a low-carbon economy, decarbonization, changes in regulation, consumer preferences, market dynamics, and adopting new technologies. Adhering to ESG principles could result in increased production costs for businesses, as they must account for environmental externalities in the cost of their products (e.g., the regulation on carbon pricing mechanisms in the form of carbon tax). Conversely, the traditional approach to cost–benefit analysis conveniently overlooks these externalities. Breakthroughs in technology could potentially reduce the cost of renewable energy, resulting in a significant decrease in the demand for fossil fuels and a transformative shift in the energy mix. The pace, magnitude, and effectiveness of integrating climate-resilient solutions, along with the readiness of financial institutions, will dictate the actual impacts on the financial services sector. Today's businesses must embrace innovation to mitigate the current high production costs and waste management expenses by transitioning from a linear to a circular business lifecycle mode.

Greenhouse gas (GHG) emissions are segregated into three categories or "scopes." Scope 1 refers to direct emissions where the company owns or controls directly the sources of emissions. Scope 2 refers to indirect emissions from the likes of purchased energy generation (i.e., electricity, steam, heat, and cooling). Scope 3 covers all other indirect emissions from the company upstream and downstream value chain which are not produced by the company itself and are not the result of activities from assets owned or controlled by them [12]. Scope 3 emissions, constituting 65–95% of the carbon impact for most companies, present a significant challenge to manage, as many of the contributing factors are beyond their direct control, originating within the supply chain.

Merely issuing marketing and declarations about accomplishing ESG goals and targeting net-zero emissions is insufficient. To effectively monitor ESG practices, companies require a strong employee culture, clear and concise policies, and established systems. Ideally, these systems should be fully integrated with one another, or better yet, consolidated onto a single management platform. This approach allows for a comprehensive understanding of the company's value chain and organizational structure. Only with this holistic view can a company make informed decisions regarding sustainability initiatives.

Integrating ESG principles into an organization and its value chain goes beyond the mere selection of measurement tools; it also involves embracing people-centric approaches, including employee mentality and values (Figure 3.3). This is where company culture becomes paramount. It is essential for employees,

Figure 3.3   Integrating ESG propositions involve embracing people-centric approaches

customers, and stakeholders to grasp the core values and reasons behind the business's commitment to sustainability. As employees gain a deeper understanding of the benefits and importance of ESG, they become better equipped to champion the company's sustainability mission and objectives.

ESG reporting represents a significant realm to oversee, particularly for large companies, posing challenges for many in terms of self-management and measurement. The ever-changing global ESG reporting standards contribute to the complexity, posing challenges for companies and investors in selecting the appropriate standards to follow. There is a persistent need for ongoing efforts to establish a comprehensive and integrated reporting framework. This endeavor should entail consultations with various stakeholders, including regulatory bodies, government agencies, companies, investors, and civil society, to improve transparency and competitiveness while safeguarding the business's viability. Collaboration through partnerships emerges as a vital element in ESG reporting. Establishing ongoing communication with stakeholders across the entire value chain, including suppliers and customers, is essential for obtaining accurate measurements and achieving ESG targets. To gain better visibility and understanding of ESG propositions, companies, employees, agencies, stakeholders, and investors need to work hand in hand, with increased communication and honest and transparent conversations on climate change and ESG factors. Forging strong partnerships enables businesses to achieve long-term sustainability goals.

Legislation and regulations play a crucial role in eradicating various forms of social inequality and promoting corporate sustainability in practice. In 2023, Germany enacted its Supply Chain Due Diligence Act, while in 2022, the European Union (EU) introduced a proposal for a Directive on corporate sustainability due diligence [11]. As compliance becomes increasingly prevalent, advancements in digitalization and technology will be essential for tracking, measuring, and upholding the rising ethical standards mandated by compliance laws. Technology serves as a unifying force, providing ESG solutions that empower businesses and their supply chains to address ethical challenges effectively.

AI is poised to play an increasingly vital role in enhancing business productivity and profitability. Breakthroughs in AI, such as advancement in robotics and the development of self-driving cars, consistently capture public attention. Concerns among the public regarding the potential displacement of jobs by AI are becoming more urgent. According to PricewaterhouseCoopers (PwC), approximately 14% of jobs in Organization for Economic Co-operation and Development (OECD) economies face a high risk of automation, while another 32% are at a high risk of partial replacement [13]. Jobs most susceptible to being "lost" are those involving predictable, routine tasks (e.g., assembly line works, record clerks) and those with minimal customer interaction (e.g., food service workers, travel booking). From an investor standpoint, it is imperative to explore the social ramifications of these rapid advancements. Mass job losses could lead to structural unemployment and heightened social inequality, potentially fueling populist and anti-globalization movements. The shift in investment from human capital to robotics and AI has the potential to fundamentally transform societies. Businesses and investors must contemplate their role in this technological transition and devise strategies to anticipate or mitigate its impacts on their workforce, sales, and markets. In addition to ESG considerations, the COVID-19 pandemic has also exacerbated disruptions to businesses, underscoring the necessity for workforces to adapt through digital transformation. Strong governmental support is essential to provide guidance and actionable roadmaps for achieving these digital transformation objectives, along with incentives and refined support schemes. Governments should also facilitate job redesign efforts and upskilling initiatives to facilitate smoother workforce transitions in industries undergoing transformation, enabling employees to transition into roles of higher value. Ernst & Young has detailed that the Singaporean Government is addressing this through its Industry 4.0 (i4.0) Human Capital Initiative, which offers practical guidance for companies in the manufacturing sector to trial i4.0 technologies and redesign jobs and human capital practices to navigate transformation efforts effectively [14].

Another significant challenge is the unequal levels of ESG readiness across companies, exacerbated by the absence of a standardized framework. This variance results in different ESG practices and disclosure standards among companies. Consequently, the quality of ESG information accessible to investors is frequently compromised, rendering it challenging for them to evaluate and analyze ESG performance accurately [10].

The Institute for Capital Market Research Malaysia (ICMR) highlighted the scarcity of talent specializing in sustainability as a common obstacle to effective

ESG implementation [10]. In developing countries, the pool of sustainable experts is relatively limited and costly. To instill a culture of sustainability within their organizations, the corporate sector must be prepared to invest in developing sustainability skills and expertise.

Advocating for the advancement and adoption of green growth and sustainable technology are integral aspects for businesses in today's landscape [15]. According to Ernst & Young, the green growth action plan encompasses three primary objectives: (1) sustainable consumption and production, which involves transitioning to net-zero emissions, advocating sustainable choices for consumers, recycling and reusing waste from production, and fostering sustainable organizational practices; (2) climate, energy, and biodiversity, which entails deploying advanced technologies to measure, reduce, and eliminate an organization's carbon footprint, implementing responsible AI solutions, and embracing energy efficiency and renewable energy (Figure 3.4); and (3) equal opportunities and social cohesion, which focuses on establishing trustworthy systems that integrate privacy, fairness, transparency, robustness, and accessibility, as well as instituting appropriate governance mechanisms [15].

In Figure 3.5, it is depicted how a robust ESG strategy correlates with value creation and impact by: (1) enabling top-line growth, (2) reducing expenses, (3) mitigating regulatory and legal interventions, (4) enhancing employee productivity, and (5) optimizing investment and capital expenditures [7].

*Figure 3.4   Green growth and sustainable technology: embracing energy
efficiency and renewable energy*

	Strong ESG proposition (examples)	Weak ESG proposition (examples)
Top-line growth	Attract B2B and B2C customers with more sustainable products Achieve better access to resources through stronger community and government relations	Lose customers through poor sustainability practices (eg, human rights, supply chain) or a perception of unsustainable/unsafe products Lose access to resources (including from operational shutdowns) as a result of poor community and labor relations
Cost reductions	Lower energy consumption Reduce water intake	Generate unnecessary waste and pay correspondingly higher waste-disposal costs Expend more in packaging costs
Regulatory and legal interventions	Achieve greater strategic freedom through deregulation Earn subsidies and government support	Suffer restrictions on advertising and point of sale Incur fines, penalties, and enforcement actions
Productivity uplift	Boost employee motivation Attract talent through greater social credibility	Deal with "social stigma," which restricts talent pool Lose talent as a result of weak purpose
Investment and asset optimization	Enhance investment returns by better allocating capital for the long term (eg, more sustainable plant and equipment) Avoid investments that may not pay off because of longer-term environmental issues	Suffer stranded assets as a result of premature write-downs Fall behind competitors that have invested to be less "energy hungry"

Figure 3.5   A strong environmental, social, and governance (ESG) proposition links to value creation and impact [7]

The demand for sustainable products is increasing, propelled by ethical and conscientious consumerism, particularly among younger consumers who are influencing consumer behavior. These consumers are actively making decisions, showing preference for products with a positive impact and services that promote environmental preservation and benefit society and the economy.

Businesses that perform well on ESG are well positioned for the future and have better opportunities of adapting their products and services to a global consumer base that is increasingly pushing for environmental protection, respect for human rights, and corporate transparency.

References

[1]   Deloitte. What is ESG? https://www2.deloitte.com/ce/en/pages/global-business-services/articles/esg-explained-1-what-is-esg.html [accessed November 2023].

[2]   United Nations. Causes and effects of climate change. https://www.un.org/en/climatechange/science/causes-effects-climate-change [accessed December 2023].

[3]   United Nations Framework Convention on Climate Change (UNFCCC). COP28 opens in Dubai with calls for accelerated action, higher ambition against the escalating climate crisis. https://unfccc.int/news/cop28-opens-in-dubai-with-calls-for-accelerated-action-higher-ambition-against-the-escalating [accessed December 2023].

[4]   Knight, C. What is the linear economy? https://www.eib.org/en/stories/linear-economy-recycling. European Investment Bank. Accessed December 2024.

[5]   Qureshi, Z. Rising inequality: A major issue of our time https://www.brookings.edu/articles/rising-inequality-a-major-issue-of-our-time/. Brookings Institution. Accessed November 2023.

[6]   Tan, Z. Y. ESG 101: The origins of ESG. https://theedgemalaysia.com/node/682788. The Edge Malaysia. Accessed December 2023.

[7]   Henisz, W., Koller, T., and Nuttall, R. *Five ways that ESG creates value.* McKinsey & Company; 2019.

[8]   Doherty, R., Pérez, L., Siegel, K., and Zucker, J. *How do ESG goals impact a company's growth performance?* https://www.mckinsey.com/capabilities/growth-marketing-and-sales/our-insights/next-in-growth/how-do-esg-goals-impact-a-companys-growth-performance. McKinsey & Company. Accessed December 2023.

[9]   Association for Supporting the SDGs for United Nations (ASDUN). SDGs and ESG. http://asdun.org/?page_id=2528&lang=en#:~:text=ESG%20%3A%20%5BEnvironmental%2C%20Social%2C%20Governance%5D&text=ESG%20has%20become%20a%20significant,ESG%20management%20is%20ever%2Dgrowing [accessed December 2023].

[10] Institute for Capital Market Research Malaysia (ICMR). Environmental, social and governance (ESG) integration in Malaysia: Navigating challenges and embracing opportunities for a sustainable future. https://www.icmr.my/wp-content/uploads/2023/08/ESG-Integration-in-Malaysia.pdf [accessed January 2024].

[11] Newman, G. Top 6 ESG issues for companies to tackle. https://www.ethics.org/top-6-esg-issues-for-companies-to-tackle/. The Ethics & Compliance Initiative (ECI). Accessed November 2023.

[12] National Grid. What are scope 1, 2 and 3 carbon emissions? https://www.nationalgrid.com/stories/energy-explained/what-are-scope-1-2-3-carbon-emissions#:~:text=Definitions%20of%20scope%201%2C%202,owned%20or%20controlled%20by%20it [accessed December 2024].

[13] PricewaterhouseCoopers (PwC). Six key challenges for financial institutions to deal with ESG risks. https://www.pwc.com/vn/en/publications/vietnam-publications/six-key-challenges-for-financial-institutions-to-deal-with-ESG-risks.html [accessed January 2024].

[14] Chiang, B., and Goh, J. Y. Governments play a crucial role in helping workforces unlock new opportunities as the COVID-19 pandemic accelerates disruptions.

https://www.ey.com/en_my/government-public-sector/how-governments-can-help-workforces-emerge-stronger-from-covid-19. Ernst & Young. Accessed November 2023.

[15] Ceccon, M., and Veser, M. The sustainable tech transformation: Paving the way for a greener future. https://www.ey.com/en_ch/sustainability/drive-the-green-transformation-enabled-by-technology. Ernst & Young. Accessed November 2023.

Chapter 4

Malaysia National Industry Environmental, Social and Governance framework

Chee Fui Wong[1]

The Malaysian government acknowledged the importance of environmental, social, and governance (ESG) as an indicator of how companies and organisations carry out their business in a sustainable approach that is respectful to the people and planet, while they generate the profits ethically. Malaysia Ministry of Investment, Trade and Industry (MITI) has launched the National Industry Environmental, Social and Governance framework (i-ESG framework) to accelerate Malaysia's aim to achieve sustainable development goals and to facilitate the transition towards sustainable practices among manufacturing companies. Malaysia MITI i-ESG framework consists of four components, namely standards, financing, capacity building, and market mechanism. These four components of the i-ESG framework are supported by six key enablers, namely stakeholder engagement, human capital and capabilities, digitalisation, technology, financing and incentives, and policies and regulations. The i-ESG framework is a roadmap for businesses to integrate ESG considerations into their operations and a tool for regulators to ensure compliance and accountability from Phase 1.0 'Just Transition' to Phase 2.0 'Accelerate ESG'. The i-ESG framework Phase 1.0 'Just Transition' (2024–26) prepares the groundwork and fosters the development of a robust ecosystem to help companies embark on their ESG journey. During Phase 1, Malaysia MITI will support the companies to start embarking the ESG into their business through self-readiness assessment programmes (such as iESGReady and iESGStart), outreach training (such as 'KenalESG'), mentoring programmes, and financing support options. The aim is to introduce ESG concepts to the companies to be ready to produce their sustainability report and eventually be ready to meet more rigorous ESG demands in Phase 2.0 'Accelerate ESG'.

4.1 ESG in Malaysia

Environmental, Social and Governance (ESG) was first introduced in 2004 by the United Nations Global Compact (UNGC) and International Financial Cooperation

[1]Department of Civil Engineering; Lee Kong Chien Faculty of Engineering and Science, Universiti Tunku Abdul Rahman Sg Long Campus, Selangor, Malaysia

Table 4.1   ESG components for sustainable business [1]

Environment	Social	Governance
• Emission management	• Labour practices	• Anti-corruption and bribery
• Energy management	• Migrant workers	• Board diversity and structure
• Waste management	• Working conditions	• Regulatory policies
• Water management	• Employee relations and diversity	• Whistleblowers
• Biodiversity	• Gender equality	• Data privacy and security
• Pollution	• Training	• Supply chain management
	• Health and safety	• Executive remuneration
	• Local community	

(IFC) in the report 'Who Cares Wins', which reported the research findings on how ESG drive the value in terms of asset and financial management [2]. The Malaysian government acknowledged the importance of ESG as an indicator of how companies and organisations carry out their business in a sustainable approach that is respectful to the people and planet, while they generate profits ethically. In Malaysia, the public are becoming more conscious about issues such as Climate Change and Environmental Impact. In terms of corporate governance, there has been a rising demand for sustainable assets and investments. Malaysia capital markets have a strong influence on how companies [and publicly listed companies (PLCs)] conduct their business with ESG and sustainable policies in place. According to a report by advisory firm PriceWaterHouseCoopers [3], 80% of PLCs have an awareness of the need for sustainable supply chains as part of their ESG governance and oversight. More importantly in terms of ESG implementation in the business operation, 66% of the PLCs in Malaysia have included ESG considerations in their procurement policies and 59% have provided guidelines for their vendors and suppliers, typically through their Supplier Code of Conduct [3]. The Malaysia National Industry Environment, Social and Governance Framework (i-ESG framework) has been defined as ESG components for companies to incorporate into their business in a sustainable approach (Table 4.1).

4.2   Malaysia ESG commitment

In terms of sustainability, Malaysia has made a global commitment to adopt the 2030 Agenda for Sustainable Development (2030 Agenda) at the United Nations General Assembly in New York on 25 September 2015 [4]. With this global commitment, Malaysia is committed to incorporating and aligning the United Nations Sustainable Development Goals (SDGs) into our national development as well as the 12th Malaysia Plan [5]. Malaysia recognised the paramount importance of SDG and ESG practices in the long-term transformation of the economy, environment, and society.

Malaysia's GHG inventory, MtCO$_2$eq (2019) from BUR4

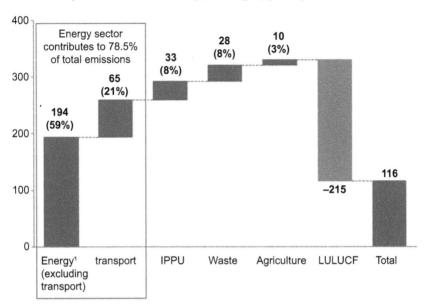

1 Refers to emissions from energy industries, manufacturing industries and construction, other sectors and non-specified energy emissions, and fugitive emissions from fuels.

Figure 4.1   Malaysia's GHG inventory in 2019 [8,9]

Malaysia has further updated its Nationally Determined Contribution (NDC) in July 2021 and has pledged to unconditionally reduce carbon and its economy-wide intensity by 45% of the gross domestic product (GDP) by 2030, as compared to the 2005 level [6].

Furthermore, Malaysia has committed to achieving net-zero GHG emissions by 2050, as planned in the Twelfth Malaysia Plan 2021–25 [4] and the National Energy Policy 2022–40 [7] that provide the foundation towards the sustainable energy transition that is fair, equitable, and inclusive for the society. The Malaysia's GHG inventory in 2019 is as shown in Figure 4.1.

4.3   National Industry Environmental, Social and Governance framework (i-ESG framework)

Malaysia Ministry of Investment, Trade and Industry (MITI) has launched the National Industry Environmental, Social and Governance framework (i-ESG framework) in 2023 to accelerate Malaysia's aim to achieve SDGs and to facilitate the transition towards sustainable practices among manufacturing companies. The objective of the i-ESG framework is to facilitate the companies in Malaysia towards the adoption and enhancement of the ESG approach in the transformation

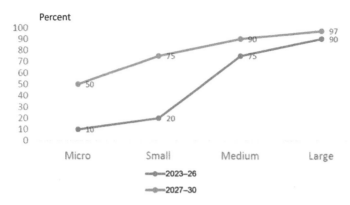

Figure 4.2   MITI estimated sustainability reporting by Malaysia companies [5]

towards achieving Malaysia's net-zero GHG goals by 2050 [5]. The i-ESG framework is a gradual adoption framework which encouraged companies to adopt the Phase 1.0 'Just Transition' model and transition into the Phase 2.0 'Accelerate ESG Practices' phase. The phased transition approach is to support the organisation's transition towards sustainable reporting that complies with the internationally accepted standard by the fourth year as estimated by the Malaysia Ministry of International Trade and Industry as shown in Figure 4.2.

Malaysia MITI i-ESG framework consists of four components, namely standards, financing, capacity building, and market mechanism. These four pillars of the i-ESG framework are supported by six key enablers, namely stakeholder engagement, human capital and capabilities, digitalisation, technology, financing and incentives, and policies and regulations.

4.4   Malaysia ESG reporting requirements

Bursa Malaysia the Malaysia exchange holding company has mandated that it is compulsory for all PLCs in Malaysia to disclose their sustainability from the financial year end (FYE) 2023 onwards as part of the listing requirements [10]. Malaysia Financial Institutions are required to comply with Bank Negara's Climate Change and Principle-based Taxonomy (CCPT) by 2024 [11].

Currently, ESG reporting is not mandated by the government. However, commencing in FYE 2023, it will be compulsory for PLCs to disclose their sustainability statement in accordance with the listing requirements of Bursa Malaysia. Meanwhile, financial institutions will be required to comply with Bank Negara's CCPT by 2024. The requirements of the sustainable reports by PLCs in Malaysia are implemented in a phased approach where main market-listed companies must disclose the common sustainability matters for FYE 2023 onwards, waste management and emissions indicators are required for FYE 2024 onwards, and the sustainability reporting aligned with Taskforce on Climate-related Financial Disclosure is only applicable for FYE 2025 onwards.

Table 4.2   *Bursa Malaysia common indicators for*
           sustainable reporting [10]

No.	Common sustainability matters
1	Anti-corruption
2	Community/society
3	Diversity
4	Energy management
5	Health and safety
6	Labour practices and standards
7	Supply chain management
8	Data privacy and security
9	Water
10	Waste management
11	Emission management

Bursa Malaysia has recognised the challenges of the different companies with diverse stakeholders, business environments, and different sectoral industries environment which may require different priorities in sustainable reporting. The Bursa Sustainable Reporting Guide (3rd edition) [10] has identified 11 common indicators that are deemed material for all PLCs in Malaysia as indicated in Table 4.2.

The sustainability reporting ecosystem in Malaysia has seen considerable growth in the past decade, from a largely nascent stage to one where all listed issuers are undertaking sustainability-related initiatives as well as producing sustainability statements or reports on an annual basis.

4.5   Green financing

Malaysia is promoting green financing through the provision of direct funding toward projects and initiatives that have a positive impact on the sustainable environment such as renewable energy, energy efficiency, and carbon capture and storage technologies. Malaysia Bank Negara has initiated the 'Low Carbon Transition Facility (LCTF)' funds to support small and medium-sized enterprises (SMEs) in adopting sustainable and low-carbon practices. Malaysia's central bank, Bank Negara Malaysia, has allocated RM2 billion in matching funds for SMEs in all sectors for funding (up to a maximum of RM10 million per SME) capital expenditures or working capital to initiate or facilitate the transition to low carbon and sustainable operation [12]. Besides that, Malaysia's central bank, Bank Negara Malaysia, has initiated the 'High Tech and Green Facility (HTG)' fund to assist SMEs and innovative start-up companies to expand their businesses and invest in strategic sectors and technologies fields (digital tech, green tech and biotech) for a sustainable and entrenched economic recovery [12]. In line with Malaysia's commitment to promote the green economy, the Malaysian government is also providing a Green Technology Tax Incentive. This includes the Green Investment Tax

Allowance (GITA) for companies seeking to acquire qualifying green technology assets or those undertaking qualifying green technology projects for business or own consumption. Besides that, the Green Income Tax Exemption (GITE) is available for qualifying green technology service provider companies [5].

4.6   ESG capacity building

The concept of ESG is relatively new in Malaysia. The National Industry Environmental, Social and Governance framework (i-ESG framework) has included capacity building as one of the key components of the ESG framework. MITI is also conducting a 'KenalESG' awareness programme to reach out to micro, small, and medium enterprises on the awareness of ESG [5]. The dissemination of information and awareness programmes is actively conducted through national forums, seminars, webinars, the Internet, social media, and other platforms. Besides the corporate sector, ESG awareness among government agencies is paramount as the public sector officers are involved in formulating the policies, procurement, and legislations that support ESG initiatives.

4.7   Market mechanism

Market mechanisms in the carbon market are usually introduced in the form of 'carbon pricing' or 'carbon trading' systems, which are formulated as incentives for the reduction of GHG gas emissions. This concept of carbon pricing quantifies a price on carbon emissions as an incentive for companies to reduce carbon emissions and to encourage organisations to invest in more sustainable technologies.

However, Malaysia does not have a carbon tax and does not levy an explicit carbon price or fuel excise taxes. In total, 0% of GHG emissions in Malaysia are subject to a positive net effective carbon rate (ECR) in 2021, unchanged since 2018 [13]. In adversary, the Malaysian government provided fossil fuel subsidies for 21.7% of emissions in 2021 as shown in Figure 4.3 [13].

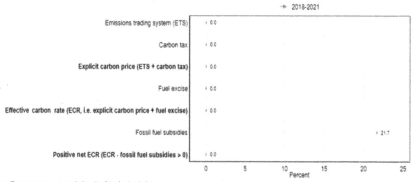

Note: Percentages are rounded to the first decimal place.

Figure 4.3   Malaysia GHG emissions subject to a positive price by instrument, 2018–21 [13]

Malaysia is considering to design the Carbon Price Index (CPI) in partnership with the World Bank to accurately gauge the economic impact of carbon pricing mechanisms and the cost implications of reducing carbon emissions across various sectors.

4.8 ESG sustainability reporting standards

To comply with Bursa Malaysia (Malaysia bourse), all PLCs in Malaysia are required to disclose the sustainability statement or sustainability report inclusive of the past 3 years' data from the FYE 2023 [10]. To comply with these requirements, the PLCs are encouraged to adopt an internationally recognised sustainable reporting format as shown in Table 4.3 [5]. However, these standards are not

Table 4.3   ESG reporting standards [5]

No.	ESG reporting standard	
1	Global Reporting Initiative (GRI)	GRI is a non-profit organisation that develops sustainability reporting guidelines for organisations around the world to use to measure and report on their environmental, social, and governance performance. The GRI's standards are used by over 10,000 organisations in over 160 countries. https://www.globalreporting.org/
2	Sustainability Accounting Standards Board (SASB)	SASB is a non-profit organisation that develops sustainability accounting standards for specific industries that companies can use to measure and report on their environmental, social, and governance performance. SASB's standards are used by over 500 companies in the United States. https://sasb.org/
3	Task Force on Climate-related Financial Disclosures (TCFD)	TCFD is an international organisation that developed recommendations for climate-related financial disclosures. The TCFD's recommendations are used by over 2000 organisations around the world. https://www.fsb-tcfd.org/
4	International Integrated Reporting Council (IIRC)	IIRC is an international organisation that developed the International Integrated Reporting Framework (IIRC Framework). The IIRC Framework is a set of principles and guidelines that organisations can use to report on their economic, environmental, and social performance. https://www.fsb-tcfd.org/
5	Climate Disclosure Standards Board (CDSB)	CDSB is an international consortium of business and environmental NGOs that are committed to advancing and aligning the global mainstream corporate reporting model to equate natural capital with financial capital. The CDSB offer companies a framework for reporting environmental information with the same rigour as financial information. https://www.cdsb.net/index.html
6	Carbon Disclosure Project (CDP)	CDP is a not-for-profit charity that runs the global disclosure system for investors, companies, cities, states and regions to manage their environmental impacts. https://www.cdp.net/en

compulsory but the standard to be adopted will be flexible to suit the company's size, industry, and other factors.

4.9   Conclusion

The Malaysia government is committed towards achieving SDGs and facilitating the transition towards sustainable practices among businesses to align with net-zero goals in 2050. Environmental, Social and Governance (ESG) is recognised as a paramount indicator of how companies and organisations carry out their business in a sustainable approach that is respectful to the people and planet, while they generate the profits ethically.

In addressing the challenges towards sustainability and aims to support businesses in adopting ESG practices, Malaysia MITI has launched the National Industry Environmental, Social and Governance framework (i-ESG framework) in 2023 to accelerate Malaysia's aim to achieve SDGs and to facilitate the transition towards sustainable practices among businesses. The i-ESG framework is a phased transition approach to support businesses' transition toward sustainable reporting that complies with international standards through a gradual adoption framework which encouraged companies to adopt the Phase 1.0 'Just Transition' model and transition into the Phase 2.0 'Accelerate ESG Practices' phase. The i-ESG Framework is a comprehensive and strategic approach to support businesses in adopting ESG practices while fostering collaboration between the public and private sectors to drive economic, environmental, and social progress.

References

[1]   Economic Planning Unit, *SDG Roadmap Phase 1: 2016-2020*, Economic Planning Unit (EPU), Malaysia, 2016.
[2]   International Financial Cooperation, *Who Cares Win 2004-2008*, International Financial Cooperation (IFC), 2004.
[3]   PwC Malaysia, *Corporate Malaysia's Journey Towards a Sustainable Supply Chain*, Capital Market Malaysia (CCM) and PwC Malaysia, April 2023.
[4]   Economic Planning Unit, *Twelfth Malaysia Plan 2021-2025 - A Prosperous, Inclusive, Sustainable Malaysia*, Economic Planning Unit (EPU), Malaysia, 2021.
[5]   MITI, *National Industry Environment, Social and Governance Framework (i-ESG framework)*, Ministry of Investment, Trade and Industry, 2023.
[6]   UNDP, *Malaysia's Update of Its First Nationally Determined Contribution*, United Nation Development Programme, July 2021.
[7]   Economic Planning Unit, *National Energy Policy 2022-2040*, Economic Planning Unit (EPU), Malaysia, 2022.
[8]   Ministry of Economy, *Malaysia Energy Transition Roadmap*, Ministry of Economy, 2023.

[9]   MNRE, *Fourth Biennial Update Report Under the United Nations Framework Convention on Climate Change*, Ministry of Natural Resources, Environment and Climate Change, Malaysia, December 2022.

[10]   Bursa Malaysia, *Sustainability Reporting Guide*, 3rd edition, Bursa Malaysia, 2023.

[11]   Bank Negara Malaysia, *Climate Change and Principle-based Taxonomy*, Bank Negara Malaysia, 2021.

[12]   Bank Negara Malaysia, *Financial Sector Blueprint 2022-2026*, Bank Negara Malaysia, 2022.

[13]   OECD, *Pricing Greenhouse Gas Emission - Country Notes (Malaysia)*, Organisation for Economic Co-operation and Development (OECD), 2022.

Chapter 5

ESG and green sustainable technology: catalysts for the next production revolution

Wai Yie Leong[1]

The global landscape of production is undergoing a transformative shift, driven by the imperative to address environmental, social, and governance (ESG) concerns. This chapter explores the intersection of ESG principles and green sustainable technology as catalysts for the next production revolution. By examining the evolving landscape of production and the role of ESG in shaping corporate strategies, we highlight how sustainable technologies are becoming pivotal in achieving both economic growth and environmental stewardship. This chapter also delves into case studies and examples to illustrate the successful integration of ESG principles and green technology in various industries.

5.1 ESG and its impact on corporate strategies

The next production revolution (NPR) is characterized by a paradigm shift in how goods and services are produced, emphasizing sustainability, efficiency, and social responsibility. ESG factors, encompassing environmental, social, and governance dimensions, have emerged as critical considerations for businesses aiming to thrive in this new era [1–3]. Green sustainable technology, driven by innovation and a commitment to sustainability, plays a central role in realizing the goals set by ESG principles, Figure 5.1.

ESG criteria have become instrumental in shaping corporate strategies across industries. The adoption of ESG principles reflects a paradigm shift in the way businesses perceive their role in society and the environment. In this section, we delve into the three dimensions of ESG and explore how each dimension influences corporate strategies.

5.1.1 Environmental dimension

The environmental dimension of ESG focuses on a company's impact on the planet and natural resources. Companies are increasingly recognizing the importance of

[1]Faculty of Engineering and Quantity Surveying, INTI International University, Malaysia

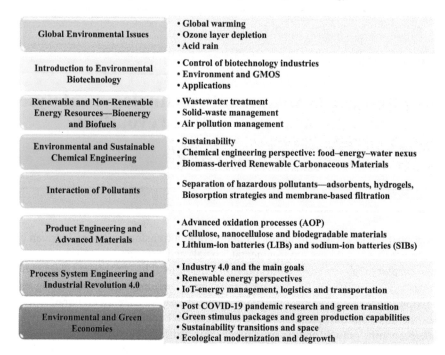

Global Environmental Issues	• Global warming • Ozone layer depletion • Acid rain
Introduction to Environmental Biotechnology	• Control of biotechnology industries • Environment and GMOS • Applications
Renewable and Non-Renewable Energy Resources—Bioenergy and Biofuels	• Wastewater treatment • Solid-waste management • Air pollution management
Environmental and Sustainable Chemical Engineering	• Sustainability • Chemical engineering perspective: food–energy–water nexus • Biomass-derived Renewable Carbonaceous Materials
Interaction of Pollutants	• Separation of hazardous pollutants—adsorbents, hydrogels, Biosorption strategies and membrane-based filtration
Product Engineering and Advanced Materials	• Advanced oxidation processes (AOP) • Cellulose, nanocellulose and biodegradable materials • Lithium-ion batteries (LIBs) and sodium-ion batteries (SIBs)
Process System Engineering and Industrial Revolution 4.0	• Industry 4.0 and the main goals • Renewable energy perspectives • IoT-energy management, logistics and transportation
Environmental and Green Economies	• Post COVID-19 pandemic research and green transition • Green stimulus packages and green production capabilities • Sustainability transitions and space • Ecological modernization and degrowth

Figure 5.1   Recent advances in green technology and Industrial Revolution 4.0 for a sustainable future [8]

environmental sustainability and are incorporating it into their strategic planning [3,4]. Key factors within this dimension include:

Carbon footprint reduction: Mitigating greenhouse gas emissions has become a cornerstone of ESG strategies. Corporations are investing in renewable energy sources, energy-efficient technologies, and sustainable practices to minimize their carbon footprint. Strategies encompass transitioning to cleaner energy sources, adopting energy-efficient processes, and implementing carbon offset initiatives.

Resource efficiency and circular economy: ESG-driven strategies emphasize the efficient use of resources and the adoption of circular economy principles. Corporations are reevaluating their supply chains, seeking sustainable sourcing of materials, and designing products with a focus on recyclability [5–7]. The shift toward a circular economy involves reducing waste, reusing materials, and promoting sustainable consumption.

Biodiversity conservation: Preserving biodiversity has gained prominence in ESG considerations. Corporate strategies are evolving to minimize negative impacts on ecosystems. Companies are incorporating biodiversity conservation into their business practices by avoiding deforestation, supporting conservation initiatives, and integrating biodiversity considerations into land-use planning.

5.1.2   Social dimension

The social dimension of ESG centers on a company's relationships with its employees, communities, and broader society. Corporations are increasingly recognizing the importance of social responsibility and are aligning their strategies with the following factors:

Stakeholder engagement: Engaging with a diverse set of stakeholders, including employees, customers, communities, and investors, is crucial for ESG strategies. Companies are incorporating stakeholder perspectives into decision-making processes, ensuring inclusivity and responsiveness to the diverse needs and expectations of different groups.

Diversity and inclusion: Promoting diversity and inclusion has become a key aspect of corporate strategies. Companies are recognizing the value of diverse perspectives and are implementing initiatives to foster inclusivity. This includes promoting diversity in leadership roles, addressing pay gaps, and creating inclusive workplaces that celebrate differences.

Labor practices and human rights: ESG-driven strategies prioritize fair and ethical labor practices. Companies are focusing on ensuring safe working conditions, fair wages, and adherence to human rights standards throughout their supply chains. This involves conducting due diligence to identify and address potential human rights risks within their operations and those of their suppliers.

5.1.3   Governance dimension

The governance dimension of ESG pertains to the structure and oversight of corporate decision-making. Companies are strengthening their governance structures to ensure transparency, accountability, and ethical conduct [8]. Key elements within this dimension include the following:

Ethical business conduct: Maintaining high ethical standards in business operations is a fundamental aspect of ESG governance. Companies are implementing robust ethical guidelines, codes of conduct, and whistleblower protection mechanisms to foster a culture of integrity and transparency.

Board diversity and independence: ESG strategies advocate for diverse and independent boards of directors. Companies are recognizing the value of diverse perspectives at the board level, which contributes to better decision-making and risk management. Independence ensures that boards can effectively oversee management and act in the best interests of shareholders.

Transparency and accountability: Transparency and accountability are paramount in ESG governance. Companies are enhancing their disclosure practices, providing stakeholders with comprehensive information about their ESG performance. This transparency builds trust and allows stakeholders to hold companies accountable for their environmental and social impacts.

The integration of ESG principles into corporate strategies represents a fundamental shift in how businesses perceive and manage their impact on the environment, society, and governance structures. Embracing ESG not only aligns businesses with societal values but also contributes to long-term resilience and sustainable growth. Companies that prioritize ESG considerations in their strategies are better positioned to navigate the challenges of the NPR while fostering positive societal and environmental outcomes.

5.2   Green sustainable technology in the next production revolution

The NPR is characterized by a transformative shift in manufacturing and production processes, driven by the imperative to achieve sustainability goals. Green sustainable technology plays a central role in shaping this revolution, enabling industries to reduce environmental impact, enhance efficiency, and meet the demands of a rapidly changing global landscape. In this section, we explore key areas of green sustainable technology that are instrumental in driving the NPR.

5.2.1   Renewable energy

One of the fundamental pillars of green sustainable technology in the NPR is the adoption of renewable energy sources. Companies are increasingly shifting away from traditional fossil fuels to embrace cleaner alternatives, contributing to a significant reduction in carbon emissions. Key components of renewable energy in the NPR include the following:

Solar and wind power: The widespread adoption of solar and wind power technologies has become a cornerstone of sustainable manufacturing. Solar panels and wind turbines are being integrated into production facilities to harness clean and abundant energy sources. This not only reduces dependence on non-renewable energy but also decreases the environmental footprint of manufacturing processes.

Energy storage solutions: To address the intermittent nature of renewable energy sources, advancements in energy storage solutions are crucial. Batteries and other energy storage technologies enable industries to store excess energy during peak production periods and utilize it during low production times or when renewable sources are unavailable. This enhances the reliability and stability of renewable energy integration in manufacturing.

5.2.2   Sustainable manufacturing

Sustainable manufacturing practices are at the core of the NPR, driven by innovative technologies that prioritize efficiency, waste reduction, and eco-friendly processes. Key components of sustainable manufacturing include the following:

3D printing and additive manufacturing: 3D printing and additive manufacturing technologies have revolutionized production processes by enabling the creation of intricate and customized designs with minimal waste. These technologies significantly reduce material consumption, energy requirements, and transportation-related emissions. They also open new possibilities for on-demand, localized production.

Advanced materials and eco-friendly processes: The development of advanced materials with reduced environmental impact is a key focus in sustainable manufacturing. Companies are investing in research and development (R&D) to create materials that are not only durable and efficient but also eco-friendly. Additionally, eco-friendly processes, such as water-based manufacturing and closed-loop systems, contribute to minimizing waste and resource consumption.

5.2.3   Smart and efficient supply chains

Green sustainable technology is transforming supply chain management, making it more efficient, transparent, and environmentally friendly. Technologies such as the Internet of Things (IoT), blockchain, and data analytics play vital roles in achieving sustainable supply chain practices, Figure 5.2. Key components of smart and efficient supply chains include the following:

IoT and blockchain integration: IoT sensors are deployed throughout the supply chain to gather real-time data on production processes, energy consumption, and transportation. This data is securely recorded and tracked using blockchain technology, enhancing transparency and traceability. The integration of these technologies helps identify inefficiencies, reduce waste, and ensure ethical sourcing.

Data analytics for optimization: Data analytics tools are employed to analyze vast amounts of information generated by production processes and supply chains. Machine learning algorithms identify patterns and inefficiencies, enabling

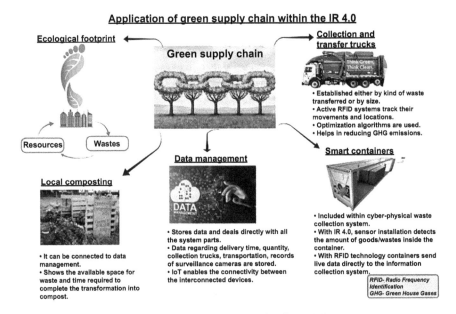

Figure 5.2   Green supply chain [8]

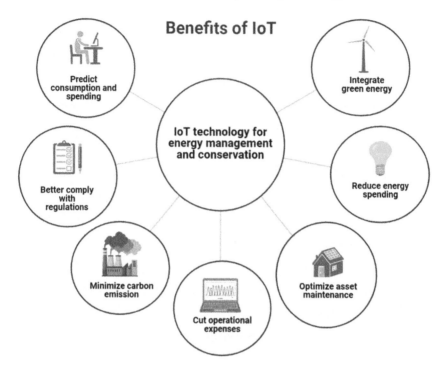

Figure 5.3   IoT in energy management and conservation [8]

companies to optimize operations, reduce energy consumption, and make data-driven decisions that align with sustainability goals.

By leveraging these green sustainable technologies, industries can not only reduce their environmental impact but also enhance operational efficiency and resilience. The integration of renewable energy, sustainable manufacturing practices, and smart supply chain technologies is pivotal in achieving the goals of the NPR, ushering in an era of environmentally responsible and economically viable production processes (Figure 5.3).

5.3   Advancing ESG and green sustainable technology in the automotive industry: a case study of Tesla, Inc.

This case study explores the integration of ESG principles and green sustainable technology in the automotive industry, with a focus on Tesla, Inc. [8,9]. As a pioneer in electric vehicles (EVs) and sustainable energy solutions, Tesla serves as an exemplary model for how ESG considerations and innovative technologies can be synergized to drive the NPR. The study delves into Tesla's commitment to ESG values, the incorporation of green sustainable technology in its products and operations, and the resulting impact on the automotive industry, Figure 5.4.

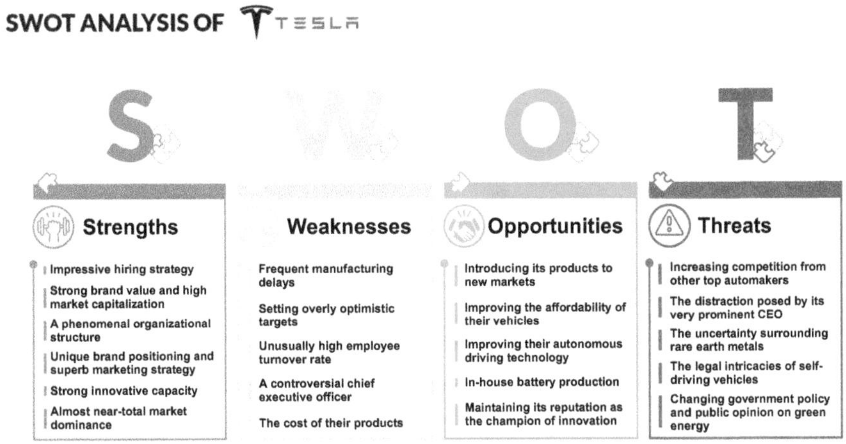

SWOT ANALYSIS OF T ᴛᴇꜱʟᴀ

Figure 5.4   *Tesla strengths, weaknesses, opportunities, and threats (SWOT)*
analysis [9]

The automotive industry is undergoing a profound transformation, marked by a shift toward sustainable practices and technologies. Tesla, Inc., founded by Elon Musk in 2003, has played a pivotal role in spearheading this transformation. The company's commitment to ESG principles and the integration of green sustainable technology in its electric vehicles exemplify the potential for positive change in the industry.

5.3.1   ESG integration at Tesla

5.3.1.1   Environmental dimension

Tesla's ESG strategy places a strong emphasis on environmental sustainability. The company's electric vehicles, powered by lithium-ion batteries, aim to reduce reliance on fossil fuels and minimize the carbon footprint associated with traditional combustion engines. Key initiatives within this dimension include the following:

Carbon footprint reduction: Tesla actively works toward reducing the carbon footprint of its manufacturing processes, supply chain, and product lifecycle. The company's Gigafactories are powered by renewable energy sources, and Tesla aims to achieve a carbon-neutral supply chain.

Resource efficiency: Tesla focuses on resource efficiency through the use of recyclable materials and sustainable sourcing practices. The company is committed to minimizing waste and employing closed-loop manufacturing processes.

5.3.1.2   Social dimension

Tesla's approach to the social dimension of ESG centers around innovation, employee well-being, and customer satisfaction:

Innovation and stakeholder engagement: Tesla engages with its stakeholders through a commitment to innovation, providing cutting-edge electric vehicles that have redefined the automotive industry. The company's emphasis on continuous improvement and responsiveness to customer feedback aligns with ESG principles.

Labor practices: Tesla has faced scrutiny regarding its labor practices, particularly worker safety and fair wages. The company has taken steps to address these concerns, implementing safety protocols and initiatives to improve workplace conditions.

5.3.1.3   Governance dimension

Tesla places a strong emphasis on governance principles to ensure transparency, accountability, and ethical conduct:

Ethical business conduct: The company has implemented robust ethical guidelines and a code of conduct to govern its business operations. Elon Musk's leadership is characterized by a commitment to transparency and integrity, aligning with good governance practices.

Board diversity and independence: Tesla's board of directors includes a diverse group of individuals, and efforts are ongoing to enhance diversity and independence. The company recognizes the value of diverse perspectives in decision-making.

5.4   Green sustainable technology in Tesla's electric vehicles

5.4.1   Electric vehicle innovation

Tesla's electric vehicles represent a paradigm shift in the automotive industry. The company's focus on electric propulsion systems, advanced battery technology, and energy-efficient design has positioned Tesla as a leader in the green sustainable technology space.

Battery technology: Tesla's advancements in battery technology, including the development of the Gigafactory to produce lithium-ion batteries at scale, contribute to the widespread adoption of electric vehicles. The company's commitment to battery recycling further aligns with sustainable practices.

Autopilot and software updates: Tesla integrates smart and sustainable technologies into its vehicles, such as Autopilot and over-the-air software updates. These innovations enhance vehicle efficiency, safety, and longevity.

5.4.2   Impact on the automotive industry

Tesla's success in integrating ESG principles and green sustainable technology has had a profound impact on the automotive industry:

Market disruption: Tesla's innovative approach to electric vehicles has disrupted traditional automakers, compelling them to accelerate their own efforts in electrification and sustainability.

Industry-wide shift: Tesla's success has prompted a broader industry-wide shift toward electric and sustainable vehicles. Competitors are investing heavily in R&D to compete in this evolving market.

5.4.3 Challenges and future considerations

Despite its successes, Tesla faces challenges related to scalability, supply chain sustainability, and regulatory scrutiny. Ongoing efforts to address these challenges will be crucial as the industry continues to evolve.

Tesla's journey serves as a compelling case study of how ESG principles and green sustainable technology can drive the NPR in the automotive industry. By prioritizing environmental sustainability, innovating in green technology, and addressing social and governance considerations, Tesla has not only transformed its business but has also catalyzed a broader shift toward a more sustainable and responsible automotive industry. The case study highlights the importance of visionary leadership, technological innovation, and a holistic approach to ESG integration in shaping the future of production and industry.

5.5 Challenges and opportunities of ESG and green sustainable technology: catalysts for the next production revolution

The integration of ESG principles and green sustainable technology into the NPR presents both challenges and opportunities for businesses [9–11]. This section explores key challenges and opportunities associated with this transformative process, Figure 5.5.

5.5.1 Challenges

Regulatory uncertainty: The lack of standardized and consistent global regulations for ESG reporting and sustainable practices poses a challenge for companies aiming to align with ESG principles.

High initial costs: Implementing green sustainable technologies often requires substantial upfront investments, posing financial challenges for companies, especially smaller enterprises.

Technological complexity: The adoption of advanced sustainable technologies may be hindered by the complexity of integrating new systems into existing production processes, requiring significant training and adjustment.

Supply chain complexity: Ensuring sustainability throughout the entire supply chain can be challenging, especially when dealing with global networks. Identifying and mitigating environmental and social risks across the supply chain requires robust monitoring and management.

Consumer awareness and adoption: Despite growing awareness, consumer adoption of sustainable products and technologies may still face resistance due to factors such as cost, habits, and perceived inconveniences.

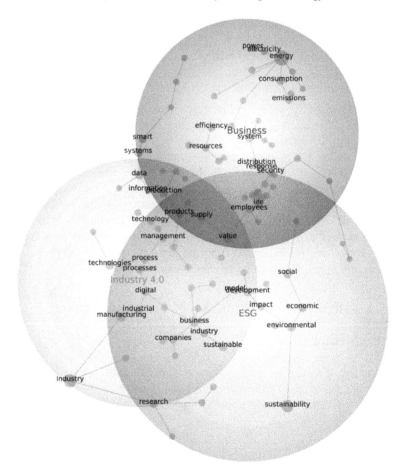

Figure 5.5   The impact factors of Industry 4.0 on ESG in the energy sector

5.5.2   Opportunities

Innovation and competitive advantage: Companies that invest in green sustainable technology can gain a competitive advantage by fostering innovation. Sustainable practices can attract environmentally conscious consumers and position companies as industry leaders.

Cost savings in the long run: While there are initial investment costs, adopting green technologies often leads to long-term cost savings through energy efficiency, waste reduction, and improved resource management.

Access to capital: Companies with robust ESG practices and sustainable technologies may find it easier to attract investments and secure favorable financing terms as investors increasingly prioritize sustainability.

Market expansion: The rising demand for sustainable products provides opportunities for companies to expand their market share and reach new customer segments interested in environmentally friendly options.

Positive brand image: Companies that successfully integrate ESG principles and green technology can build a positive brand image, enhancing customer loyalty and attracting socially conscious consumers.

5.6 Regulatory landscape and its influence on ESG adoption and green technology development

The regulatory landscape plays a significant role in shaping the adoption of ESG principles and the development of green technology [12–15], Figure 5.6. Here's how regulations influence these aspects:

Mandatory reporting requirements: Many jurisdictions have implemented mandatory reporting requirements for ESG factors, requiring companies to disclose information on their environmental impact, social policies, and governance

THE NEW GREEN DEAL GOALS AND STRATEGIES FOR
SUSTAINABLE AND ECO-FRIENDLY TRANSITION

BIO-ENERGY	BIO-DEFENSE	BIO-CITIES

→ Microalgae biomass.
→ Nanocellulose and other wood products.
- Bio-based construction materials.
→ Use of organic waste in biorefineries.
→ Bamboo-based textiles.
→ Bioplastics.

BIO-ENERGY	BIO-DEFENSE	BIO-CITIES
→ Biofuel production from domestic waste. → Sustainable trade. → Bio-based integrated power production. → Reduce carbon footprint and GHG emissions. → Carbon pricing. → Solar energy.	→ One health policy. → Proper disposal of medical waste. → Biopharmaceuticals. → Adoption of healthier diets by consuming plant-based meats and proteins. → Cooking healthier meals and proper storage.	→ Urban resilience → Regenerative urban development. → Circular economy. → Urban mobility—car pooling, public transport. → Green housing—indoor health and well-being. → Green spaces—parks.

AGRICULTURE	BIODIVERSITY	BIO-ECONOMICS	GREEN HOUSING
→ Foster organic farming. → Hydroponics. → Multitrophic aquaculture. → Sustainable forest management. → Microalgae production. → Promote traditional livestock farming.	→ Agroecology and regional bioeconomy. → More areas for national parks and wildlife sanctuaries. → Reforestation. → Replenish fisheries, conserve biodiversity. → Re-establish organic carbon and microbiota in soil and land.	→ Bio-based local public transport. → Circular and sustainable economy. → Green infrastructure projects. → Low carbon economy and green entrepreneurship. → Green artificial intelligence, smart cities.	→ Sustainable housing. → Eco-friendly and safe quality choice of building materials. → Green indoor environment. → Good air ventilation facilities. → Green open spaces. → Good hygiene and sanitation.

Figure 5.6   The Green New Deal, goals, and strategies for sustainable and eco-friendly transition

practices. These regulations compel companies to integrate ESG considerations into their operations and decision-making processes.

Financial disclosure regulations: Regulatory bodies may require companies to disclose the financial implications of ESG risks and opportunities. This encourages companies to assess and manage ESG-related risks, leading to more sustainable business practices and investment decisions.

Incentives and subsidies: Governments often provide incentives and subsidies to promote the development and adoption of green technologies. These incentives can take the form of tax credits, grants, or subsidies for R&D in sustainable technologies. Such measures encourage innovation and investment in green technology solutions.

Environmental standards and targets: Regulatory agencies set environmental standards and targets to limit pollution, conserve natural resources, and mitigate climate change. Compliance with these standards may necessitate the adoption of green technologies and sustainable practices, driving innovation and investment in environmentally friendly solutions.

Carbon pricing mechanisms: Carbon pricing mechanisms, such as carbon taxes or cap-and-trade systems, create financial incentives for companies to reduce their carbon emissions. By internalizing the cost of carbon pollution, these mechanisms encourage businesses to invest in cleaner technologies and implement emissions reduction measures.

Product labeling and certification: Regulatory bodies may establish product labeling and certification schemes to inform consumers about the environmental and social attributes of products. Compliance with these schemes often requires companies to adhere to specific ESG criteria and standards, influencing their product development and supply chain practices.

Stakeholder engagement requirements: Some regulations mandate stakeholder engagement processes, requiring companies to consult with communities, Indigenous groups, and other stakeholders affected by their operations. This fosters transparency, accountability, and social responsibility, driving companies to consider the social and environmental impacts of their activities.

International agreements and treaties: International agreements and treaties, such as the Paris Agreement on climate change, set global targets and commitments for addressing environmental challenges. Signatory countries may enact domestic regulations to fulfill their obligations under these agreements, creating a supportive regulatory environment for ESG adoption and green technology development [16,19].

The regulatory landscape exerts a significant influence on ESG adoption and green technology development by setting standards, providing incentives, and shaping market dynamics, Figure 5.7. Compliance with regulations often becomes a driving force for companies to integrate sustainability into their business strategies and invest in innovative green solutions [14,15,17,18].

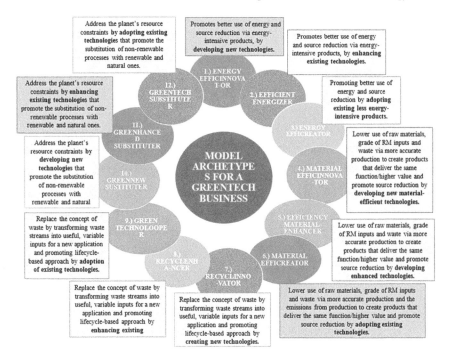

Figure 5.7   The model archetypes for a green technology (GT) business

5.7   Conclusions

As businesses navigate the NPR, the integration of ESG principles and green sustainable technology is imperative for long-term success. This chapter underscores the interdependence of ESG and sustainable technologies, providing a roadmap for businesses to contribute to economic development while preserving the environment and promoting social well-being. The examples presented demonstrate that the synergy between ESG and green technology is not only feasible but also essential for a sustainable and prosperous future.

The integration of ESG principles with green sustainable technology stands as a pivotal force propelling the forthcoming production revolution. This synthesis of ESG considerations and sustainable technologies not only holds the promise of fostering corporate resilience but also emerges as a cornerstone for sustainable development in the industrial landscape.

ESG principles have become central to corporate strategy, with companies increasingly recognizing their significance in driving long-term value creation. By incorporating ESG factors into decision-making frameworks, organizations can navigate environmental challenges, enhance social responsibility, and uphold robust governance practices. Moreover, empirical evidence underscores the positive correlation between strong ESG performance and financial returns, underlining the imperative for businesses to embrace sustainability as a strategic imperative.

Concurrently, the advent of green sustainable technology heralds a transformative era in industrial production. Innovations across various sectors, including renewable energy, clean transportation, and circular economy solutions, offer pathways toward decarbonization and resource efficiency. Through the adoption of green technologies, businesses can mitigate their environmental footprint, reduce operational costs, and capitalize on emerging market opportunities driven by sustainability trends.

Crucially, the convergence of ESG principles and green technology amplifies their collective impact, fostering synergies that drive systemic change. Companies that align ESG goals with sustainable technology strategies are better positioned to capitalize on market shifts, mitigate risks, and unlock new avenues for growth. Case studies underscore the efficacy of this integrated approach, showcasing how organizations leverage sustainability as a source of competitive advantage while contributing to broader societal goals.

However, the realization of this vision is not without its challenges. Regulatory frameworks, while instrumental in fostering ESG adoption and green technology development, must evolve to address emerging complexities and ensure a level playing field. Moreover, barriers such as capital constraints, technological limitations, and resistance to change pose significant hurdles to widespread adoption.

Nonetheless, amidst these challenges lie opportunities for collaboration, innovation, and collective action. Stakeholders across the public and private sectors can leverage partnerships, investment incentives, and knowledge-sharing platforms to accelerate progress toward a more sustainable future. By harnessing the transformative potential of ESG and green sustainable technology, businesses can catalyze the NPR while advancing the imperative of global sustainability.

References

[1]   Elkington, J. (1997). *Cannibals with Forks: The Triple Bottom Line of 21st Century Business*. New Society Publishers.

[2]   Eccles, R. G., Ioannou, I., and Serafeim, G. (2011). The impact of corporate sustainability on organizational processes and performance. *Management Science*, 60(11), 2835–2857.

[3]   World Economic Forum. (2019). *Fourth Industrial Revolution: Beacons of Technology and Innovation in Manufacturing*. White Paper. Geneva, Switzerland: World Economic Forum. https://www3.weforum.org/docs/WEF_4IR_Beacons_of_Technology_and_Innovation_in_Manufacturing_report_2019.pdf.

[4]   Schaltegger, S., and Burritt, R. (2018). Business cases and corporate engagement with sustainability: Differentiating ethical motivations. *Journal of Business Ethics*, 147(2), 241–259.

[5]   Porter, M. E., and Kramer, M. R. (2011). Creating shared value. *Harvard Business Review*, 89(1/2), 62–77.

[6]   Global Reporting Initiative (GRI). (2016). *GRI Standards*.

[7]   International Integrated Reporting Council (IIRC). (2013). The International IR Framework.

[8] Bradu, P., Biswas, A., Nair, C. *et al.* Recent advances in green technology and Industrial Revolution 4.0 for a sustainable future. *Environmental Science and Pollution Research* 30, 124488–124519 (2023). https://doi.org/10.1007/s11356-022-20024-4.

[9] D. Pereira, *Tesla SWOT Analysis*, The Business Model Analyst, 2023.

[10] W. Y. Leong, Y. Z. Leong, and W. S. Leong, "ESG imperatives for healthcare sustainability," In *IEEE 6th Eurasia Conference on Biomedical Engineering, Healthcare and Sustainability (IEEE ECBIOS 2024)*, Taiwan, 2024.

[11] W. Y. Leong, Y. Z. Leong, and W. S. Leong, "Green building initiatives in ASEAN countries," In *Asia Meeting on Environment and Electrical Engineering (EEE-AM)*, Hanoi, Vietnam, 2023, pp. 1–6, doi: 10.1109/EEE-AM58328.2023.10395261.

[12] W. Y. Leong, "Investigating and enhancing energy management policy and strategies in ASEAN," In *Asia Meeting on Environment and Electrical Engineering (EEE-AM)*, Hanoi, Vietnam, 2023, pp. 1–6, doi: 10.1109/EEE-AM58328.2023.10395060.

[13] W. Y. Leong, Y. Z. Leong and W. S. Leong, "Accelerating the energy transition via 3D printing," In *Asia Meeting on Environment and Electrical Engineering (EEE-AM)*, Hanoi, Vietnam, 2023, pp. 1–5, doi: 10.1109/EEE-AM58328.2023.10395773.

[14] Leong W. Y., and Kumar R. 5G intelligent transportation systems for smart cities. In *Convergence of IoT, Blockchain, and Computational Intelligence in Smart Cities* 2023 (pp. 1–25). CRC Press.

[15] Leong W. Y., Heng L. S., and Leong W. Y. "Malaysia renewable energy policy and its impact on regional countries". In *Proc. 12th International Conference on Renewable Power Generation (RPG 2023)* 2023 October 14 (Vol. 2023, pp. 7–13). IET.

[16] Jain V. and Yie L. W. *Convergence of IoT, Blockchain, and Computational Intelligence in Smart Cities*. Kumar R, Teyarachakul S, editors. CRC Press; 2023.

[17] W. Y. Leong, Y. Z. Leong and W. S. Leong, Green Policies and Programmes in ASEAN, The 9th scientific conference on "Applying new Technology in Green Buildings, August 2024, Da Nang City, Vietnam.

[18] W. Y. Leong, Y. Z. Leong and W. S. Leong, Green Communication Systems: Towards Sustainable Networking, The 5th International Conference on Information Science, Parallel and Distributed Systems (ISPDS 2024), 2024.

[19] W. Y. Leong, Y. Z. Leong and W. S. Leong, Blockchain Technology in Next Generation Energy Management System, The 2024 7th International Conference on Green Technology and Sustainable Development (GTSD), 2024, Ho Chi Minh City.

Chapter 6

Medical equipment engineering in ESG and green sustainable technology

Sumit Majumder[1]

6.1 Introduction

Humankind in today's world faces a plethora of unprecedented challenges across environmental, public health, financial, and social domains. From the rapid deterioration of the environment and climate situation driven by greenhouse gas emissions to the persistent threat of severe public health crises such as Ebola, H1N1 flu, and the recent COVID-19 situation, the world continues to face significant challenges in terms of environmental and human well-being. Simultaneously, economic instability and widening social disparities [1] further compound these pressing concerns, highlighting the urgent need for concerted action and innovative solutions to address environmental and social issues. In response, the United Nations introduced the Sustainable Development Goals (SDGs) in September 2015, aiming to tackle environmental, social, and economic challenges over the next 15 years [2].

Nevertheless, these challenges have reshaped investment and operational strategies, driving organizations to adopt environmental, social, and governance (ESG) principles, which not only promote responsible and ethical practices, and accountability but also enhance business sustainability and performance and drive innovations in sustainable finance models that align with market needs, demands, and stakeholder expectations [3]. The healthcare sector, being one of the large contributors to the global economy with nearly 10% of global GDP in 2018 [4], must integrate ESG and green sustainable practices to address these emerging challenges effectively.

The impact of ESG and green sustainable technologies in medical equipment engineering and healthcare cannot be overstated. As global consciousness regarding environmental impact, social responsibility, and ethical governance grows, these principles can potentially shape the entire healthcare ecosystem that transcends conventional boundaries. ESG principles focus on an organization's

[1]Department of Biomedical Engineering, Chittagong University of Engineering and Technology, Bangladesh

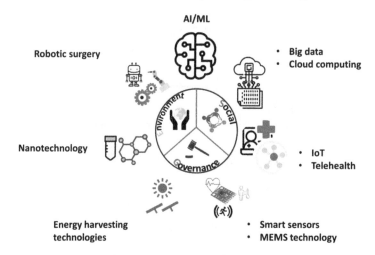

Figure 6.1   Key technologies impacting ESG and sustainability goals

ecological impact, societal contributions, and commitment to transparency and ethical practices, guiding it toward a more responsible, accountable, and sustainable future whereas green technologies serve as the driving force behind this evolution. Furthermore, with the evolution of advanced technologies (see Figure 6.1) such as artificial intelligence (AI) [5,6] robotics and automation [7,8], the Internet of Things (IoT) [9,10] smart sensors [11,12], and Big Data analytics [13–15], the healthcare landscape is undergoing a transformative shift. These innovations have the potential to enhance the precision, efficiency, accessibility, and inclusivity of healthcare while minimizing environmental footprint and resource utilization and ensuring transparency and ethical governance.

In this chapter, we explore how advanced technologies may contribute to achieving ESG and sustainability goals in healthcare and medical equipment engineering domains. We will discuss specific areas where medical equipment engineering intersects with ESG and green sustainable technology, exploring how these advancements can potentially reshape the healthcare landscape and foster patient-centric, ethical, and environmentally responsible medical practices. We will highlight how leading players in various domains of the healthcare sector are spearheading ESG and sustainable initiatives. Subsequently, we will discuss key challenges and considerations associated with implementing ESG practices. Finally, the chapter will conclude by outlining the future possibilities in this sector.

6.2   Key technologies impacting ESG and sustainability goals in healthcare

6.2.1   Artificial intelligence and machine learning

AI and machine learning (ML) technologies can potentially shape the future of healthcare and medical device engineering, offering unique avenues to address

ESG concerns. AI can enable sustainable production of medical devices [16] by optimizing manufacturing processes, potentially reducing waste, landfill, energy consumption, and overall environmental impact. Furthermore, AI and ML technologies can use data from the devices and sensors to predict equipment malfunctions and failures. This enables proactive maintenance measures, ensuring device reliability, extended lifespan, and contributing to resource efficiency.

The International Medical Device Regulators Forum (IMDRF) coined the term "Software as a Medical Device" (SaMD), which has later been adopted by the Food and Drug Administration (FDA) to refer to any software used for medical purposes without relying on a hardware medical device [5,17]. According to the FDA, any AI/ML-based software that is intended to "treat, diagnose, cure, mitigate, or prevent disease or other conditions" is considered a medical device [18]. Therefore, AI can enable developing cost-effective and accessible SaMDs tailored to diverse patient needs, thus addressing social disparities in healthcare access, and promoting inclusivity. Additionally, AI-powered remote monitoring solutions have the potential to enhance patient care beyond traditional healthcare settings, particularly benefiting remote or underserved communities.

In addition, AI/ML technologies can assist medical device manufacturers in navigating complex regulatory landscapes to facilitate adherence to evolving standards and ensure devices meet ethical and legal requirements. Furthermore, AI technologies can be used to establish robust data security protocols [19,20], which are essential for addressing governance concerns related to patient data. By automating and enhancing data encryption processes, these technologies can help safeguard against unauthorized access, thereby enhancing data security in medical devices. Moreover, AI algorithms can analyze normal user behavior and detect anomalies and potential threats to data breaches.

Besides, AI and ML technologies can be used to customize medical devices based on patients' unique health profiles, which provide opportunities to improve personalized medicine and tailor effective therapeutic strategies [21,22], while ensuring compliance with ethical standards. This approach potentially paves the way for innovation while minimizing resource consumption, waste, and energy usage. Moreover, AI can optimize supply chain processes, leading to efficient procurement and reduced environmental footprints associated with resource extraction and transportation (Figure 6.1) [23,24].

6.2.2 *Big data analytics and cloud computing*

Big data analytics can potentially play a crucial role in optimizing resource utilization and reducing energy consumption and waste generation in healthcare facilities. By analyzing data on energy usage, waste production, and resource allocation, hospitals can minimize their environmental footprint. The healthcare industry can benefit from emerging technologies like data science, Big Data analytics, and cloud computing to lower operating costs [25]. In addition, the evolution of cloud computing and energy-efficient data centers offers an opportunity for cost-effective green IT adoption in hospitals. Technologies such as virtual servers, virtual data

storage, and private cloud computing can reduce equipment and system management costs, thus enhancing savings and promoting sustainability efforts [25].
Additionally, data science and analytics can facilitate collaboration within the
supply chain, thus improving coordination between departments and assisting in
decision-making for environmental initiatives [14].

Besides, Big Data analytics has the potential to drive social innovation in
healthcare, by leveraging data from wearables and other sources, including clinical
data and behavioral patterns [26,27], to enhance the quality of life and social
relations. Big data analytics enables the analysis of large datasets, which can help
individuals manage their health, address healthcare challenges, and facilitate personalized medicine with tailored treatment plans [26]. It also can promote preventive care through predictive analytics by enabling early identification and
intervention in the case of potential health issues before they escalate [28,29].
Furthermore, with the capability of analyzing large datasets, Big Data analytics can
facilitate improved public health management by identifying trends, patterns, and
risk factors within communities [13,15].

Big data analytics can enhance governance in healthcare by enabling better
decision-making processes and ensuring compliance with regulatory requirements
[30]. By analyzing data associated with healthcare practices, outcomes, and regulatory compliance, organizations can identify areas for improvement, mitigate
risks, and ensure accountability. The abundant and diverse healthcare data can be
leveraged by Big Data analytics to create effective policies, improve management,
and deliver more efficient services to patients [28]. Moreover, organizations can
make informed decisions to streamline operations, reduce patient wait times, and
optimize resource utilization by analyzing historical data, patient flow patterns, and
demand forecasting, ultimately leading to enhanced patient experiences and cost
savings. Additionally, by analyzing large volumes of claims data using advanced
techniques like anomaly detection and social network analysis [31,32], Big Data
analytics can identify suspicious patterns, potential fraud schemes, and high-risk
areas, thus reducing financial losses and strengthening the integrity of healthcare
systems [33,34].

6.2.3   IoT, telemedicine, and remote patient monitoring

The IoT technology enables real-time monitoring of equipment and facilities,
which helps optimize resource utilization like energy, water, and medical supplies.
This ultimately leads to a reduction in environmental impact in terms of waste and
energy consumption, thus contributing to sustainability efforts [35]. Additionally,
IoT has the potential to enable remote health monitoring, diagnostics, and telemedicine [10,36], which in turn can reduce the carbon footprint by minimizing
transportation-related greenhouse gas (GHG) emissions [37,38] for patients and
healthcare providers.

Moreover, telemedicine and remote patient monitoring improve healthcare
access, particularly for underserved or remote communities, thus reducing disparities in healthcare and medical services [10,36]. IoT allows patients to actively

engage in their healthcare management by continuously monitoring vital signs and health metrics, thus promoting autonomy and control [39,40].

However, IoT devices used in healthcare must adhere to strict and robust governance frameworks that include stringent data security and privacy standards to safeguard patient information from cyber threats and unauthorized access [41,42]. It is also crucial to comply with regulatory requirements to meet ethical and legal standards and ensure transparency, accountability, and responsible use of technology in healthcare [9,42].

6.2.4 Smart sensors and MEMS technology

Smart sensors are modern sensing devices that come with built-in processing capabilities, communication interfaces, and sometimes even artificial intelligence algorithms [43]. Unlike traditional sensors, smart sensors not only detect and measure physical parameters but also possess the ability to process and analyze data locally. They can also communicate with other devices or systems, often wirelessly, to transmit data or receive commands. Smart sensors are integrated into wearable devices, such as fitness trackers and smartwatches, to continuously monitor vital signs like heart rate [12], blood pressure, and oxygen levels [39].

In addition, micro-electro-mechanical systems (MEMS) sensors find extensive use in healthcare for motion sensing, orientation detection, and environmental monitoring. MEMS sensors like accelerometers are used to measure movement, while pressure sensors can be used to monitor heart rate [44]. Together, smart sensors and MEMS technologies enable remote patient monitoring, telehealth consultations, and personalized treatment plans. This not only enables early detection of health issues but also promotes more accessible and efficient healthcare services. Moreover, smart environmental sensors can provide real-time environmental data, including temperature, humidity, and air quality, within healthcare facilities. This helps optimize energy usage and reduce carbon emissions, thus contributing to environmental sustainability.

6.2.5 Energy harvesting technologies

Integrating energy harvesting technologies (EHTs) into healthcare infrastructure and monitoring technology can have a significant impact on the ESG factors within the industry. EHT utilizes renewable energy sources such as solar, thermal, or kinetic energy, thereby reducing the environmental impact of healthcare facilities. By generating clean energy onsite, healthcare facilities can lower their dependence on fossil fuels and carbon emissions and mitigate environmental degradation [45]. Additionally, by reducing energy costs and operational expenses, healthcare institutions can allocate resources more effectively toward patient care, infrastructure development, and staff training, while enhancing financial sustainability and transparency in governance.

Moreover, EHT can be integrated into wearables to harness ambient energy from renewable sources [46]. EHT-powered wearables are less dependent on traditional portable power sources such as disposable batteries, thus reducing the environmental footprint and electronic waste. This promotes environmental sustainability and also helps to reduce the cost of wearables in the long run.

Energy harvesting technologies allow for decentralizing energy production and reducing reliance on centralized power grids [47,48]. As a result, healthcare facilities can operate independently and efficiently, even in off-grid locations. This improves access to healthcare services, particularly in remote areas with limited infrastructure or unreliable energy supply, promoting social equity and inclusivity. When coupled with modern communication technologies, EHT-powered wearables enable long-term telemonitoring of people's health without frequent battery replacements, thus making healthcare more accessible to individuals from underserved and remote communities [49,50].

6.2.6   Nanotechnology

Nanotechnology has significant potential in healthcare for advanced health monitoring, diagnostics, drug delivery, and therapeutics. Nanotechnology enables the fabrication of lightweight, wearable, and low-cost sensors that can detect various biochemicals in body fluids, such as electrolytes and metabolites, making precise monitoring of health parameters possible [51]. Nanotechnology also improves drug solubility [52], bioavailability, and targeted delivery to specific cells or tissues [53], which ultimately improves treatment efficacy while minimizing side effects. Furthermore, nanoparticles provide ample surface area for the functionalization of reporter groups and surface ligands, which makes it possible to enable site-specific localization and stimuli-responsive behavior. This, in turn, facilitates the development of advanced high-contrast imaging techniques for early disease detection and precise monitoring of treatment responses [54,55].

Nanotechnology has made it possible to fabricate smaller and more efficient biochemical sensors and electronic devices. This has led to a reduction in material consumption and waste generation [56]. Furthermore, it facilitates the development of lightweight and energy-efficient [51] portable medical devices, wearable sensors, energy harvesters [57], and drug delivery systems, thereby contributing to energy conservation and a reduced carbon footprint. These innovations in portable medical devices and wearable sensors enable remote health monitoring [10,36,51], potentially reducing health-related travel and associated emissions. Additionally, the green synthesis of nanoparticles [58] allows for eco-friendly manufacturing of biosensors, which helps minimize waste and harmful chemicals thereby promoting environmental sustainability by reducing pollution.

Nanotechnology enables the development of portable and cost-effective medical devices and sensors and thus enhances healthcare accessibility, especially in underserved and remote areas [10,36]. Nanomaterials-based drug delivery systems and diagnostic tools improve treatment outcomes and early disease detection, which helps to enhance patient care and quality of life. By providing affordable and accessible healthcare solutions, nanotechnology can help reduce healthcare disparities.

6.2.7   Robotics and automation in surgery

Robot-assisted surgeries were reported to have a significant environmental impact due to their high energy requirements for operation and maintenance [59,60], as well as the resulting increase in greenhouse gas emissions [7] and waste production [59,61], thus resulting in a higher carbon footprint [59]. However, the use of remanufactured single-use medical devices in robotic surgery can potentially reduce complex waste [7]. In addition, robotic tools allow surgeons to perform surgery with greater precision, which minimizes tissue damage and subsequently reduces medical waste [62]. It is also crucial to design energy-efficient systems to mitigate energy consumption. Moreover, by streamlining surgical procedures, using less GHG-emitting substances such as sevoflurane, and propofol for anesthesia, reducing surgical duration through robotic assistance, and reusing equipment, the carbon footprint associated with robot-assisted surgeries can be reduced [7].

Robotic-assisted surgery has shown great potential in enhancing precision, control, and safety during surgical procedures. This can lead to better patient outcomes such as reduced post-surgical complications, shorter hospital stays, and faster recovery times [63]. This, in turn, contributes to improved patient satisfaction and quality of life, while making procedures more efficient and cost-effective [64]. Importantly, robotic surgery can help bridge healthcare disparities for underserved populations and thus improve healthcare accessibility by enabling advanced surgical care offered remotely [65,66].

Nevertheless, specialized training and certification requirements are crucial to ensure standardized surgical techniques, which can potentially minimize outcome variability across healthcare facilities [67,68]. Additionally, robotic surgery procedures must undergo rigorous regulatory oversight to ensure patient safety and efficacy. Furthermore, the adoption of robotics in surgery raises ethical concerns regarding patient autonomy, informed consent, and equitable access to care [62]. Therefore, a transparent governance framework is essential to harness the benefits of robotic surgery while minimizing potential risks and addressing ethical concerns.

6.3   ESG practice in healthcare industry

6.3.1   Remote monitoring and diagnosis

Advancements in modern communication, computing, and sensor technology have paved the way for remote health monitoring and diagnosis [10,36,69,70]. These sustainable alternatives align seamlessly with the ESG principles. These transformative solutions can potentially integrate AI and the IoT to enhance diagnostic capabilities. AI algorithms can analyze extensive patient data, providing timely insights and personalized healthcare recommendations. This can improve diagnostic accuracy and contribute to a more sustainable healthcare system by reducing unnecessary tests and treatments.

Traditional healthcare typically requires patients to visit healthcare facilities frequently, which contributes to increased carbon emissions due to transportation and energy consumption. However, remote health monitoring allows patients to receive healthcare services from the comfort of their homes, potentially making them more interested in receiving healthcare remotely. For instance, at the onset of the COVID-19 pandemic, there was a big increase in using telehealth services at the US Department of Veterans Affairs (VA) [71]. Remote health monitoring thus reduces the need for unnecessary travel, which helps to lower carbon footprints [72,73], and thus aligns with ESG goals.

It also promotes inclusivity by overcoming geographical barriers. This ensures individuals in remote or underserved areas have access to quality healthcare. This approach is helpful in addressing social equity concerns, democratizing healthcare services, and fostering a more inclusive system, especially for the elderly or those with mobility challenges. In the United Kingdom, the National Health Service (NHS) has initiated programs to deliver healthcare remotely over the digital platform as a part of the Regional Scale Programme [74]. In the United States, similar telehealth services are provided by initiatives such as "Remote Area Medical" (RAM) clinics [75], and Florence Health, which particularly focuses on managing chronic conditions from a distance [76]. This approach not only reduces the burden on healthcare resources but also harmonizes with sustainability goals. Furthermore, these technologies allow for real-time data collection, which enables efficient resource allocation, prioritizing high-risk patients and proactively addressing potential health issues. This enhances overall care quality and contributes to responsible resource management.

By facilitating continuous health monitoring, remote health monitoring enables early diagnosis. This method of healthcare delivery promotes a preventive approach and helps in alleviating the burden on healthcare systems.

Companies such as Philips have developed programs like "Philips Virtual Care Management" that use remote monitoring for chronic disease management, promoting a sustainable model centered on preventive care [77,78]. This program empowers patients with chronic conditions to monitor their health remotely, thus reducing the need for frequent hospital visits.

Remote health monitoring and diagnosis represent a paradigm shift in healthcare delivery, which offers sustainable alternatives aligned with ESG goals. By embracing these technologies, healthcare providers can improve patient outcomes, and increase efficiency, while also mitigating environmental impact, fostering social inclusivity, and promoting a more sustainable and responsible approach to healthcare.

6.3.2 *Personalized medicine*

Personalized medicine refers to the customization of medical treatment based on the attributes of each patient. This approach relies on the understanding of an individual's molecular and genetic profile, which can influence their susceptibility to specific diseases [79]. By integrating ESG principles and sustainability practices into the development of personalized medicine it is possible not only to improve patient outcomes but also to effectively address ethical, social, and environmental concerns.

Traditionally, pharmaceutical manufacturing processes are resource-intensive and generate significant waste. However, personalized medicine holds promise to offer a more environmentally friendly solution due to its targeted approach tailored to individual patient needs. Nevertheless, it is important to proactively address and minimize the environmental impacts associated with personalized medicine manufacturing [80]. In response to a growing awareness, ESG initiatives are pushing the pharmaceutical industry toward more sustainable practices. For instance, Roche, a leading healthcare company, has committed to reducing its environmental footprint [81] by adopting eco-friendly strategies such as phasing out potent GHGs such as halogenated hydrocarbons, utilizing heat reuse for building heating, implementing energy-saving action plans, and incorporating sustainable energy sources like solar panels, windmills, or hydropower plants. These efforts aim to actively contribute to fostering a more sustainable pharmaceutical sector.

ESG principles also highlight the importance of social responsibility and advocate for accessible healthcare for all. Personalized medicine, which focuses on individual patient needs, inherently promotes equity and inclusivity. This approach can potentially reduce healthcare costs by using individual genetic or biological information to improve and diagnose diseases earlier. Furthermore, it can facilitate the implementation of cost-effective preventive measures, mitigating disease risk and promoting inclusive, sustainable, and efficient use of therapeutic options [82].

For example, The All of Us Research Program, funded by the NIH in the United States, collects diverse health data from one million participants, with the aim of establishing a foundation for leveraging diversity in precision medicine research [83]. The participants include racial and ethnic minorities, those with limited medical access, individuals under 18 or over 65, with an annual household income at or below 200% of the federal poverty level, those with cognitive or physical disabilities, less than a high school education, intersex individuals, sexual or gender minorities, or those in rural areas. This effort aims to address health disparities and ensure social equity in healthcare.

Governance in the domain of personalized medicine involves ethical considerations and the establishment of regulatory frameworks regarding data privacy and the responsible use of genetic information [84]. An illustrative case is 23andMe, a direct-to-consumer genetic testing company, where a security breach in 2023 allowed hackers to access 14,000 user accounts, revealing their detailed health information [85]. Nevertheless, 23andWe, the research wing of 23andMe, conducts research on genetic data derived from customer samples, along with follow-up surveys about health and lifestyle [86]. However, considering recent security concerns and past controversies regarding genetic samples in research [87], the extent to which customers are informed about the nature of this research and the dissemination of results remains unclear. Additionally, governance includes considerations such as the origin of manufacturing equipment, adequacy of the producers' remuneration, their working conditions, and proper disposal practices.

Sustainability in healthcare goes beyond environmental impact and involves resource optimization and ensuring efficiency. Personalized medicine, which offers customized treatments based on genetic profiles, contributes to efficient resource utilization. Foundation Medicine [88], a company that specializes in genomic profiling, assists oncologists in identifying targeted therapies. This approach minimizes unnecessary treatments, reduces the burden on healthcare resources, and thus aligns with sustainability goals.

Furthermore, personalized medicine's focus on preventive healthcare aligns seamlessly with sustainability objectives. By identifying genetic predispositions to diseases, individuals can take proactive measures to maintain their health. Color Health [89], for instance, offers genetic testing for hereditary conditions, empowering individuals to make informed decisions about their health. This preventive approach contributes to long-term sustainability by reducing the prevalence of chronic diseases and associated healthcare costs.

6.3.3   Precision surgery

Precision surgery refers to the customization of the surgical procedures for each individual patient by taking into account their clinical stage, molecular features, environmental factors, and lifestyle [90], while completely avoiding surgical procedures on individuals who are unlikely to succeed [91,92]. This has the potential to revolutionize contemporary oncology and oncological surgery, which extends beyond mere technical expertise and requires a deep understanding of the molecular and phenotypic characteristics of cancer cells [90].

The substantial progress made in precision surgery can be attributed to the gradual adoption of minimally invasive techniques, which are facilitated by image-guided surgery [93,94]. This allows for accurate anatomical targeting, ensures safety margins, and facilitates complete excision while minimizing damage to healthy tissues [94]. Additionally, mass spectrometry imaging [95] coupled with AI algorithms [96] can play a pivotal role in predicting cancer development and surgical margins, guiding optimal timing of surgery, and assessing risk for precise stratification [97]. AI-based image recognition techniques can also assist in identifying minor lesions, making them useful in dermatology [98] and endoscopy [99]. Ongoing technological progress might ultimately enable predicting the outcomes of particular surgical procedures, such as tissue coagulation or suturing [93].

Integrating ESG principles with precision surgery can potentially create a transformative healthcare environment that takes into consideration sustainability, ethical considerations, and social responsibility alongside medical advancements.

Precision Surgery is a medical approach that uses advanced technologies such as AI and robotics to enable minimally invasive surgical procedures. This approach helps reduce the environmental impact associated with traditional surgical procedures, aligning with sustainability goals. For example, the da Vinci robotic surgical systems by Intuitive (Sunnyvale, California, USA) enable surgeons to perform minimally invasive procedures with enhanced precision [100]. Through precise control of incisions, this technology contributes to reducing postoperative waste and shorter recovery times, aligning with ESG's environmental goals. Another example is the Mako SmartRobotics^{TM} by Stryker (Portage, MI, USA) which facilitates precise bone preparations, preserving healthy tissue and minimizing the overall impact on the patient's body in total knee arthroplasty (TKA) surgeries [101].

By facilitating minimally invasive procedures, Precision surgery can help alleviate both physical and financial burdens on patients. In addition, tailoring interventions to individual clinical and molecular features ensures personalized, targeted treatments which can potentially reduce the need for extensive and costly procedures. For example, The Leksell Gamma Knife® Perfexion™ by Elekta (Stockholm, Sweden) enables non-invasive precision surgery by delivering targeted radiation to treat brain conditions and thereby minimizing damage to surrounding healthy tissue. Furthermore, Precision surgery can potentially result in shorter hospital stays, quicker recovery, and fewer complications, thus enhancing healthcare efficiency, sustainability, and accessibility. For instance, the Aravind Eye Care System in India leverages advanced surgical instruments to provide high-quality eye surgeries, including cataract procedures [102], irrespective of their patients' socio-economic backgrounds, thus promoting accessibility and social equality.

Governance in precision surgery involves establishing principles, rules, and oversight mechanisms to regulate the implementation of precision surgical practices. This includes developing ethical standards, protocols, and guidelines to ensure the responsible application of precision surgery techniques. Governance also addresses issues related to patient consent, data privacy, and the responsible use of advanced technologies such as artificial intelligence and robotics in surgical procedures [103]. Additionally, to establish regulatory standards that promote the safety, quality, and accessibility of precision surgery while aligning with broader healthcare goals and ethical considerations, collaboration with regulatory bodies and professional organizations may also be required.

6.3.4   Predictive maintenance

Predictive maintenance is a proactive approach that exploits data, analytics, and AI and ML techniques to predict equipment failures, defects, and operational anomalies. This allows for minimizing accidental downtime, reducing maintenance costs, avoiding unnecessary repairs and replacements, and maximizing the overall lifespan of equipment. This, in turn, contributes to reducing the overall environmental impact of industries.

Leading predictive maintenance platforms such as ABB's Ability™ [104], GE Digital's SmartSignal [105], Siemens' Senseye Predictive Maintenance [106], and IBM's Maximo Predict [107] leverage IoT and predictive analytics to monitor the performance of industrial equipment, enabling proactive maintenance and contributing to long-term sustainability. Similar tools can be developed for monitoring large medical devices and equipment such as CT scanners and MRI machines at doctors' offices and hospitals. This can help minimize unplanned service interruptions and extend their lifespan, thus addressing ESG's environmental goals by minimizing waste and promoting resource efficiency.

While the primary focus of predictive maintenance in healthcare is to optimize equipment performance for seamless patient care, it also addresses vital social considerations [108]. Predictive maintenance enhances equipment reliability, reducing the risk of failures during critical medical procedures and improving patient safety and quality of care. This, in turn, enhances patient trust and confidence in healthcare services. Additionally, predictive maintenance fosters a predictable, efficient, and well-maintained work environment. It not only improves the well-being of healthcare professionals in the workplace but also allows them to focus on patient care rather than dealing with unexpected equipment issues. Furthermore, predictive maintenance contributes to realizing sustainable healthcare systems by ensuring efficient use of resources. This enables organizations to allocate funds to critical areas, such as improving patient outcomes and investing in innovative technologies.

Governance plays a crucial role in predictive maintenance which involves establishing policies, and frameworks for effective, transparent, and responsible practices. Predictive maintenance relies on analyzing extensive data from sensors, and IoT devices, which makes it important to invest in robust infrastructure aligned with governance principles. Furthermore, it is crucial to implement validation and quality assurance processes to ensure data reliability and integrity. In addition, robust data management frameworks and encryption mechanisms are essential to protect sensitive equipment and operational data and ensure regulatory compliance. Comprehensive cybersecurity measures are also critical for safeguarding predictive maintenance data from threats such as cyber-attacks and data breaches.

Besides, employee resistance can be a potential obstacle while implementing predictive maintenance effectively, which is driven by concerns about job security, skepticism about new technologies, or a lack of understanding. It is essential to communicate program objectives clearly and engage with stakeholders at all levels. In addition, employee training programs that focus on their technical skills and data literacy can foster a culture of continuous learning and improvement. Organizations should also incorporate performance metrics and feedback mechanisms to gather insights from frontline employees to drive continuous improvement and innovation.

6.3.5 Improved patient experience

Integrating ESG principles into healthcare practices has the potential not only to transform the industry's operational landscape but also to elevate the patient's experience. Sustainable and ethical practices in healthcare facilities foster an environment that promotes inclusivity, patient well-being, and satisfaction.

One of the key aspects of ESG practices in healthcare is associated with their environmental impact, highlighting the importance of eco-friendly infrastructure. For example, Cleveland Clinic has taken strides toward sustainability by implementing practices such as green building initiatives [109], achieving ENERGY STAR® certification [110], and recycling plastic. Furthermore, the clinic collaborates with local partners to develop recycling markets to divert waste from landfills, contributing to eco-friendly practices. In addition, the clinic reduces its carbon footprint by adopting energy-efficient designs and environmentally friendly technologies, ultimately enhancing patient well-being through promoting healing environments [111].

Promoting social inclusivity is pivotal to enabling ESG practices in healthcare. For instance, Kaiser Permanente® (Oakland, California, United States), a leading non-profit healthcare provider in the United States, incorporates cultural competence to deliver equitable and culturally responsive care to address health disparities [112]. They promote health and equity by addressing societal gaps through impact investments, grants, and community partnerships. They prioritize health equity by screening their members for social needs, connecting them to resources, and expanding community healthcare access through safety net partnerships and financial assistance. Besides, Singapore General Hospital (SGH) has taken initiatives to provide more inclusive healthcare services by leveraging emerging technologies like AI and ML, focusing on telemedicine and precise patient health analytics [113].

ESG principles play a crucial role in promoting ethical, transparent, trustworthy, and patient-centric governance in healthcare. One such example is the Mayo Clinic, which considers shared decision-making (SDM) and patient engagement as vital components of its healthcare strategy, resulting in an improved patient experience and a collaborative and respectful healthcare environment [114,115]. Ethical governance also mandates safeguarding patient confidentiality and privacy, ensuring informed consent, and maintaining financial transparency [116]. Healthcare professionals must adhere to high ethical standards, demonstrating honesty, integrity, and transparency in all their interactions.

Moreover, sustainability practices in healthcare must extend to resource efficiency, which directly impacts the patient's experience. SGH, for example, leverages smart technologies and data analytics to streamline workflows, ultimately ensuring timely and effective diagnosis [113] and care delivery [117]. This resource-efficient approach contributes to a positive patient experience by minimizing wait times and optimizing the use of healthcare resources.

6.4   Concerns and challenges

The adoption of ESG principles into medical equipment engineering and healthcare holds promise for transformative advancements in patient care and sustainability. However, this endeavor is accompanied by a myriad of challenges that must be addressed to realize its full potential. Researchers and industries must work together to navigate through various considerations to ensure ethical, responsible, and sustainable innovation.

6.4.1   Investment in innovation

As global awareness of environmental issues increases, sustainability emerges as a significant catalyst for corporate expansion and progress [118]. However, incorporating advanced technologies like artificial intelligence, robotics, and the IoT into healthcare, while maintaining ESG and sustainable practices, demands significant investment in research and development of sustainable practices and technologies, such as renewable energy sources, waste management systems, eco-friendly materials, and green manufacturing processes. This calls for a careful balance between innovation and financial aspects. Radical advancements across various fronts, including technologies, products, processes, business models, and environmental innovations, are imperative to transition current market-driven practices toward a more sustainable approach. However, sustainable innovation strategies often struggle with the inherent complexities and uncertainties associated with ESG requirements, presenting a challenge compared to conventional market-driven innovation [119].

6.4.2 Standardization of ESG frameworks

Medical equipment manufacturers face a persistent challenge in navigating the complex landscape of regulations and standards to ensure the accuracy and efficacy of their devices. Additionally, ESG requirements demand that innovations not only adhere to regulations but also contribute to sustainability goals. This emphasizes the need for medical equipment manufacturers to consider both regulatory requirements and sustainability objectives in their innovation processes, requiring a nuanced understanding of the legal framework governing the industry. However, the current ESG reporting landscape lacks consensus on terminology and definitions and is crowded with various frameworks, with the Global Reporting Initiative (GRI) and the Sustainable Accounting Standards Board's standards (SASB) being the most widely used of them, each highlighting different performance metrics such as environment, social aspects, governance, carbon emissions, energy usage, waste management, and water conservation [120,121]. In addition, disclosure requirements undergo frequent changes, making the reporting process more complicated. This challenge is exacerbated by the absence of consensus on terminology and definitions. A thorough understanding and assessment of these frameworks are essential for organizations to choose the most suitable one. A unified standardized framework for ESG disclosure is, therefore, essential to streamline reporting processes and ensure consistency across the healthcare industry. Nevertheless, The International Sustainability Standards Board (ISSB) has agreed to release global guidelines that aim to harmonize corporate environmental regulatory disclosures from 2024 onwards [122].

6.4.3 Data management

ESG considerations involve various aspects that are crucial for a company's sustainability and ethical performance, including leadership, diversity, and inclusivity in board composition, executive compensation, financial transparency, regulatory compliance, and ethical business practices. Gathering extensive amounts of financial and non-financial data from different organizational areas is essential to assess these factors. However, this data is often scattered across different disconnected systems of the organization, requiring manual processing of a large volume of data, potentially leading to delays and errors. This also makes it difficult to quantitatively evaluate the impact of ESG activities on the organization's financial outcome. Therefore, efficient data management systems and analytical tools are essential to manage and analyze large volumes of data required for ESG planning and reporting.

6.4.4 Data privacy and security

The integration of advanced technologies in healthcare often involves collecting and analyzing sensitive health data, highlighting the need to address concerns about data privacy and security. A breach in security could lead to misuse of personal data, posing severe risks to users, including potentially fatal outcomes. Therefore, it is crucial for medical equipment engineering companies to prioritize safeguarding patient information. Users should also have greater control over their data usage and be informed about security policies and practices, including details about any data breaches. Developing advanced algorithms and adopting stringent cybersecurity measures are essential to enhance data security and user control, ensuring alignment with ethical data handling and governance standards as part of ESG practices.

6.4.5 Supply chain

The complex structure of global supply chains presents significant challenges for both medical device manufacturers and healthcare service providers as they strive to meet ESG standards. These challenges include various aspects, such as ensuring the ethical and responsible sourcing of materials and components, reducing the environmental impact of production processes, and promoting fair labor practices across the supply chain. Addressing these challenges requires organizations to navigate intricate logistical and ethical considerations associated with ESG practices while simultaneously ensuring the reliability and quality of their medical equipment.

6.4.6 End-of-life product management

Implementing a lifecycle management plan is crucial for a smooth phase-out of a product and to prevent disruptions within an organization. By understanding the lifecycles of product components and end-of-life (EOL) dates, organizations can anticipate when parts may be discontinued and plan transitions accordingly. However, without careful planning, EOL issues can lead to service disruptions, security breaches, and compliance failures. Providing EOL notifications well in advance allows customers to prepare for the transition and maximize the lifespan of the products. Additionally, the disposal and recycling of medical equipment at the end of its lifecycle presents sustainability challenges. It is essential to develop responsible strategies for medical and electronic waste disposal and implement recycling programs. Medical equipment manufacturers must consider the environmental impact of their products beyond their primary use to ensure compliance with ESG principles through responsible EOL product management.

6.4.7 Lack of understanding and resistance to change

Sustainable healthcare is a relatively new concept for many healthcare professionals, leaders, and patients. Introducing sustainable practices may involve altering established procedures, workflows, and cultures within the organization, which can be challenging. In addition, limited awareness and education about the benefits of sustainable practices often lead to resistance to change, hindering adoption. This resistance can stem from various stakeholders, including healthcare professionals and administrators, who may perceive sustainability practices as unfamiliar, demanding, and costly. Overcoming this resistance and fostering a culture of sustainability within healthcare organizations and medical device manufacturers calls for a significant cultural shift. This shift requires leadership commitment, educational initiatives for employees and stakeholders, comprehensive training programs highlighting the benefits of sustainable healthcare, and organizational restructuring. Ensuring that employees are well-versed in the principles of ESG is crucial for successfully implementing sustainable practices. Nevertheless, building awareness about the importance of sustainability and prioritizing responsible practices remains an ongoing challenge.

6.4.8   *Lack of quantitative and comparable ESG measure*

As mentioned earlier, many organizations are involved in ESG practices, but not all have formal definitions, identified key performance indicators (KPIs), or systems in place to monitor them. The lack of a structured method of evaluating ESG practices poses a challenge, especially because many indicators are not easily quantifiable. Quantitative ESG metrics typically include measurable factors such as carbon emissions, water usage, employee turnover rates, and executive compensation. On the other hand, qualitative ESG metrics encompass aspects like a company's commitment to diversity, equity, and inclusion (DEI), its labor practices, and its impact on local communities. While these qualitative metrics offer valuable insights into a company's culture and values, they are difficult to quantify.

Furthermore, data related to ESG activities is often scattered across different disconnected systems within organizations. This fragmentation makes it challenging to quantitatively evaluate the impact of ESG activities on the organization's financial outcomes. Therefore, it is crucial to develop methods for measuring ESG activities and their impact quantitatively.

Currently, there are over 600 agencies, including Bloomberg ESG Data Services, Dow Jones Sustainability Index Family, and Thomson Reuters ESG Scores, that issue ESG scores [123]. However, these agencies use different approaches to measure ESG ratings, resulting in wide variations in the ratings. Moreover, the existing method of measuring ESG falls short of capturing the intricate, systemic nature of social and environmental systems, as well as that of business organizations themselves [124]. A robust standardized scoring system is therefore necessary for companies to assess, compare and improve their performance.

6.5   Future perspectives and conclusion

The integration of ESG principles into medical equipment engineering and the adoption of green sustainable technology offer promising opportunities for the healthcare industry. Prioritizing sustainability and responsible practices enables medical equipment manufacturers and healthcare providers to enhance patient care while minimizing environmental impact and contributing to societal well-being.

Moreover, beyond its social and environmental benefits, an ESG-driven business approach also proves financially beneficial [125]. While some manufacturers are hesitant to adopt ESG practices due to concerns about increased costs, studies have shown a positive correlation between ESG performance, equity returns, and risk reduction, indicating potential financial savings [126]. Sustainable strategies, including investments in renewable energy, waste reduction, green manufacturing, and efficient transportation methods, can yield significant long-term cost savings. Moreover, businesses can capitalize on tax incentives and manage operational risks related to ESG initiatives [127,128].

As technology continues to evolve, it is important to develop stringent regulatory frameworks to ensure that medical equipment meets sustainability standards and ensure that healthcare innovations adhere to these ESG principles. This involves establishing standardized frameworks for ESG reporting, standardized quantitative measures of evaluating ESG initiatives, improving methods for measuring the outcomes of sustainability initiatives, and promoting a culture of sustainability across organizations. Researchers need to explore bio-based materials, recycled plastics, and other renewable resources to replace traditional materials in medical devices.

There is a growing concern regarding the environmental impact stemming from the energy requirements of medical equipment and healthcare facilities. Integrating energy harvesting and energy-saving technologies, as well as low-power electronic components, into medical sensors and devices can effectively minimize energy consumption. Additionally, healthcare facilities can adopt green building technologies and utilize renewable energy sources to reduce energy consumption and decrease their carbon footprints.

Furthermore, the incorporation of digital technologies such as the IoT, AI, and Big Data analytics holds promise in promoting sustainable practices in medical equipment engineering and healthcare. These technologies can potentially enable telehealth, predictive analytics for resource optimization, mitigate environmental impact, and promote sustainable innovation. However, it is extremely important to regulate the use of these technologies to promote user safety, transparency, accountability, and mutual trust. Besides, the growing use of these technologies may potentially lead to a new global energy challenge. For example, it is estimated that data centers hosting cloud services are projected to consume 20% of global electricity and emit up to 5.5% of the world's carbon emissions by 2025 [103]. Moreover, a considerable portion of the energy utilized is converted into heat, resulting in operational issues such as decreased system reliability, shortened device lifespan, and increased cooling demands. Hence, more research is needed to prioritize energy efficiency and environmental sustainability to effectively tackle the repercussions of substantial energy consumption, while also ensuring superior quality of service.

Besides, it is important to implement effective product lifecycle management strategies to achieve ESG goals. It is essential to design products with EOL strategies in consideration right from the outset. Manufacturers can adopt a circular economy model, prioritizing durability and repairability to extend product lifespan and promote reuse, recycling, and responsible material disposal. This model aims to minimize waste and resource consumption by maximizing the value of materials and products through reuse and recycling, thereby enhancing environmental sustainability. Additionally, investing in research on bio-based materials, recycled plastics, and other renewable resources can pave the way for potential alternatives to traditional materials. By leveraging materials such as plant-derived polymers and nanoparticles, reliance on limited resources can be reduced while minimizing environmental impact. However, the use of nanomaterials in healthcare raises concerns regarding human exposure and adverse effects. These materials can enter the body through various pathways, affecting organs and tissues and potentially leading to cytotoxic effects and diseases. The properties of nanomaterials, including size and concentration, influence their interactions with biological systems [129]. Understanding and addressing these risks is vital for the safe utilization of nanotechnology in healthcare. Therefore, stringent regulatory standards and compliance monitoring are essential to ensure the safe use of nanomaterials.

Engaging a diverse industry stakeholder, from manufacturers and healthcare providers to policymakers, regulators, and consumers, is essential for the successful integration of ESG practices and for fostering innovation in green sustainable healthcare technology. Establishing a culture of open communication and collaboration, along with ensuring alignment of interests and priorities among these diverse stakeholders, is paramount for achieving these goals. Furthermore, it is imperative to establish partnerships and promote knowledge-sharing to accelerate the development and adoption of sustainable medical technologies. Investing in education and training programs is crucial not only to raise awareness about ESG and sustainable practices but also to cultivate a skilled workforce capable of driving innovation in green sustainable healthcare technologies. This can be achieved by integrating ESG and sustainability principles into general education, health science, and engineering curricula, while also providing regular training for the professionals in the healthcare and medical device industry. Increased consumer awareness can potentially lead to a greater demand for sustainable healthcare solutions. Consequently, manufacturers may need to prioritize transparency and communication regarding the environmental and social impact of their products to meet evolving consumer expectations and market demands.

As discussed in this chapter, integrating ESG principles into medical equipment engineering offers substantial benefits. There is no denying that numerous challenges lie ahead. However, these challenges also present new opportunities for innovation. Embracing ESG practices, sustainability, and green technology can lead us toward a more resilient, equitable, and environmentally conscious healthcare system.

References

[1]   Z. Javed, M. H. Maqsood, T. Yahya, *et al.*, "Race, Racism, and Cardio-vascular Health: Applying a Social Determinants of Health Framework to Racial/Ethnic Disparities in Cardiovascular Disease," *Circ. Cardiovasc. Qual. Outcomes*, vol. 15, no. 1, p. E007917, 2022.

[2]   United Nations, "The 17 Goals," *Department of Economic and Social Affairs Sustainable Development*, 2015. [Online]. Available: https://sdgs.un.org/goals.

[3]   A. Sepetis, F. Rizos, G. Pierrakos, H. Karanikas, and D. Schallmo, "A Sustainable Model for Healthcare Systems: The Innovative Approach of ESG and Digital Transformation," *Healthcare*, vol. 12, no. 2, 2024.

[4]   World Health Organization, "Global spending on health: Weathering the storm," 2020.

[5]   J. Boubker, "When Medical Devices Have a Mind of Their Own: The Challenges of Regulating Artificial Intelligence," *Am. J. Law Med.*, vol. 47, no. 4, pp. 427–454, 2021.

[6]   M. Naghshvarianjahromi, S. Majumder, S. Kumar, N. Naghshvarianjahromi, and M. J. Deen, "Natural Brain-Inspired Intelligence for Screening in Healthcare Applications," *IEEE Access*, vol. 9, pp. 67957–67973, 2021.

[7]   A. Papadopoulou, N. S. Kumar, A. Vanhoestenberghe, and N. K. Francis, "Environmental Sustainability in Robotic and Laparoscopic Surgery: Systematic Review," *Br. J. Surg.*, vol. 109, no. 10, pp. 921–932, 2022.

[8]   K. Reddy, P. Gharde, H. Tayade, M. Patil, L. S. Reddy, and D. Surya, "Advancements in Robotic Surgery: A Comprehensive Overview of Current Utilizations and Upcoming Frontiers," *Cureus*, 2023.

[9]   R. A. Rayan, C. Tsagkaris, and R. B. Iryna, "The Internet of Things for Healthcare: Applications, Selected Cases and Challenges," *Stud. Comput. Intell.*, vol. 933, pp. 1–15, 2021.

[10] S. Majumder, E. Aghayi, M. Noferesti, *et al.*, "Smart Homes for Elderly Healthcare—Recent Advances and Research Challenges," *Sensors (Switzerland)*, vol. 17, no. 11, p. 2496, 2017.

[11] S. Majumder, T. Mondal, and M. J. Deen, "A Simple, Low-Cost and Efficient Gait Analyzer for Wearable Healthcare Applications," *IEEE Sens. J.*, vol. 19, no. 6, pp. 2320–2329, 2019.

[12] S. Majumder, L. Chen, O. Marinov, C. H. Chen, T. Mondal, and M. Jamal Deen, "Noncontact Wearable Wireless ECG Systems for Long-Term Monitoring," *IEEE Rev. Biomed. Eng.*, vol. 11, pp. 306–321, 2018.

[13] N. L. Bragazzi, H. Dai, G. Damiani, M. Behzadifar, M. Martini, and J. Wu, "How Big Data and Artificial Intelligence can Help Better Manage the Covid-19 Pandemic," *Int. J. Environ. Res. Public Health*, vol. 17, no. 9, 2020.

[14] S. Benzidia, O. Bentahar, J. Husson, and N. Makaoui, "Big Data Analytics Capability in Healthcare Operations and Supply Chain Management: The

Role of Green Process Innovation," *Ann. Oper. Res.*, vol. 333, no. 2–3, pp. 1077–1101, 2024.

[15]   S. J. Mooney and V. Pejaver, "Big Data in Public Health: Terminology, Machine Learning, and Privacy," *Annu. Rev. Public Health*, vol. 39, pp. 95–112, 2018.

[16]   H. W. Lo, "A Data-Driven Decision Support System for Sustainable Supplier Evaluation in the Industry 5.0 era: A Case Study for Medical Equipment Manufacturing," *Adv. Eng. Informatics*, vol. 56, 2023.

[17]   International Medical Device Regulators Forum, "Software as a Medical Device (SaMD): Key Definitions," *Int. Med. Device Regul. Forum*, pp. 1–9, 2013.

[18]   FDA, "Developing a Software Precertification Program: A working Model," *U.S Food Drug Adm.*, pp. 1–58, 2019.

[19]   J. Hua Li, "Cyber Security Meets Artificial Intelligence: A Survey," *Front. Inf. Technol. Electron. Eng.*, vol. 19, no. 12, pp. 1462–1474, 2018.

[20]   F. Oumaima, Z. Karim, E. G. Abdellatif, and B. Mohammed, "A Survey on Blockchain and Artificial Intelligence Technologies for Enhancing Security and Privacy in Smart Environments," *IEEE Access*, 2022.

[21]   N. J. Schork, "Artificial Intelligence and Personalized Medicine," *Cancer Treat. Res.*, vol. 178, pp. 265–283, 2019.

[22]   K. B. Johnson, W. Q. Wei, D. Weeraratne, *et al.*, "Precision Medicine, AI, and the Future of Personalized Health Care," *Clin. Transl. Sci.*, vol. 14, no. 1, pp. 86–93, 2021.

[23]   M. Pournader, H. Ghaderi, A. Hassanzadegan, and B. Fahimnia, "Artificial Intelligence Applications in Supply Chain Management," *Int. J. Prod. Econ.*, vol. 241, 2021.

[24]   Y. Riahi, T. Saikouk, A. Gunasekaran, and I. Badraoui, "Artificial Intelligence Applications in Supply Chain: A Descriptive Bibliometric Analysis and Future Research Directions," *Expert Syst. Appl.*, vol. 173, 2021.

[25]   N. S. Godbole and J. Lamb, "Using Data Science & Big Data Analytics to Make Healthcare Green," *2015 12th Int. Conf. Expo Emerg. Technol. a Smarter World, CEWIT 2015*, 2015.

[26]   K. Batko, "Digital Social Innovation Based on Big Data Analytics for Health and Well-Being of Society," *J. Big Data*, vol. 10, no. 1, 2023.

[27]   L. van Niekerk, L. Manderson, and D. Balabanova, "The Application of Social Innovation in Healthcare: A Scoping Review," *Infect. Dis. Poverty*, vol. 10, no. 1, 2021.

[28]   D.W. Bates, S. Saria, L. Ohno-Machado, A. Shah, and G. Escobar, "Big Data in Health Care: Using Analytics to Identify and Manage High-Risk and High-Cost Patients," *Health Aff.*, vol. 33, no. 7, pp. 1123–1131, 2014.

[29]   M. I. Razzak, M. Imran, and G. Xu, "Big Data Analytics for Preventive Medicine," *Neural Comput. Appl.*, vol. 32, no. 9, pp. 4417–4451, 2020.

[30]   R. Saini and R. K. Kanna, "Big Data Analytics: Understanding Its Capabilities and Potential Benefits for Healthcare Organizations," *2023 3rd Int. Conf. Adv. Electron. Commun. Eng. AECE 2023*, pp. 733–738, 2023.

[31]   V. Chandola, S. R. Sukumar, and J. Schryver, "Knowledge Discovery From Massive Healthcare Claims Data," *Proc. ACM SIGKDD Int. Conf. Knowl. Discov. Data Min.*, vol. Part F128815, pp. 1312–1320, 2013.

[32]   A. J. Mary and S. P. A. Claret, "Design and Development of Big Data-Based Model for Detecting Fraud in Healthcare Insurance Industry," *Soft Comput.*, vol. 27, no. 12, pp. 8357–8369, 2023.

[33]   A. Verma, A. Taneja, and A. Arora, "Fraud Detection and Frequent Pattern Matching in Insurance Claims Using Data Mining Techniques," *2017 10th Int. Conf. Contemp. Comput. IC3 2017*, vol. 2018, pp. 1–7, 2017.

[34]   J. Koreff, M. Weisner, and S. G. Sutton, "Data Analytics (AB) use in Healthcare Fraud Audits," *Int. J. Account. Inf. Syst.*, vol. 42, 2021.

[35]   N. H. Motlagh, M. Mohammadrezaei, J. Hunt, and B. Zakeri, "Internet of Things (IoT) and the Energy Sector," *Energies*, vol. 13, no. 2, 2020.

[36]   S. Majumder, T. Mondal, and M. J. Deen, "Wearable Sensors for Remote Health Monitoring," *Sensors (Switzerland)*, vol. 17, no. 1, p. 130, 2017.

[37]   A. Purohit, J. Smith, and A. Hibble, "Does Telemedicine Reduce the Carbon Footprint of Healthcare? A Systematic Review," *Futur. Healthc. J.*, vol. 8, no. 1, pp. e85–e91, 2021.

[38]   S. Rodler, L. S. Ramacciotti, M. Maas, *et al.*, "The Impact of Telemedicine in Reducing the Carbon Footprint in Health Care: A Systematic Review and Cumulative Analysis of 68 Million Clinical Consultations," *Eur. Urol. Focus*, vol. 9, no. 6, pp. 873–887, 2023.

[39]   W. Jiang, S. Majumder, S. Kumar, *et al.*, "A Wearable Tele-Health System Towards Monitoring COVID-19 and Chronic Diseases," *IEEE Rev. Biomed. Eng.*, vol. 15, pp. 61–84, 2022.

[40]   F. Ye, S. Majumder, W. Jiang, *et al.*, "A Framework for Infectious Disease Monitoring With Automated Contact Tracing - A Case Study of COVID-19," *IEEE Internet Things J.*, vol. 10, no. 1, pp. 144–165, 2023.

[41]   A. Riahi Sfar, E. Natalizio, Y. Challal, and Z. Chtourou, "A Roadmap for Security Challenges in the Internet of Things," *Digit. Commun. Networks*, vol. 4, no. 2, pp. 118–137, 2018.

[42]   A. Karale, "The Challenges of IoT Addressing Security, Ethics, Privacy, and Laws," *Internet of Things (Netherlands)*, vol. 15, 2021.

[43]   C. M. Kyung, H. Yasuura, Y. Liu, and Y. L. Lin, *Smart Sensors and Systems: Innovations for Medical, Environmental, and IoT Applications*. 2016.

[44]   M. Kaisti, J. Leppänen, O. Lahdenoja, *et al.*, "Wearable Pressure Sensor Array for Health Monitoring," *Comput. Cardiol.*, vol. 44, pp. 1–4, 2017.

[45]   L. Olatomiwa, R. Blanchard, S. Mekhilef, and D. Akinyele, "Hybrid Renewable Energy Supply for Rural Healthcare Facilities: An Approach to Quality Healthcare Delivery," *Sustain. Energy Technol. Assessments*, vol. 30, pp. 121–138, 2018.

[46]   G. Rong, Y. Zheng, and M. Sawan, "Energy Solutions for Wearable Sensors: A Review," *Sensors*, vol. 21, no. 11, 2021.

[47] M. Moner-Girona, G. Kakoulaki, G. Falchetta, D. J. Weiss, and N. Taylor, "Achieving Universal Electrification of Rural Healthcare Facilities in sub-Saharan Africa with Decentralized Renewable Energy Technologies," *Joule*, vol. 5, no. 10, pp. 2687–2714, 2021.

[48] P. Alstone, D. Gershenson, and D. M. Kammen, "Decentralized Energy Systems for Clean Electricity Access," *Nat. Clim. Chang.*, vol. 5, no. 4, pp. 305–314, 2015.

[49] M. M. H. Shuvo, T. Titirsha, N. Amin, and S. K. Islam, "Energy Harvesting in Implantable and Wearable Medical Devices for Enduring Precision Healthcare," *Energies*, vol. 15, no. 20, 2022.

[50] H. Heidari, O. Onireti, R. Das, and M. Imran, "Energy Harvesting and Power Management for IoT Devices in the 5G Era," *IEEE Commun. Mag.*, vol. 59, no. 9, pp. 91–97, 2021.

[51] S. Majumder, A. K. Roy, W. Jiang, T. Mondal, and M. Jamal Deen, "Smart Textiles to Enable In-Home Health Care: State of the Art, Challenges, and Future Perspectives," *IEEE J. Flex. Electron.*, pp. 1–1, 2023.

[52] S. Kumar, N. Dilbaghi, R. Saharan, and G. Bhanjana, "Nanotechnology as Emerging Tool for Enhancing Solubility of Poorly Water-Soluble Drugs," *Bionanoscience*, vol. 2, no. 4, pp. 227–250, 2012.

[53] M. Napagoda and S. Witharana, "Nanotechnology in Drug Delivery," *Nanotechnol. Mod. Med.*, pp. 47–73, 2022.

[54] A. Singh and M. M. Amiji, "Application of Nanotechnology in Medical Diagnosis and Imaging," *Curr. Opin. Biotechnol.*, vol. 74, pp. 241–246, 2022.

[55] J. Nam, N. Won, J. Bang, *et al.*, "Surface Engineering of Inorganic Nanoparticles for Imaging and Therapy," *Adv. Drug Deliv. Rev.*, vol. 65, no. 5, pp. 622–648, 2013.

[56] M. Thakur, A. Sharma, M. Chandel, and D. Pathania, "Modern Applications and Current Status of Green Nanotechnology in Environmental Industry," *Green Funct. Nanomater. Environ. Appl.*, pp. 259–281, 2021.

[57] A. K. Roy, W. Ghann, S. Rabi, *et al.*, "Hydrogen Peroxide Assisted Synthesis of Fluorescent Carbon Nanoparticles From Teak Leaves for dye-Sensitized Solar Cells," *RSC Sustain.*, 2024.

[58] S. H. Khan, "Green Nanotechnology for the Environment and Sustainable Development," pp. 13–46, 2020.

[59] D. L. Woods, T. McAndrew, N. Nevadunsky, *et al.*, "Carbon Footprint of Robotically-Assisted Laparoscopy, Laparoscopy and Laparotomy: A Comparison," *Int. J. Med. Robot. Comput. Assist. Surg.*, vol. 11, no. 4, pp. 406–412, 2015.

[60] K. S. Chan, H. Y. Lo, and V. G. Shelat, "Carbon Footprints in Minimally Invasive surgery: Good Patient Outcomes, but Costly for the Environment," *World J. Gastrointest. Surg.*, vol. 15, no. 7, pp. 1277–1285, 2023.

[61] S. Sasse, A. Bleasdale, J. Niemeier, J. Zaslavsky, J. Q. Huang, and C. Thiel, "Waste Audit of Robotic Gynecologic Surgery (1228)," *Gynecol. Oncol.*, vol. 176, p. S147, 2023.

[62] S. O'Sullivan, N. Nevejans, C. Allen, *et al.*, "Legal, Regulatory, and Ethical Frameworks for Development of Standards in Artificial Intelligence (AI) and Autonomous Robotic Surgery," *Int. J. Med. Robot. Comput. Assist. Surg.*, vol. 15, no. 1, 2019.

[63] N. Nadjmi, "Definition, History, and Indications of Robotic Surgery in Oral and Maxillofacial Surgery," *Emerg. Technol. Oral Maxillofac. Surg.*, pp. 239–265, 2023.

[64] A. Zemmar, A. M. Lozano, and B. J. Nelson, "The Rise of Robots in Surgical Environments During COVID-19," *Nat. Mach. Intell.*, vol. 2, no. 10, pp. 566–572, 2020.

[65] M. L. Jin, M. M. Brown, D. Patwa, A. Nirmalan, and P. A. Edwards, "Telemedicine, Telementoring, and Telesurgery for Surgical Practices," *Curr. Probl. Surg.*, vol. 58, no. 12, 2021.

[66] N. Hachach-Haram, "A Digital Doorway to Global Surgery," *Digit. Surg.*, pp. 351–360, 2021.

[67] R. Chen, P. Rodrigues Armijo, C. Krause, K. C. Siu, and D. Oleynikov, "A Comprehensive Review of Robotic Surgery Curriculum and Training for Residents, Fellows, and Postgraduate Surgical Education," *Surg. Endosc.*, vol. 34, no. 1, pp. 361–367, 2020.

[68] I. H. A. Chen, A. Ghazi, A. Sridhar, *et al.*, "Evolving Robotic Surgery Training and Improving Patient Safety, with the Integration of Novel Technologies," *World J. Urol.*, vol. 39, no. 8, pp. 2883–2893, 2021.

[69] S. Majumder and M. J. Deen, "Smartphone Sensors for Health Monitoring and Diagnosis," *Sensors (Switzerland)*, vol. 19, no. 9. 2019.

[70] S. Majumder and M. Jamal Deen, "Wearable IMU-Based System for Real-Time Monitoring of Lower-Limb Joints," *IEEE Sens. J.*, vol. 21, no. 6, pp. 8267–8275, 2021.

[71] C. Der-Martirosian, T. Wyte-Lake, M. Balut, *et al.*, "Implementation of Telehealth Services at the US Department of Veterans Affairs During the COVID-19 Pandemic: Mixed Methods Study," *JMIR Form. Res.*, vol. 5, no. 9, 2021.

[72] A. Holmner, J. Rocklöv, N. Ng, and M. Nilsson, "Climate Change and eHealth: A Promising Strategy for Health Sector Mitigation and Adaptation," *Glob. Health Action*, vol. 5, 2012.

[73] B. Duane, M. B. Lee, S. White, R. Stancliffe, and I. Steinbach, "An Estimated Carbon Footprint of NHS Primary Dental Care Within England. How can Dentistry be More Environmentally Sustainable?," *Br. Dent. J.*, vol. 223, no. 8, pp. 589–593, 2017.

[74] "Supporting care with remote monitoring." [Online]. Available: https:// transform.england.nhs.uk/covid-19-response/technology-nhs/supporting-the-innovation-collaboratives-to-expand-their-remote-monitoring-plans/ [accessed: 28-Mar-2034].

[75] "About RAM - Remote Area Medical." [Online]. Available: https://www. ramusa.org/about-ram/ [accessed: 28-Mar-2024].

[76] "Florence Health." [Online]. Available: https://florencehealth.com/ [accessed: 28-Mar-2024].

[77]   "Philips Virtual Care Management." [Online]. Available: https://www.gobio.com/philips-virtual-care-management/ [accessed: 28-Mar-2024].

[78]   "Philips Debuts Remote Patient Management Solution," 2023. [Online]. Available: https://www.dicardiology.com/content/philips-debuts-remote-patient-management-solution [accessed: 28-Mar-2024].

[79]   S. Avicenna, "Personalized Medicine: A Tailor Made Medicine," *Avicenna J. Med. Biotechnol.*, vol. 6, no. 4, p. 191, 2014.

[80]   "Personalized Medicine Sector Needs to Consider Environmental Impact," *Genetic Engineering & Biotechnology News*, 2023. https://www.genengnews.com/insights/personalized-medicine-sector-needs-to-consider-environmental-impact/.

[81]   F. Hoffmann-La Roche Ltd, "Minimise the Environmental Footprint." [Online]. Available: https://www.roche.com/about/sustainability/fighting-climate-change [accessed: 28-Mar-2024].

[82]   C. Carlsten, M. Brauer, F. Brinkman, *et al.*, "Genes, the Environment and Personalized Medicine," *EMBO Rep.*, vol. 15, no. 7, pp. 736–739, 2014.

[83]   B. M. Mapes, C. S. Foster, S. V. Kusnoor, *et al.*, "Diversity and Inclusion for the All of Us Research Program: A Scoping Review," *PLoS One*, vol. 15, no. 7, 2020.

[84]   P. Trein and J. Wagner, "Governing Personalized Health: A Scoping Review," *Front. Genet.*, vol. 12, 2021.

[85]   M. DeGeurin, "Hackers got Nearly 7 million People's Data from 23andMe. The firm Blamed Users in 'Very Dumb' Move," *Guardian News & Media Limited*, 2024.

[86]   S. L. Tobin, M. K. Cho, S. S. J. Lee, *et al.*, "Customers or Research Participants?: Guidance for Research Practices in Commercialization of Personal Genomics," *Genet. Med.*, vol. 14, no. 10, pp. 833–835, 2012.

[87]   B. A. Tarini, "Storage and use of Residual Newborn Screening Blood Spots: A Public Policy Emergency," *Genet. Med.*, vol. 13, no. 7, pp. 619–620, 2011.

[88]   "Foundation Medicine." [Online]. Available: https://www.foundationmedicine.com/ [accessed: 28-Mar-2024].

[89]   "Genomics - Color Health." [Online]. Available: https://www.color.com/genomics [accessed: 28-Mar-2024].

[90]   G. Cannone, G. M. Comacchio, G. Pasello, *et al.*, "Precision Surgery in NSCLC," *Cancers (Basel).*, vol. 15, no. 5, 2023.

[91]   B. Cady, "Basic Principles in Surgical Oncology," *Arch. Surg.*, vol. 132, no. 4, pp. 338–346, 1997.

[92]   M. E. Lidsky and M. I. D' Angelica, "An outlook on Precision Surgery," *Eur. J. Surg. Oncol.*, vol. 43, no. 5, pp. 853–855, 2017.

[93]   U. Boggi, "Precision Surgery," *Updates Surg.*, vol. 75, no. 1, pp. 3–5, 2023.

[94]   P. Dell'Oglio, E. Mazzone, T. Buckle, *et al.*, "Precision Surgery: The Role of Intra-Operative Real-Time Image Guidance – Outcomes From a Multi-disciplinary European Consensus Conference.," *Am. J. Nucl. Med. Mol. Imaging*, vol. 12, no. 2, pp. 74–80, 2022.

[95]   A. M. L. Santilli, K. Ren, R. Oleschuk, *et al.*, "Application of Intraoperative Mass Spectrometry and Data Analytics for Oncological Margin Detection, A Review," *IEEE Trans. Biomed. Eng.*, vol. 69, no. 7, pp. 2220–2232, 2022.

[96]   N. Ogrinc, P. Saudemont, Z. Takats, M. Salzet, and I. Fournier, "Cancer Surgery 2.0: Guidance by Real-Time Molecular Technologies," *Trends Mol. Med.*, vol. 27, no. 6, pp. 602–615, 2021.

[97]   D. Schlanger, F. Graur, C. Popa, E. Moiş, and N. Al Hajjar, "The Role of Artificial Intelligence in Pancreatic Surgery: A Systematic Review," *Updates Surg.*, vol. 74, no. 2, pp. 417–429, 2022.

[98]   Z. Li, K. C. Koban, T. L. Schenck, R. E. Giunta, Q. Li, and Y. Sun, "Artificial Intelligence in Dermatology Image Analysis: Current Developments and Future Trends," *J. Clin. Med.*, vol. 11, no. 22, 2022.

[99]   Y. Okagawa, S. Abe, M. Yamada, I. Oda, and Y. Saito, "Artificial Intelligence in Endoscopy," *Dig. Dis. Sci.*, vol. 67, no. 5, pp. 1553–1572, 2022.

[100]   Intuitive Surgical, "Intuitive da Vinci." [Online]. Available: https://www. intuitive.com/en-us/products-and-services/da-vinci [accessed: 28-Mar-2024].

[101]   stryker, "Mako® Total Knee arthroplasty."

[102]   Aravind Eye Care System, "Cataract – Aravind Eye Care System." [Online]. Available: https://aravind.org/clinics/cataract/ [accessed: 28-Mar-2024].

[103]   K. Lam, F. M. Iqbal, S. Purkayastha, and J. M. Kinross, "Investigating the Ethical and Data Governance Issues of Artificial Intelligence in Surgery: Protocol for a Delphi study," *JMIR Res. Protoc.*, vol. 10, no. 2, 2021.

[104]   ABB, "Empowering Insight With Digital Solutions." [Online]. Available: https://global.abb/topic/ability/en/about [accessed: 28-Mar-2024].

[105]   G.E. Vernova, "Equipment Downtime Prevention." [Online]. Available: https://www.ge.com/digital/applications/asset-performance-management-apm-software/equipment-downtime-prevention-ge-smartsignal [accessed: 28-Mar-2024].

[106]   Siemens, "Senseye Predictive Maintenance." [Online]. Available: https:// www.siemens.com/global/en/products/services/digital-enterprise-services/ analytics-artificial-intelligence-services/predictive-services/senseye-predictive-maintenance.html [accessed: 28-Mar-2024].

[107]   IBM, "Predictive Maintenance – Maximo Application Suite." [Online]. Available: https://www.ibm.com/products/maximo/predictive-maintenance [accessed: 28-Mar-2024].

[108]   Sensemore, "Challenges and Considerations in Implementing Predictive Maintenance." [Online]. Available: https://sensemore.io/challenges-in-implementing-predictive-maintenance/ [accessed: 28-Mar-2024].

[109]   Cleveland Clinic, "Healthy Buildings." [Online]. Available: https://my. clevelandclinic.org/about/community/sustainability/sustainability-global-citizenship/environment/healthy-buildings [accessed: 28-Mar-2024].

[110]   ENERGY STAR, "The Simple Choice for Saving Energy." [Online]. Available: https://www.energystar.gov/ [accessed: 28-Mar-2024].

[111] Cleveland Clinic, "Our Operations." [Online]. Available: https://my.cleve-landclinic.org/about/community/sustainability/our-operations [accessed: 28-Mar-2024].

[112] Kaiser Permanente, "High-Quality, Equitable Care." [Online]. Available: https://about.kaiserpermanente.org/commitments-and-impact/equity-inclu-sion-and-diversity/high-quality-equitable-care [accessed: 28-Mar-2024].

[113] "SGH Deploys Prediction-Based AI Tool to Better Assess Patient's Surgery Risk," 2023. [Online]. Available: https://www.sgh.com.sg/news/patient-care/sgh-deploys-prediction-based-ai-tool-to-better-assess-patients-surgery-risk [accessed: 28-Mar-2024].

[114] V. M. Montori, M. M. Ruissen, I. G. Hargraves, J. P. Brito, and M. Kunneman, "Shared Decision-Making as a Method of Care," *BMJ Evidence-Based Med.*, vol. 28, no. 4, pp. 213–217, 2023.

[115] A. M. Stiggelbout, T. Van der Weijden, M. P. T. De Wit, *et al.*, "Shared Decision Making: Really Putting Patients at the Centre of Healthcare," *BMJ*, vol. 344, no. 7842, 2012.

[116] J. Béranger, "Ethical Governance and Responsibility in Digital Medicine: The Case of Artificial Intelligence," *Digit. Revolut. Heal.*, vol. 2, pp. 169–190, 2021.

[117] Shabana Begum, "Smart Tech one way to Boost Healthcare System: Masagos," 2021. [Online]. Available: https://www.sgh.com.sg/news/tomorrows-medicine/smart-tech-one-way-to-boost-healthcare-system-masagos [accessed: 28-Mar-2024].

[118] L. Bouten, "Sustainable Entrepreneurship and Sustainability Innovation: Categories and Interactions," *Soc. Environ. Account. J.*, vol. 32, no. 2, pp. 110–111, 2012.

[119] H. Vredenburg and J. Hall, "The Challenges of Innovating for Sustainable Development," *MIT Sloan management review*, 2003. [Online]. Available: https://sloanreview.mit.edu/article/the-challenges-of-innovating-for-sus-tainable-development/ [accessed: 28-Mar-2024].

[120] IBM, "What are ESG Frameworks?" [Online]. Available: https://www.ibm.com/topics/esg-frameworks. [Accessed: 28-Mar-2024].

[121] S. Bose, "Evolution of ESG Reporting Frameworks," *Values Work Sustain. Invest. ESG Report.*, pp. 13–33, 2020.

[122] D. Holger, "SN236 - Global Sustainability Disclosure Standards for Com-panies Set to Take Effect in 2024." [Online]. Available: https://www.wsj.com/articles/global-sustainability-disclosure-standards-for-companies-set-for-2024-749f68be.

[123] K. Farnham, "ESG Scores and Ratings: What They are, why They Matter," *Diligent Corporation*, 2023. [Online]. Available: https://www.diligent.com/resources/blog/esg-risk-scores [accessed: 28-Mar-2024].

[124] J. Howard-Grenville, "ESG Impact Is Hard to Measure — But It's Not Impossible," *Harvard Business Review*, 22-Jan-2021.

[125] G. Friede, T. Busch, and A. Bassen, "ESG and Financial Performance: Aggregated Evidence From More than 2000 Empirical Studies," *J. Sustain. Financ. Invest.*, vol. 5, no. 4, pp. 210–233, 2015.

[126] N. Ahmad, A. Mobarek, and N. N. Roni, "Revisiting the Impact of ESG on Financial Performance of FTSE350 UK Firms: Static and Dynamic Panel Data Analysis," *Cogent Bus. Manag.*, vol. 8, no. 1, 2021.

[127] N. Zhu, Y. Zhou, S. Zhang, and J. Yan, "Tax Incentives and Environmental, Social, and Governance Performance: Empirical Evidence from China," *Environ. Sci. Pollut. Res.*, vol. 30, no. 19, pp. 54899–54913, 2023.

[128] Z. Zaporowska and M. Szczepański, "The Application of Environmental, Social and Governance Standards in Operational Risk Management in SSC in Poland," *Sustainability*, vol. 16, no. 6, p. 2413, 2024.

[129] S. Singh and H. S. Nalwa, "Nanotechnology and Health Safety – Toxicity and Risk Assessments of Nanostructured Materials on Human health," *J. Nanosci. Nanotechnol.*, vol. 7, no. 9, pp. 3048–3070, 2007.

Chapter 7

A framework of ESG and green sustainability technology for physiotherapy

Surya Vishnuram[1], Vinodhkumar Ramalingam[1] and Prathap Suganthirababu[1]

7.1 Introduction

Sustainability has emerged as a paramount concern across various sectors of society, and healthcare is no exception. The integration of sustainability principles into healthcare practices not only addresses the pressing environmental challenges facing our planet but also plays a crucial role in delivering improved patient care and fostering social responsibility [1]. In this chapter, we delve into the intersection of sustainability and physiotherapy, exploring how environmental, social, and governance (ESG) principles are reshaping the landscape of healthcare, with a particular focus on the field of physiotherapy.

7.1.1 The importance of sustainability in healthcare

Healthcare systems worldwide are facing increasing scrutiny due to their environmental footprint, which includes resource consumption, waste generation, and greenhouse gas emissions [2]. In an era characterized by climate change, resource depletion, and social inequities, healthcare providers and institutions are recognizing the imperative of adopting sustainable practices. Sustainable healthcare aims to minimize its environmental impact while maximizing positive social outcomes, all while maintaining high standards of patient care [3].

7.1.2 The relevance of ESG principles

ESG principles have gained prominence in the corporate world as a framework for evaluating the ESG practices of businesses. Now, these principles are finding their place in healthcare, including physiotherapy. Environmental factors encompass issues such as energy efficiency, waste reduction, and the use of eco-friendly materials in clinical settings [4]. Social factors involve the fair treatment of patients, diversity and inclusion, and community engagement. Governance, on the

[1]Saveetha College of Physiotherapy, SIMATS, India

other hand, deals with transparency, accountability, and ethical decision-making within healthcare organizations. The incorporation of ESG principles in healthcare not only aligns with global sustainability goals but also enhances the quality of care and builds trust among stakeholders [5].

7.1.3   Overview of the chapter structure

This chapter is structured to provide a comprehensive understanding of the fusion between ESG principles and green sustainability technology within the realm of physiotherapy. We will begin by examining the fundamental principles of ESG and their applicability to healthcare. Subsequently, we will explore the integration of green sustainability technology, emphasizing its relevance in physiotherapy practices. Throughout the chapter, we will delve into sustainable clinic operations, treatment protocols, and patient engagement strategies. We will also discuss the methods of assessing the environmental and social impact of sustainable physiotherapy practices. Finally, we will acknowledge the challenges faced in adopting sustainable approaches and offer a glimpse into the future of eco-conscious physiotherapy.

7.2   The intersection of ESG and physiotherapy

7.2.1   Explanation of ESG principles

ESG principles serve as a guiding framework for assessing the sustainability and ethical practices of organizations as shown in Figure 7.1. In the context of physiotherapy and healthcare, understanding these principles is crucial [6].

Environmental (E): The "E" in ESG relates to environmental sustainability. This principle emphasizes the reduction of environmental impact, including aspects such as energy consumption, waste management, and carbon emissions. In physiotherapy, ESG-driven practices may involve eco-friendly clinic designs, energy-efficient equipment, and responsible waste disposal [7].

Figure 7.1   ESG framework

Social (S): The "S" in ESG pertains to social responsibility. In healthcare, this means ensuring equitable access to care, promoting diversity and inclusion, and engaging with local communities. Physiotherapy clinics can embody social responsibility by offering affordable services, providing accessible facilities for individuals with disabilities, and actively participating in community health initiatives [8].

Governance (G): Governance refers to the ethical and transparent management of healthcare organizations. This principle ensures that healthcare institutions maintain high standards of ethics, accountability, and patient data security. Within physiotherapy, governance principles dictate ethical treatment protocols, patient confidentiality, and the responsible use of patient data [9].

7.2.2   How ESG applies to healthcare and physiotherapy

ESG principles are highly applicable to the healthcare sector, including physiotherapy, due to several reasons:

Patient-centric focus: ESG principles align with patient-centric care. By emphasizing equitable access, quality of care, and ethical governance, ESG ensures that patients receive the best possible treatments and services [10].

Economic and regulatory factors: As healthcare costs rise and regulations become more stringent, organizations that adopt sustainable practices often find cost savings and regulatory compliance easier to achieve. Sustainable physiotherapy practices can lead to long-term financial benefits [11].

Public trust and reputation: Patients and communities increasingly expect healthcare providers to be socially responsible and environmentally conscious. Organizations that prioritize ESG principles build trust and gain a competitive edge [12].

7.2.3   Case studies showcasing successful ESG integration in healthcare

Green hospital initiatives: Numerous hospitals and healthcare systems globally have embraced ESG principles. For example, the Cleveland Clinic in the United States implemented energy-efficient lighting, reduced waste through recycling programs, and integrated sustainable practices into its operations. These efforts led to cost savings and a reduced environmental footprint [13].

Patient-centered sustainability: The National Health Service (NHS) in the United Kingdom launched the "Greener NHS" campaign, which aims to reduce the carbon footprint of healthcare services. This initiative includes sustainability-focused training for healthcare professionals and patient engagement strategies, demonstrating the importance of social and environmental factors in patient care [14].

Sustainable physiotherapy clinics: Physiotherapy clinics are also contributing to sustainable healthcare. Some clinics are adopting paperless record-keeping systems to reduce paper waste and minimize their environmental impact. Others are implementing energy-efficient equipment and using eco-friendly materials in their facilities, showcasing how ESG can be integrated into everyday physiotherapy practices [15].

These case studies highlight the tangible benefits of integrating ESG principles into healthcare and physiotherapy, including cost savings, improved patient outcomes, and a positive impact on the environment and society. In the subsequent sections of this chapter, we will delve deeper into the specific applications of ESG and green sustainability technology in physiotherapy settings.

7.3   Green sustainability technology in physiotherapy

Green technology, often referred to as "clean technology" or "sustainable technology," encompasses innovative solutions designed to minimize environmental impact while maximizing resource efficiency. In healthcare, including physiotherapy, the adoption of green technology holds significant promise for improving patient care and contributing to environmental sustainability.

Green technology in physiotherapy aligns with the broader movement toward sustainable healthcare practices. Its integration not only reduces the environmental footprint of healthcare facilities but also enhances patient outcomes and operational efficiency [16].

7.3.1   Benefits of adopting green technology in physiotherapy

Environmental impact reduction: Green technology in physiotherapy reduces energy consumption, minimizes waste production, and lowers greenhouse gas emissions. This reduction in environmental impact contributes to the overall sustainability of healthcare operations.

Cost saving: Energy-efficient equipment and sustainable facility designs can lead to substantial cost savings over time. Lower utility bills, reduced resource consumption, and decreased waste disposal expenses contribute to a more economically efficient practice.

Improved patient experience: Green technology often results in quieter, cleaner, and more comfortable clinical environments. Patients benefit from improved air quality, reduced noise pollution, and the knowledge that they are receiving care in a socially responsible and eco-friendly setting.

Enhanced reputation: Physiotherapy clinics that prioritize green technology demonstrate a commitment to environmental responsibility and community well-being. This commitment can enhance the clinic's reputation and attract environmentally conscious patients.

7.3.2   Examples of green technology applications in physiotherapy clinics

Energy-efficient equipment: Physiotherapy clinics can invest in energy-efficient rehabilitation equipment such as treadmills, stationary bikes, and therapeutic modalities. These devices consume less electricity and may incorporate smart features to optimize usage.

Renewable energy sources: Some clinics integrate renewable energy sources, such as solar panels or wind turbines, to power their facilities. These sources generate clean energy, reduce dependence on fossil fuels, and lower energy costs.

Smart thermostat and lighting system: Automated heating, ventilation, and air conditioning (HVAC) systems, along with smart lighting controls, help clinics optimize energy usage. These systems adjust temperature and lighting based on occupancy and daylight, reducing energy waste.

Telehealth and digital record-keeping: Telehealth technologies enable remote physiotherapy consultations, reducing the need for patients to travel and associated emissions. Additionally, digital record-keeping systems reduce paper usage and improve data accessibility.

Sustainable facility design: When building or renovating clinics, sustainable design principles can be applied. This includes using eco-friendly construction materials, incorporating natural lighting, and implementing efficient insulation to reduce heating and cooling needs.

Recycling and waste management: Establishing recycling programs within physiotherapy clinics encourages proper disposal of waste materials, including medical supplies and packaging, diverting them from landfills.

Green rehabilitation practices: Physiotherapists can incorporate eco-conscious rehabilitation practices into patient treatment plans. For example, recommending outdoor therapy sessions in natural settings promotes both physical and mental well-being while reducing the need for energy-consuming indoor facilities.

These examples illustrate the diverse ways in which green sustainability technology can be integrated into physiotherapy clinics. The adoption of such technologies not only aligns with ESG principles but also serves as a testament to a clinic's commitment to environmental stewardship and holistic patient care (Figure 7.2). In the subsequent sections of this chapter, we will explore sustainable

Figure 7.2   Outdoor therapy sessions in natural setting

clinic operations and treatment protocols that complement the integration of green technology [17,18].

7.4   Sustainable clinic operations

7.4.1   *Sustainable facility design and construction*

Eco-friendly building materials: When designing or renovating physiotherapy clinics, selecting environmentally friendly building materials can significantly reduce the facility's environmental impact. Examples include using recycled materials, low-volatile organic compound (VOC) paint, and sustainable wood products.

Energy-efficient building design: Sustainable clinic design incorporates energy-efficient features such as large windows for natural lighting, proper insulation, and orientation to optimize solar gain. These design elements reduce the need for artificial lighting and heating, leading to lower energy consumption.

Water conservation: Implementing water-saving fixtures, such as low-flow faucets and toilets, minimizes water wastage within the clinic. Rainwater harvesting systems can also be installed to reuse rainwater for non-potable purposes.

Green roofs and gardens: Rooftop gardens or green roofs not only improve insulation but also provide an attractive and sustainable environment. These features enhance air quality and reduce heat absorption, contributing to energy efficiency.

7.4.2   *Waste reduction and recycling practices in physiotherapy clinics*

Recycling stations: Set up clearly labeled recycling stations throughout the clinic to encourage proper disposal of recyclables, such as paper, cardboard, and plastic. Educate staff and patients about recycling practices.

Reusable supplies: Whenever possible, opt for reusable medical supplies and equipment. For instance, consider using washable, sterilizable items rather than single-use disposable products.

Eco-friendly packaging: Select suppliers that prioritize eco-friendly packaging for medical supplies and equipment. Minimize unnecessary packaging materials and choose recyclable or biodegradable options.

Electronic records: Transition to electronic health records (EHRs) and digital record-keeping systems to reduce paper consumption. EHRs improve data accessibility, streamline administrative tasks, and minimize paper waste.

Medication and hazardous waste disposal: Adhere to proper disposal protocols for medical waste and hazardous materials. Work with licensed waste disposal companies to ensure safe and environmentally responsible disposal practices.

Patient education: Engage patients in sustainable practices by educating them on proper waste disposal within the clinic. Encourage patients to bring their reusable water bottles and reduce unnecessary waste generation.

By implementing sustainable facility design, energy-efficient equipment, and effective waste reduction and recycling practices, physiotherapy clinics can significantly reduce their environmental footprint and promote eco-conscious healthcare. These sustainable operations not only align with ESG principles but also contribute to cost savings, enhanced patient experiences, and a greener future for healthcare facilities. In the subsequent section, we will explore the incorporation of sustainable treatment protocols within physiotherapy practices [19,20].

7.5 Sustainable treatment protocols

7.5.1 Incorporating ESG principles into patient care

Equitable access: ESG principles advocate for equitable access to healthcare. Physiotherapy clinics should ensure that all patients, regardless of socioeconomic status or physical abilities, have equal access to treatment. This includes providing affordable options, accommodating disabilities, and addressing language and cultural barriers.

Ethical governance: Ethical governance practices should extend to patient care. Clinics should uphold transparency, informed consent, and respect for patient autonomy. Clear communication and ethical decision-making contribute to a patient-centric approach aligned with ESG.

Social responsibility: Engage in community outreach and social responsibility initiatives. Physiotherapy clinics can organize health education workshops, support local health programs, or collaborate with nonprofit organizations to address healthcare disparities and social determinants of health.

7.5.2 Sustainable rehabilitation techniques

Natural rehabilitation environments: Embrace the healing potential of nature by incorporating outdoor therapy sessions when feasible. Natural settings provide a therapeutic backdrop that promotes patient well-being and mental health while reducing the need for energy-consuming indoor facilities.

Ergonomic equipment: Choose rehabilitation equipment and tools that promote patient comfort and minimize energy consumption. Ergonomically designed equipment reduces the risk of injury and enhances the patient's rehabilitation experience.

Active rehabilitation: Encourage active patient involvement in the rehabilitation process. Empowering patients to take charge of their health not only improves outcomes but also reduces the environmental impact associated with passive treatments.

Sustainable techniques: Explore rehabilitation techniques that have a lower environmental footprint, such as hydrotherapy can use less water than traditional baths, and manual therapies can be energy-efficient compared to electronic modalities.

7.5.3 The role of telehealth in reducing environmental impact

Reduced travel: Telehealth eliminates the need for patients to travel to the clinic, thereby reducing carbon emissions associated with transportation, This is particularly significant for patients who live far from healthcare facilities.

Energy efficiency: Telehealth appointments consume significantly less energy than in-person visits, which often require lighting, heating, cooling, and medical equipment operation in clinical settings.

Digital records: Telehealth relies on EHRs, minimizing paper usage and the need for physical storage space. Digital records improve data accessibility and reduce administrative waste.

Remote monitoring: Telehealth can include remote monitoring of patients' progress and vital signs. This reduces the need for frequent in-person check-ups, saving time, resources, and energy.

Flexible scheduling: Telehealth offers flexible scheduling options, reducing wait times and congestion in clinic waiting rooms. This contributes to a more efficient use of healthcare resources.

Integrating ESG principles into patient care, adopting sustainable rehabilitation techniques, and leveraging telehealth for eco-friendly healthcare delivery not only align with sustainability goals but also enhance the quality and accessibility of physiotherapy services. In the subsequent section, we will explore strategies for patient education and engagement in eco-conscious physiotherapy practices [21,22].

7.6   Patient education and engagement

7.6.1   Raising awareness among patients about ESG and sustainability

Information dissemination: Provide patients with educational materials and resources that explain the concept of ESG and its relevance to healthcare. This can include pamphlets, brochures, or digital content accessible through the clinic's website or patient portal.

In-person discussion: Take advantage of face-to-face interactions with patients to discuss the clinic's commitment to sustainability and how it aligns with their values. Clinicians can explain how sustainable practices benefit both patient care and the environment.

Social media and online platforms: Utilize social media channels and online platforms to share sustainability updates and tips with patients. Regular posts on the clinic's social media profiles can raise awareness and engage patients in sustainability discussions.

7.6.2   Strategies for involving patients in eco-friendly physiotherapy practices

Patient-centered goals: Collaborate with patients to set eco-friendly rehabilitation goals. For instance, encourage patients to incorporate walking or cycling into their daily routines as a sustainable means of transportation to the clinic.

Home exercise programs: Develop home exercise programs that utilize minimal equipment and emphasize sustainable, cost-effective exercises that patients can perform in their own environment.

Reuse and recycling: Educate patients about the importance of reusing and recycling equipment and supplies. Encourage them to return or recycle items such as orthopedic braces, resistance bands, or elastic tape responsibly.

Digital resources: Provide patients with digital resources, such as exercise videos and guides, accessible via a patient portal or mobile app. Digital resources reduce the need for printed materials.

Engagement in sustainability initiatives: Invite patients to participate in sustainability initiatives, such as community clean-up events or health and wellness programs with an environmental focus. Involvement in such activities promotes eco-friendly practices beyond the clinic.

7.6.3 Measuring patient outcomes and satisfaction

Surveys and feedback: Implement patient satisfaction surveys that include questions related to sustainability and eco-friendly practices. Gathering feedback from patients helps gauge their perception of the clinic's commitment to sustainability.

Outcome measures: Assess patient outcomes related to physiotherapy interventions. Evaluate whether sustainable rehabilitation techniques and eco-friendly practices have a positive impact on patient recovery and overall well-being.

Patient reviews and testimonials: Encourage patients to share their experiences with sustainable physiotherapy practices through online reviews or testimonials. Positive reviews can serve as endorsements of the clinic's commitment to eco-conscious care.

Long-term follow-up: Monitor patient progress and satisfaction in the long term to determine if sustainable physiotherapy practices contribute to improved outcomes and sustained patient engagement.

Comparative analysis: Compare patient outcomes and satisfaction between clinics that prioritize sustainability and those that do not. Such comparative analyses can provide insights into the effectiveness of eco-friendly practices in healthcare.

By raising awareness, involving patients in eco-friendly physiotherapy practices, and measuring patient outcomes and satisfaction, clinics can foster a culture of sustainability that not only benefits the environment but also enhances patient engagement and care quality. In the subsequent section, we will delve into the methods of assessing the environmental and social impact of sustainable physiotherapy practices [23,24].

By using relevant metrics and key performance indicators, tracking and reporting on environmental and social performance, and ensuring compliance with regulatory requirements and industry standards, physiotherapy clinics can effectively assess their sustainability efforts. This comprehensive approach not only promotes eco-conscious practices but also enhances the clinic's reputation, patient trust, and contribution to a more sustainable healthcare system. In the concluding section, we will acknowledge the challenges and opportunities in adopting sustainable approaches and offer insights into the future of eco-conscious physiotherapy [25,26].

7.7 Conclusion

In conclusion, sustainable physiotherapy faces challenges related to costs, awareness, and regulatory complexities, but it holds promise for a more eco-conscious and patient-centered future. Emerging technologies and innovations are set to revolutionize physiotherapy practices, making them more efficient, accessible, and personalized. As healthcare providers continue to embrace ESG principles and green technology, they will contribute to a greener and healthier world while delivering high-quality patient care. To healthcare professionals in physiotherapy and beyond, we extend an encouragement to embrace sustainability wholeheartedly. By incorporating ESG principles and green technology into your practices, you are not only advancing your profession but also making a profound impact on the health of our planet and the well-being of your patients. Sustainability is not just a trend; it is a responsibility we all share as healthcare providers.

References

[1]   World Health Organization. *Environmental Health in Emergencies and Disasters: A Practical Guide*. World Health Organization; 2015.

[2]   United Nations. *Transforming our World: The 2030 Agenda for Sustainable Development*. United Nations; 2015.

[3]   American Physical Therapy Association. *APTA's Environmental Sustainability Guide for Physical Therapy Practice*. American Physical Therapy Association; 2020.

[4]   Cleveland Clinic. *Cleveland Clinic Greening the Clinic*. [Accessed on October 1, 2023]. Available from: https://my.clevelandclinic.org/-/scassets/files/org/about/greening-the-clinic/gtc-overview-2018.pdf; 2023.

[5]   NHS England. *Greener NHS*. [Accessed on October 1, 2023]. Available from: https://www.england.nhs.uk/greenernhs/; 2023.

[6]   Healthcare Without Harm. *Healthier Hospitals Initiative*. [Accessed on October 1, 2023]. Available from: https://noharm-global.org/issues/us-canada/healthier-hospitals; 2023.

[7]   Smith J, Johnson A, Brown K, *et al.* Sustainable physiotherapy: A review of current practices and future considerations. *Journal of Physiotherapy* 2022; 68(1):42–50.

[8]   Green PT, *EcoHealth Rehabilitation Network*. *Eco-Friendly Practices in Physiotherapy: A Practical Guide*. EcoHealth Publishing; 2021.

[9]   Jones R, Patel B, and Smith C. Sustainable rehabilitation: Integrating ESG principles into physiotherapy. *Physiotherapy Sustainability Journal*. 2023;5 (2):78–88.

[10] Global Green Healthcare. *Sustainable Healthcare Practices: A Handbook for Clinics*. Global Green Healthcare Publications; 2020.

[11]   International Council of Nurses. *Sustainability and Healthcare: A Nursing Perspective*. International Council of Nurses; 2019.

[12]   American College of Healthcare Architects. *Sustainable Design in Healthcare Facilities: Key Trends and Best Practices*. American College of Healthcare Architects; 2021.

[13]   European Society of Physiotherapy. *ESG Principles in Physiotherapy: Guidelines for Clinicians*. European Society of Physiotherapy; 2022.

[14]   Health and Sustainability Alliance. *Sustainable Healthcare: A Global Perspective*. Health and Sustainability Alliance Publications; 2018.

[15]   Johnson E, *Green Technology in Physiotherapy: Case Studies and Applications*. Sustainable Healthcare Institute; 2022.

[16]   Green PT, *Sustainable Clinic Operations: Strategies for Green Facility Management*. Sustainable Rehabilitation Network; 2021.

[17]   Wai Yie L. *Medical Equipment Engineering*. Vol. 54, Healthcare Technologies; 2023, pp. 29–37.

[18]   World Physiotherapy. *Tele-Rehabilitation: The Future of Physiotherapy Care*. World Physiotherapy; 2023.

[19]   Green PT, *Sustainable Rehabilitation Techniques: A Handbook for Physiotherapists*. Sustainable Rehabilitation Network; 2021.

[20]   Telehealth Alliance. *The Role of Telehealth in Sustainable Healthcare*. Telehealth Alliance; 2022.

[21]   American Telemedicine Association. *Telehealth and Environmental Sustainability: A White Paper*. American Telemedicine Association; 2021.

[22]   Green PT, *Sustainable Patient Engagement: Strategies for Eco-Friendly Physiotherapy Practices*. Sustainable Rehabilitation Network; 2022.

[23]   International Journal of Sustainable Healthcare. Special issue: Sustainable physiotherapy practices. *International Journal of Sustainable Healthcare*. 2021;20(3).

[24]   Green PT, *Measuring Patient Outcomes and Satisfaction in Sustainable Physiotherapy*. Sustainable Rehabilitation Network; 2023.

[25]   Sustainable Healthcare Foundation. *Sustainable Healthcare: A Handbook for Healthcare Professionals*. Sustainable Healthcare Foundation; 2020.

[26]   United Nations Framework Convention on Climate Change. *Healthcare and Climate Change: A Global Assessment*. United Nations Framework Convention on Climate Change; 2019.

Chapter 8

Physiotherapy – a green sustainable alternative healthcare

Vinodhkumar Ramalingam[1], Jagatheesan Alagesan[2] and Surya Vishnuram[1]

8.1 Introduction

Physiotherapy has an extensive background in healthcare that has transformed over the years. Physiotherapy's foundations can be traced back to ancient cultures [1]. In ancient Greece, Rome, China, and other cultures, various forms of manual therapy, exercise, and hydrotherapy were used to cure injuries and physical diseases [2,3]. The contemporary field of physiotherapy began to emerge in the late nineteenth and early twentieth centuries. The profession expanded in popularity, especially during World Wars I and II [4,5]. The demand for physiotherapy services grew in response to the crucial need to rehabilitate injured soldiers. This time period was considered critical in establishing the physiotherapy profession as a recognized profession globally. In addition, physiotherapy associations and professional organizations were founded to standardize physiotherapy practice and education. In that regard, notable examples are the American Physical Therapy Association (APTA) in the United States and the Chartered Society of Physiotherapy (CSP) in the United Kingdom [6,7].

Besides, treatment methods have advanced significantly in the field of physiotherapy over time, which consists of ultrasonography, electrotherapy, manual therapy, exercise treatment, and more. These methods help patients recover from a variety of ailments, reduce discomfort, and increase mobility. Additionally, physiotherapy has expanded its scope, which includes orthopaedic, neurological, paediatric, and sports physiotherapy. These specializations address the requirements of particular patient populations, with a rising focus on physiotherapy evidence-based practice [8]. A deeper comprehension of the efficacy of various treatment modalities has been made possible through research and clinical trials. Additionally, wearable technology, electrical stimulation devices, and ultrasound equipment have been used as a result of the integration of technology into physical

[1]Saveetha College of Physiotherapy, SIMATS, India
[2]School of Paramedical, Allied and Health Care Sciences, Mohan Babu University, India

therapy. Emerging cutting-edge tools for physiotherapy include virtual reality (VR) and telerehabilitation [9,10]. Through the promotion of wellness initiatives, injury prevention, and healthy lifestyles, physiotherapy has increased its role in preventive care. Physiotherapists assist patients in avoiding injuries and preserving their physical health [11]. With the ability to provide rehabilitation services for a broad range of illnesses and age groups, it is an essential component of healthcare systems. Physiotherapy has evolved from a conventional profession to one that is evidence-based and science-based and plays a vital part in the healthcare system [12]. Physiotherapists are highly appreciated nowadays for their skills in assisting patients with their physical well-being, chronic condition management, and injury recovery. Therefore, when practising, physiotherapy professionals need to keep sustainable physiotherapy in mind in the present industry revolution.

8.2   Industrial revolution in healthcare

The industrial revolution refers to a mix of physical, biological, and digital technologies that are transforming various economic sectors, including physiotherapy in the healthcare industry [13]. Automation, artificial intelligence (AI), the Internet of Things (IoT), and the growing interconnectedness of systems and devices are some of the major characteristics that define this revolution in healthcare [14]. It has the potential to completely transform the sustainability, accessibility, and delivery of healthcare services.

A myriad of technological innovations, including wearable technology, AI and machine learning, 3D printing, big data and data analytics, VR, augmented reality (AR), and telehealth and telemedicine, have the potential to completely transform healthcare practices during the industrial revolution with no limitation in physiotherapy practice.

Wearable technology: Smartwatches and fitness trackers are two examples of wearable technology that have proliferated. With the help of these gadgets, people may proactively manage their health by keeping an eye on their vital signs, physical activity levels, and other health-related data. Additionally, these devices are useful for individual illness prevention and remote patient monitoring [15].

AI and machine learning: AI is being used to evaluate enormous volumes of medical data, offering insights into treatment options, patterns of disease, and patient outcomes (Figure 8.1). AI-powered diagnostic systems have the potential to decrease the need for pointless testing and interventions by improving the precision and effectiveness of medical assessments [16].

3D printing: This cutting-edge technology is transforming the production of prosthetic limbs, medical equipment, and even human tissues. Given that 3D printing can be more accurate and resource-efficient than traditional production techniques, this could lead to a decrease in waste [17].

Big data and data analytics: From genomics to electronic health records (EHRs), the healthcare sector is gathering enormous volumes of data. In addition to helping to personalize care and find trends, data analytics can also increase the

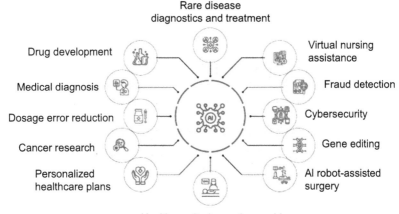

Figure 8.1   Applications of AI in healthcare

general effectiveness of healthcare systems [18]. VR and AR are being used in patient education, therapy, and medical training. Immerse technologies have the potential to improve healthcare delivery while conserving resources [19].

Telemedicine and telehealth: Patients can now obtain physiotherapy or medical treatments and consultations from a distance thanks to the growing popularity of telehealth services. This change has the added benefit of being more environmentally friendly because it lessens the need for patients to physically travel to healthcare facilities, which lowers the industry's carbon footprint [10,20].

8.3   Impact on healthcare by industrial revolution

Access to care, preventive and personalized care, efficiency and resource management, patient engagement, and environmental responsibility are just a few of the ways that the healthcare industry is being significantly impacted by the Fourth Industrial Revolution [21].

Access to treatment: Telehealth services are removing obstacles based on geography, making it possible for people living in rural or underdeveloped areas to receive medical treatment. By doing this, long-distance travel's damaging environmental effects are lessened, and on the other hand, accessibility to healthcare services is also improved [22].

Preventive and personalized care: Cutting-edge technologies like wearables and AI-powered analytics make individualized and preventive care possible. Patients who actively manage their health with the adaptation of advanced technology-supported devices in their regular therapy sessions may find that they require fewer, more resource-intensive therapies [23].

Efficiency and resource management: The healthcare industry benefits from the integration of AI, data analytics, and 3D printing as it lowers waste, simplifies procedures, and optimizes the use of resources. This may result in healthcare practices that are more sustainable [24].

Patient engagement: Through increased engagement with patients, VR and AR technologies are improving the efficacy of therapy and education. Patients who are actively involved in their healthcare are more likely to adopt sustainable lifestyle choices [25].

Environmental responsibility: Reducing needless examinations, operations, and travel related to online consultations will help lower the healthcare industry's carbon footprint. Healthcare operations are progressively incorporating sustainable practices [26].

The Fourth Industrial Revolution is radically altering the healthcare environment by utilizing technology to enhance sustainability, efficiency, and accessibility [27]. These developments could completely change the way healthcare is provided, making it more patient-focused and ecologically conscious. As a field of healthcare, physiotherapy practice may adapt to these developments and conform to Industry 4.0's sustainable healthcare principles. The following sections of this chapter will address the need for sustainable alternatives in healthcare practice.

8.4   Need for sustainable alternatives in healthcare

Numerous interrelated factors, including public health, patient-centred care, climate change, resource scarcity, economic viability, environmental impact, regulatory and ethical considerations, long-term viability, resilience and preparedness, and corporate and social responsibility, have made the need for sustainable alternatives in healthcare, which includes physiotherapy services, increasingly essential [28].

Environmental impact: The healthcare industry's degradation of the environment is quite high. In the past and present, the healthcare industry produces a lot of waste, uses a lot of energy and water, and pollutes the air since it uses fossil fuels for medical equipment and transportation [27].

Climate change: The energy-intensive operations in the healthcare sector are a major contributor to the industry's carbon footprint. In addition to being a concern to global health, climate change increases the frequency of diseases and extreme weather events, which calls for a more sustainable strategy [29].

Scarcity of resources: A lot of drugs and medical equipment depend on finite, non-renewable resources. Healthcare expenses may increase and the supply of essential medical supplies may become rarer as these resources grow more constrained [30].

Economic viability: Healthcare systems may face financial strain as a result of unsustainable healthcare practices. Saving money and making the healthcare industry more financially sustainable can be achieved by cutting waste and increasing resource efficiency [31].

Public health: Environmental deterioration, unsustainable practices, and climate change can all result in health catastrophes. For example, zoonotic disease outbreaks and respiratory illnesses can be attributed to habitat degradation and air pollution [32]. These dangers can be reduced by using sustainable healthcare practices.

Patient-centred care: People are becoming more aware of how their healthcare decisions may affect the environment. When presented with the opportunity, they are more likely to choose sustainable health practices, therefore they are looking for providers who take sustainability into account [33].

Considerations related to ethics and regulations: The healthcare sector is moving towards more environmentally friendly procedures, considering ethics and regulations. As a result, these factors cover the need for appropriate trash disposal as well as the effects of medications on the environment [34].

Long-term viability: Healthcare providers play a vital role in ensuring long-term sustainability. They guarantee the continued provision of healthcare for future generations by adopting green and sustainable practices [35].

Resilience and preparedness: Disasters and alterations in the environment, such as severe weather, can cause havoc in the healthcare system. Resilience and readiness in the face of such difficulties can be enhanced through sustainable healthcare practices [36].

Corporate and social responsibility: Sustainability and corporate social responsibility are becoming central to the missions of many healthcare organizations. This not only improves individual standing but also supports more general social objectives related to sustainability [37].

Healthcare is changing to become more sustainably run in response to the above-discussed considerations. This shift includes green alternatives for medical treatment and rehabilitation, like sustainable physiotherapy, as well as eco-friendly building design, energy-efficient procedures, waste reduction, sustainable sourcing, telehealth services, and patient education on sustainability [38]. The fight for sustainable healthcare is a component of a broader international effort to combat climate change, save natural resources, safeguard public health, and make sure that, despite environmental difficulties, healthcare is still accessible and reasonably priced. The transition to healthier and more egalitarian healthcare practices, like physiotherapy, is essential. The following sections of this chapter will address the crucial role of physiotherapy in the context of sustainable healthcare.

8.5 Crucial role of physiotherapy in the context of sustainable healthcare

Sustainable healthcare assists people in maintaining and enhancing their physical well-being, and physiotherapists frequently participate in preventive healthcare [29]. Physiotherapists inspire patients to take proactive steps to prevent injuries and eliminate the need for more resource-intensive medical treatments by encouraging physical activity or exercise, appropriate body mechanics and posture education, and injury prevention strategies.

Non-invasive approach: To treat musculoskeletal problems and encourage recovery, physical modalities, manual therapy, and therapeutic exercises are some of the non-invasive approaches used in physiotherapy. This method frequently eliminates the need for intrusive medical procedures or surgery, which can have an impact on the environment and resources [27].

Rehabilitation: Physiotherapists are essential to the rehabilitation process when surgery is required for a patient's condition. By assisting patients in regaining their strength, mobility, and functionality, they speed up their recovery and reduce the need for lengthy hospital stays.

Telerehabilitation: During the COVID-19 pandemic, telerehabilitation was very popular, and more people are using telerehabilitation services even after the pandemic. Even physiotherapy follow-up services use digital technology to monitor and treat patients from a distance by the physiotherapist in virtual mode. This strategy lessens the impact that patient travel has on the environment while also improving accessibility to physiotherapy services [10].

Patient education: An essential part of physical therapy is patient education. Physiotherapists instruct patients on how to lead sustainable lifestyles, the advantages of exercises and self-care practices, and their ailments. Patients take an active role in their care, which promotes sustainability over time.

Sustainable rehabilitation programs: Physiotherapists frequently concentrate on early intervention and long-term rehabilitation initiatives. Their goal is to enhance the long-term health and well-being of their patients by treating the underlying causes of their ailments and imparting long-lasting self-management skills [27].

Efficiency and resource management: Physiotherapy clinics are, on a growing basis, implementing sustainable practices. This includes using energy-efficient Heating Ventilation and Air Conditioning (HVAC) and lighting systems, using EHRs to cut down on paper usage, and designing clinics using environmentally friendly materials. These actions result in cost reductions and more effective resource management [39].

Green rehabilitation equipment: It is important to choose the right materials and equipment for rehabilitation. Physiotherapists can reduce their environmental impact by choosing sustainable and eco-friendly equipment. This could include equipment that is long-lasting and composed of sustainable materials [39,40].

Environmental awareness: Physiotherapists can help patients become more conscious of their surroundings. They can advise patients on sustainable living habits and eco-friendly workout regimens, as well as educate them on the effects healthcare decisions have on the environment [27,41].

Decreased medication usage: Physiotherapy services use non-pharmaceutical therapies for patients with acute and chronic illnesses [26]. On the other hand, physiotherapy interventions result in less medication usage and disposal, which has a negative environmental impact. Further, physiotherapy can occasionally assist in managing diseases without the need for medication.

Physiotherapy adds to the larger movement of sustainable healthcare in the community by utilizing technology and incorporating sustainable practices [27]. It is an essential part of the transition to sustainable healthcare in the context of the Fourth Industrial Revolution (Industry 4.0) since it not only offers patients effective care but also adheres to environmental responsibility and resource conservation [21]. The following sections of this chapter will address the environmental and social impacts of traditional healthcare.

8.6 Environmental and social impacts of traditional healthcare

Physiotherapy experts play an important role in the community's health and well-being, primarily by implementing traditional healthcare. Traditional healthcare, although important for enhancing health and well-being, can have detrimental social and environmental impacts that should be considered when moving towards sustainable healthcare.

8.6.1 Impacts on the environment

Waste generation: Conventional healthcare providers produce a significant amount of medical trash, which includes infectious waste and potentially dangerous items like sharps (dry needles). Inappropriate disposal may contaminate water and soil (Figure 8.2).

Energy consumption: Clinics and hospitals use a lot of energy in their operations. Fossil fuel consumption for heating and power increases greenhouse gas emissions, which worsen climate change.

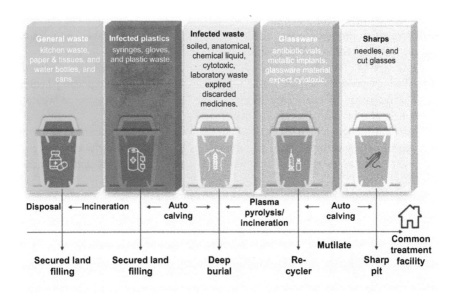

Figure 8.2   Biomedical waste management

Water usage: For a variety of uses, such as patient care, sanitation, and landscaping, healthcare facilities need a lot of water. Overuse of water can put a strain on available supplies, especially in areas where water is scarce.

Chemical use: Chemicals used in healthcare are not adequately handled, they may end up in the environment and have an adverse effect on ecosystems. Water contamination may result from improper disposal.

Resource consumption: Non-renewable resources are needed in the manufacturing of physiotherapy supplies, and equipment. This may result in a shortage of resources and increased healthcare expenses [42].

8.6.2   Impacts on society

Health inequities: Because of social, geographic, or economic constraints, certain communities or groups may have limited access to care [43]. This is a problem that is occasionally perpetuated by traditional healthcare systems.

Rising costs of healthcare: Increasing healthcare expenditures are a result of traditional healthcare's negative environmental effects, such as waste management and energy use. These expenses may put a burden on healthcare systems and constrain underprivileged populations' access.

Air pollution: Emissions from emergency generators, boilers, and automobile traffic can all be a factor in hospital and clinic air pollution in metropolitan locations. The health of people in general is negatively impacted by air pollution, which can result in respiratory illnesses and other issues.

Public health risks: By exposing populations to hazardous materials and infectious pathogens, ineffective waste management and disposal procedures in healthcare services can create a risk to public health.

Ecosystem disruption: The effects of healthcare on the environment can have an adverse effect on biodiversity and wildlife in the area. For instance, poor waste management can taint waterways and endanger aquatic life.

Resilience challenges: Traditional healthcare practices contribute to climate change, which presents serious obstacles to the infrastructure supporting healthcare. The provision of healthcare services may be impacted by extreme weather conditions and other climate-related disturbances.

Community engagement: It is possible that traditional healthcare systems do not always interact with the local community to learn about their wants and preferences, which causes a gap in communication between patients and healthcare professionals.

Health literacy: Inadequate attempts to advance health literacy and education may leave patients with incomplete knowledge of their problems and available treatments, which could have a negative impact on their health.

Sustainable healthcare practices are becoming more and more important in healthcare systems across the globe in order to address these social and environmental implications [44]. This entails implementing energy-efficient technology, minimizing waste via recycling and appropriate disposal, advancing health equity, enhancing community involvement, and placing a strong emphasis on health education and prevention. Sustainable healthcare aims to lessen these effects while offering everyone in society access to high-quality care. Considering the impacts of traditional healthcare, the following section will address sustainable alternative healthcare.

8.7   Sustainable alternative healthcare

In many respects, sustainable healthcare is in line with the ideas of Industry 4.0. It uses innovation and technology to advance efficiency, environmental responsibility, and patient-centred care. The following are the ways that Industry 4.0 and sustainable healthcare are compatible:

Patient-centred care in digital and telehealth environments (eHealth): Personalized and patient-centred care is encouraged by Industry 4.0, and eHealth makes this possible by enabling patients to obtain medical treatments from a distance [43].

Decreased carbon footprint: By eliminating the need for patients to physically travel to medical facilities, telehealth helps to reduce the carbon footprint related to patient transportation.

8.7.1   Wearable technology and the Internet of Things

Remote monitoring: IoT sensors and wearable devices can be used to continuously monitor patient health, which can help identify problems early and avoid readmissions to the hospital. In addition, with the support of a patient data system, the physiotherapist can generate individualized treatment plans, and care can be provided more effectively while using fewer resources [14].

Efficiency and preventive care: Enhancing the productivity of physiotherapy services and cutting down on waste may optimize healthcare operations efficiently. On the other hand, by identifying at-risk populations, predictive analytics makes it possible to implement cost-effective interventions and preventive care techniques that are achieved by the involvement of individual physiotherapists and institutions with long-term strategies that are accountable for application in clinical practice (Figure 8.3).

8.7.2   Artificial intelligence

Treatment personalization: AI-powered algorithms have the ability to tailor treatment plans, improving patient outcomes and cutting down on resource waste.

3D printing: Customization is made possible through 3D printing, which lowers waste and enhances patient outcomes by enabling the production of personalized medical items.

Localized production: The environmental impact of moving medical equipment and supplies can be lessened by using on-site 3D printing [14,25].

8.7.3   Smart facilities and green building practices

Energy and resource efficiency: In line with Industry 4.0's energy-efficient guidelines, sustainable healthcare facilities use renewable energy sources, energy-efficient

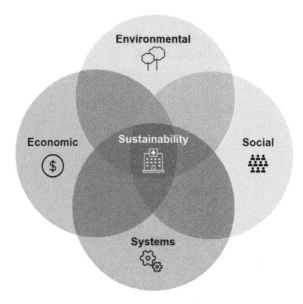

Figure 8.3   Sustainable healthcare system

HVAC systems, and energy-efficient lighting. To minimize their environmental impact, smart facilities use data to optimize the usage of resources like water and electricity.

Patient engagement: In line with Industry 4.0's emphasis on patient empowerment, telerehabilitation can help patients take an active role in their own care and encourage positive behavioural changes.

Reduced travel: Patients' carbon footprint is decreased when telerehabilitation eliminates the need for them to go to physical therapy facilities.

Engagement and education: By enhancing patient outcomes and encouraging improved patient compliance, VR/AR technology can enhance the educational and therapeutic aspects of healthcare.

Remote education: VR and AR allow for remote healthcare education and training, which eliminates the need for travel and expensive training initiatives.

Paperless records and electronic health records (EHRs): Digital health records reduce the need for paper by streamlining administrative procedures and enhancing healthcare efficiency. EHRs improve patient data accessibility, enabling preventive health initiatives and individualized treatment [45].

Sustainable healthcare practices align with Industry 4.0's goals, which include improving productivity, fostering customization, and lessening the environmental effects of healthcare and other industries. Sustainable healthcare aims to deliver high-quality treatment while reducing waste and environmental damage by utilizing technology, data, and creativity. For the healthcare system to be sustainable and successful, healthcare providers must be in line with the ideas of Industry 4.0 [46].

The following section in this chapter will address the key part of sustainability, which is environmental physiotherapy.

8.8 Environmental physiotherapy

The physiotherapy profession plays a significant role in promoting environmental responsibility through various means. Here are some ways in which physiotherapists and the profession contribute to sustainability and eco-friendly practices:

Green clinics: By using ecologically friendly procedures, physical therapy clinics can set an example for others to follow. This entails the use of eco-friendly building materials, sustainable waste management techniques, and energy-efficient HVAC and lighting systems. In addition to lessening the clinic's environmental impact, green clinic design offers patients a more comfortable and healthy setting.

Waste reduction: Physiotherapists can encourage their clinics to reduce trash. This entails recycling, managing medical waste properly, and reducing paper use by implementing EHRs and digital documentation.

Sustainable procurement: Physiotherapists have the opportunity to choose environmentally friendly and sustainable options when choosing tools, materials, and equipment for rehabilitation [47]. Choosing products composed of recyclable or biodegradable materials may fall within this category.

Telerehabilitation: By utilizing remote monitoring and telerehabilitation, physiotherapists can help patients avoid having to travel to clinics [10]. By doing this, the transportation sector's carbon impact is reduced and accessibility is enhanced.

Patient education: Physiotherapists have the ability to inform their patients about the advantages of eco-friendly exercise regimens as well as the effects that healthcare decisions have on the environment. Sustainable practices are more likely to be embraced by patients who care about the environment.

Energy efficiency: Clinics can lessen the environmental effect of everyday commuting by installing energy-efficient lighting, heating, and cooling systems as well as encouraging employees and patients to walk, cycle, or take public transit.

Environmental advocacy: By taking part in relevant campaigns, working with environmental organizations, and influencing healthcare, physiotherapists, and professional associations can promote sustainability in healthcare.

Green rehabilitation equipment: Physiotherapists can use long-lasting, robust rehabilitation equipment composed of sustainable materials, which minimizes the need for frequent replacements [40,47].

Public awareness: Physiotherapists can promote both individual health and environmental responsibility by running public awareness campaigns about the health advantages of an active lifestyle and sustainable living practices.

Research and innovation: The development of environmentally friendly rehabilitation methods and technologies that lessen the impact of healthcare delivery might be the focus of physiotherapy research.

Continuing education: Physiotherapists may be required to complete continuing education credits that address sustainability and environmental responsibility as part of their curriculum from professional organizations and educational institutions [29].

The physiotherapy profession can help advance the larger movement in favour of sustainable healthcare by incorporating these practices. By encouraging environmentally friendly procedures in their offices and informing patients and the general public about the connection between individual health and environmental well-being, physical therapists may make a significant difference in the community [27,29]. This advances patient treatment while simultaneously supporting the global mission of protecting the environment.

8.9   Conclusion

This chapter summarizes that sustainable healthcare is an alternative to traditional healthcare that reduces carbon footprints. The future direction of sustainable physiotherapy is likely to encompass a range of innovative practices and approaches that focus on reducing the environmental footprint of physiotherapy services while delivering effective patient care. Incorporating green physiotherapy care, telerehabilitation, patient-centred sustainability, and community engagements into the future of sustainable physiotherapy will likely be marked by a strong commitment to environmental responsibility, energy efficiency, and eco-friendly practices. It will not only contribute to minimizing the environmental impact of healthcare but also align with broader global goals of sustainability, climate action, and resource conservation. As patients and healthcare providers become more conscious of the need for eco-friendly healthcare, sustainable physiotherapy practices will continue to evolve and expand.

References

[1]   Roberts PA. *The Practice of Physiotherapy: Theoretical and Contextual Reflections.* Sheffield Hallam University (United Kingdom); 2000.
[2]   Mahlovanyy AV, Hrynovets VS, Kuninets OB, *et al. Bases of Physical Rehabilitation in Medicine*; 2019.
[3]   Pettman E. A history of manipulative therapy. *Journal of Manual & Manipulative Therapy* 2007;15(3):165–174.
[4]   Ottosson A. One history or many herstories? Gender politics and the history of physiotherapy's origins in the nineteenth and early twentieth century. *Women's History Review.* 2016;25(2):296–319.
[5]   Martyr P. *The Professional Development of Rehabilitation in Australia, 1870–1981*; 1994.
[6]   American Physical Therapy Association. *Physical Therapist Clinical Performance Instrument.* Alexandria, VA: American Physical Therapy Association; 1997;29.

[7]   Barclay J. *In Good Hands: The History of the Chartered Society of Physiotherapy, 1894–1994*; 1994.

[8]   Christensen N, Black L, Gilliland S, Huhn K, and Wainwright S. The role of movement in physical therapist clinical reasoning. *Physical Therapy*. 2023: pzad085.

[9]   Valverde-Martínez MÁ, López-Liria R, Martínez-Cal J, Benzo-Iglesias MJ, Torres-Álamo L, and Rocamora-Pérez P. Telerehabilitation, a viable option in patients with persistent post-COVID syndrome: A systematic review. *Healthcare* 2023 (11, 2, 187).

[10] Annadurai B, Jagatheesan GPA, and Ramalingam V. Telerehabilitation in physiotherapy. In *Medical Equipment Engineering: Design, Manufacture and Applications*; 2023.

[11] Saks M. The evolution of the medical and health professions in the National Health Services. In *National Health Services of Western Europe: Challenges, Reforms and Future Perspectives*; 2023 Aug 14.

[12] Alsalem TN and Almuhaid TM. Physiotherapist as health promotion practitioners in primary health care (PHC) in Riyadh City. *Advances in Clinical and Medical Research* 2023;4(1):1–48.

[13] Karatas M, Eriskin L, Deveci M, Pamucar D, and Garg H. Big data for healthcare industry 4.0: Applications, challenges and future perspectives. *Expert Systems with Applications*. 2022;200:116912.

[14] Thakare V, Khire G, and Kumbhar M. Artificial intelligence (AI) and Internet of Things (IoT) in healthcare: Opportunities and challenges. *ECS Transactions*. 2022;107(1):7941.

[15] Charan GS, Khurana MS, and Kalia R. Wearable technology: How healthcare is changing forever. *Journal of Chitwan Medical College*. 2023;13(3):111–3.

[16] Rajpurkar P, Chen E, Banerjee O, and Topol EJ. AI in health and medicine. *Nature Medicine*. 2022;28(1):31–8.

[17] Javaid M, Haleem A, Singh RP, and Suman R. 3D printing applications for healthcare research and development. *Global Health Journal*. 2022;6(4): 217–26.

[18] Estel K, Scherer J, Dahl H, Wolber E, Forsat ND, and Back DA. Potential of digitalization within physiotherapy: A comparative survey. *BMC Health Services Research*. 2022;22(1):496.

[19] Lucena-Anton D, Fernandez-Lopez JC, Pacheco-Serrano AI, Garcia-Munoz C, and Moral-Munoz JA. Virtual and augmented reality versus traditional methods for teaching physiotherapy: A systematic review. *European Journal of Investigation in Health, Psychology and Education*. 2022;12(12):1780–92.

[20] Kumar S, Sharma A, and Rishi P. Importance and uses of telemedicine in physiotherapeutic healthcare system: A scoping systemic review. *Data Engineering for Smart Systems: Proceedings of SSIC 2021*. 2022:411–22.

[21] Bradu P, Biswas A, Nair C, *et al.* Recent advances in green technology and Industrial Revolution 4.0 for a sustainable future. *Environmental Science and Pollution Research*. 2022:1–32.

[22] Gujral K, Scott JY, Ambady L, *et al.* A primary care telehealth pilot program to improve access: Associations with patients' health care utilization and costs. *Telemedicine and e-Health.* 2022;28(5):643–53.

[23] Bhatt P, Liu J, Gong Y, Wang J, and Guo Y. Emerging artificial intelligence–empowered mHealth: Scoping review. *JMIR mHealth and uHealth.* 2022;10(6):e35053.

[24] Manickam P, Mariappan SA, Murugesan SM, *et al.* Artificial intelligence (AI) and internet of medical things (IoMT) assisted biomedical systems for intelligent healthcare. *Biosensors.* 2022;12(8):562.

[25] Akshaya AV and Vigneshwaran S. Artificial intelligence is changing health and eHealth care. *EAI Endorsed Transactions on Smart Cities.* 2022;6(3).

[26] Banerjee S and Maric F. Mitigating the environmental impact of NSAIDs-physiotherapy as a contribution to one health and the SDGs. *European Journal of Physiotherapy.* 2023;25(1):51–5.

[27] Palstam A, Sehdev S, Barna S, Andersson M, and Liebenberg N. Sustainability in physiotherapy and rehabilitation. *Orthopaedics and Trauma.* 2022; 36(5):279–83.

[28] Haleem A, Javaid M, Singh RP, and Suman R. Medical 4.0 technologies for healthcare: Features, capabilities, and applications. *Internet of Things and Cyber-Physical Systems.* 2022;2:12–30.

[29] Stanhope J, Maric F, Rothmore P, and Weinstein P. Physiotherapy and ecosystem services: Improving the health of our patients, the population, and the environment. *Physiotherapy Theory and Practice.* 2023;39(2):227–240.

[30] Shahabi S, Skempes D, Mojgani P, Bagheri Lankarani K, and Heydari ST. Stewardship of physiotherapy services in Iran: Common pitfalls and policy solutions. *Physiotherapy Theory and Practice.* 2022;38(12):2086–2099.

[31] Rennie K, Taylor C, Corriero AC, *et al.* The current accuracy, cost-effectiveness, and uses of musculoskeletal telehealth and telerehabilitation services. *Current Sports Medicine Reports.* 2022;21(7):247–60.

[32] Al Huraimel K, Alhosani M, Gopalani H, Kunhabdulla S, and Stietiya MH. Elucidating the role of environmental management of forests, air quality, solid waste and wastewater on the dissemination of SARS-CoV-2. *Hygiene and Environmental Health Advances.* 2022;3:100006.

[33] Morera-Balaguer J, Martínez-González MC, Río-Medina S, *et al.* The influence of the environment on the patient-centered therapeutic relationship in physical therapy: A qualitative study. *Archives of Public Health.* 2023;81 (1):92.

[34] Merry L, Castiglione SA, Rouleau G, *et al.* Continuing professional development (CPD) system development, implementation, evaluation and sustainability for healthcare professionals in low-and lower-middle-income countries: A rapid scoping review. *BMC Medical Education.* 2023;23(1):498.

[35] Zaman K, Anser MK, Awan U, *et al.* Transportation-induced carbon emissions jeopardize healthcare logistics sustainability: Toward a healthier today and a better tomorrow. *Logistics.* 2022;6(2):27.

[36] Thomas A and Suresh M. Readiness for sustainable-resilience in healthcare organisations during Covid-19 era. *International Journal of Organizational Analysis*. 2022;31(1):91–123.
[37] Allard HM. *Corporate Social Responsibility and Financial Performance in the Healthcare Industry* (Doctoral dissertation, Northcentral University); 2018.
[38] Prabadevi B, Deepa N, Victor N, *et al. Metaverse for Industry 5.0 in NextG Communications: Potential Applications and Future Challenges*. arXiv pre-print arXiv: 2308.02677; 2023.
[39] William MA, Elharidi AM, Hanafy AA, Attia A, and Elhelw M. Energy-efficient retrofitting strategies for healthcare facilities in hot-humid climate: Parametric and economical analysis. *Alexandria Engineering Journal*. 2020; 59(6):4549–62.
[40] Silva BV, Holm-Nielsen JB, Sadrizadeh S, Teles MP, Kiani-Moghaddam M, and Arabkoohsar A. Sustainable, green, or smart? Pathways for energy-efficient healthcare buildings. *Sustainable Cities and Society*. 2023:105013.
[41] Maric F, Chance-Larsen K, Chevan J, *et al*. A progress report on planetary health, environmental and sustainability education in physiotherapy–Editorial. *European Journal of Physiotherapy*. 2021;23(4):201–202.
[42] Alves RR and Rosa IM. Biodiversity, traditional medicine and public health: Where do they meet? *Journal of Ethnobiology and Ethnomedicine*. 2007; 3:1–9.
[43] Myers, S. (2017). Planetary health: Protecting human health on a rapidly changing planet. *The Lancet*, 390, 2860–2868.
[44] Albers Mohrman S, Shani AB, and McCracken A. Chapter 1 organizing for sustainable health care: The emerging global challenge. In *Organizing for Sustainable Health Care* 2012 Jul 25 (pp. 1–39). Emerald Group Publishing Limited.
[45] Rameshwar R, Solanki A, Nayyar A, and Mahapatra B. Green and smart buildings: A key to sustainable global solutions. In *Green Building Management and Smart Automation* 2020 (pp. 146–163). Hershey, PA: IGI Global.
[46] Sangwan SR and Bhatia MP. Sustainable development in industry 4.0. In *A Roadmap to Industry 4.0: Smart Production, Sharp Business and Sustainable Development* 2020 (pp. 39–56). Cham: Springer.
[47] McLean R, Wagstaff K, Cooke A, *et al*. Green Team Competition 2023 Impact Report for South Warwickshire University NHS Foundation Trust.

Chapter 9

Trends for additive manufacturing in the ESG and green sustainable technology era

R. Kamalakannan[1], P. Sivakumar[2], Dadapeer Doddamani[2], Tan Koon Tatt[3] and S. Nagarajan[4]

The addition of environmental, social, and governance (ESG) investigation into the method of investment has increased in prominence. Investing professionals view gaining a deeper understanding of the firms they invest in as a primary incentive for incorporating ESG problems into their financial research. Within the ESG paradigm, additive manufacturing, or 3D printing, is going through a number of changes. Additive manufacturing offers special potential to address sustainability, social responsibility, and ethical governance practices – concerns that organisations are prioritising more and more. Despite traditional manufacturing processes, which regularly remove extra material, additive manufacturing enables precise and controlled material deposition, minimising waste. Additive manufacturing saves down on material usage and improves resource efficiency by using just the minimum quantity of material to create a product.

9.1 Sustainable in today's additive manufacturing

The requirement to reduce CO_2 emissions and reduce the impact on the environment grows for every industry and technology. In the next 10 years or so, businesses in almost every industry will pledge to reach carbon neutrality or net-zero CO_2 emissions. Additive manufacturing (AM) must help ensure a more sustainable future. Reducing CO_2 emissions is only one component of sustainability; other crucial concerns include recycling, the use of renewable raw materials, waste minimisation, and many other facets of the environment, social, and governance scopes [1]. Therefore, the decrease in CO_2 emissions is just used as an example in this article to highlight AM's future environmental impact and to further advance

[1]Department of Mechanical Engineering, Thiagarajar College of Engineering, Madurai, India
[2]Mechanical and Industrial Engineering, University of Technology and Applied Sciences, Oman
[3]School of Technology & Engineering Science, Wawasan Open University, Penang, Malaysia
[4]Faculty of Mechanical and Industrial Engineering, Bahirdar Institute of Technology, Bahirdar University, Ethiopia

AM as a next-generation technology. At first look, AM appears to be the ideal technology for environmentally friendly production that has no effect on the environment. Building components one at a time and using just the material really needed for each element, without wasting any, seems like the best approach to conserving resources. Nonetheless, it is possible that the reality falls short of this grand goal [2]. For nearly all AM techniques, materials that have previously undergone an additional processing step are necessary as indicated in Figure 9.1. The best significant polymer material classes like resins, filaments, and polymer powders have undergone prior thermal or chemical conversions, using energy and adding a debit entry to their carbon dioxide emissions ledger. In a similar vein, metal AM materials like wires and powders need to be pulled into wires or atomised into powders in the form of particles. Earlier the material can be employed in an AM machine, a significant amount of energy is required for the gas atomisation of powders. A heated, fast-moving inert gas stream melts metal ingots and disperses them into spherical powders with a specific particle size. To maintain the ideal distribution of particle sizes, these particles must next be sieved, which results in a significant amount of waste powder. Furthermore, extended processing durations are necessary for additive manufacturing; these times are typically a few hours for polymers, but they can reach a week for metal powder bed fusion items. In addition, processing requires the use of power to heat the construction room, an inert gas environment, and machine operation for printing.

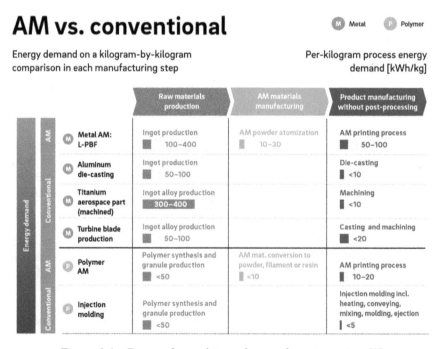

Figure 9.1   Energy demand in each manufacturing setup [3]

9.1.1   Life cycle analysis

An AM-produced item has a bigger environmental footprint at the beginning of its life – that is, before it enters the usage and recycling phases – than a traditionally manufactured part that is made on a kilogramme and process step basis [4]. The AM industry has not often compared AM components to traditional production routes or published full life cycle assessments (LCAs) for AM parts (see Figure 9.2). However, to demonstrate if AM really has a lower environmental effect than a traditional component, a fair LCA is essential. But there's a catch: if AM can make items that traditional manufacturing methods cannot, how can you compare the two manufacturing paths? In fact, AM is advancing the fastest in these 'impossible' domains. Processing AM components leads to an energy disadvantage when comparing 1 kg of material in each process step of the manufacturing chain, as shown in Figures 9.1 and 9.2. For the titanium aerospace bracket example, the ratio is different.

This part shows how the material reduction from AM more than offsets the energy-intensive AM material and AM manufacturing stages, even before design optimisation. This is due to the fact that AM processes produce less waste material than milling processes as shown in Figure 9.3. Then, by allowing for more intricate designs that save weight and only use solid materials where absolutely necessary to maintain mechanical qualities and assure part operation, parts may be optimised. The titanium aerospace brackets of today are usually machined. This production technique may be replaced with metal additive manufacturing laser powder bed fusion, which reduces the part's overall consumption of energy by approximately 75% during the stages of raw material production, manufacture, and end of life. The same bracket's shape may then be optimised to decrease mass by 66%, which will result in an additional 65% of energy savings

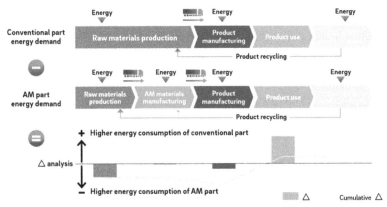

Figure 9.2   Life cycle analysis [3]

AM saves energy for an aerospace bracket

Benefits of AM: less material needed
and weight reduction

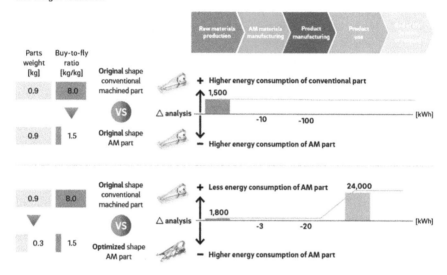

Figure 9.3   Additive manufacturing saves energy for an aerospace bracket [3]

during the manufacturing and raw material production stages. Most crucially, though, during the airlift operation in the product consumption phase, reducing the bulk of the bracket can result in considerable energy (and consequently fuel) savings of around 24 MWh. Similar energy-saving impacts are anticipated from additive manufacturing in areas like improved gas turbine vane cooling, which increases gas turbine efficiency and saves fuel once again. The extra expense that AM often entails in comparison to conventionally made components is also justified by this 'going beyond' the bounds of the conventional. The benefits of AM fabrication are already being felt by many sectors, as the technology allows for the production of increasingly sophisticated items [5]. Consequently, more transparent LCAs that accurately compare additive manufacturing use cases with traditional ones must be produced by the AM industry and its clients. To achieve this, Berger has created a four-step roadmap for AM as a supportable in terms of sustainable manufacturing techniques.

9.1.2   Road map towards greener additive manufacturing

The road map towards greener additive manufacturing has the following four-step approach as indicated below:

1. Increase transparency on the environmental effect of the materials, equipment, and processes used in additive manufacturing
2. Create a database for LCAs specifically for use in
3. Consider the effects on the environment before printing
4. Take steps to lessen additive manufacturing's negative environmental effects.

9.1.3   Improvement of ESG based on additive manufacturing

The total emissions should be considered each time a manufacturing method is used to make a product and additive manufacturing is one of the options [6]. In this case, a fair life cycle analysis of the component produced using additive manufacturing or traditional methods shows the influence, particularly on the Environment in the ESG rating. Examining the part's life cycle from raw material to recycling is crucial in this case. Additionally, 3D printing frequently plays a major role in facilitating local manufacturing, addressing short transit routes and quick delivery, which reduces transportation-related emissions. Our investigation demonstrates how improbable this is. Here, two factors are crucial: first, using fewer raw materials from AM directly lowers material prices and emissions at the same time. Second, the benefits of additive manufacturing often stand out when the item is being used. When an AM component outperforms a conventional one, it is probably also superior in terms of emissions, particularly if it is a moving element on land or in the air. The rationale behind both the AM emissions case and the AM business case is the same.

9.2   Sustainable manufacturing with ESG

It entails using industrial techniques that strike a balance between social and environmental well-being and economic prosperity. A rising number of firms are switching from traditional to sustainable production as they become aware of the growing demand for 'green' operations [7]. This is due to the fact that conventional manufacturing processes frequently employ inefficient methods like milling or cutting, demand a lot of electricity, and frequently depend on non-ecofriendly materials. Sustainable manufacturing is the use of production methods that minimise waste, pollution, and harmful emissions to protect the environment and its people (as shown in Figure 9.4).

9.2.1   Benefits of green sustainable technology

Reduced waste and energy consumption: This issue is resolved by sustainable manufacturing, which incorporates energy-saving technology into every stage of the production process [8]. For instance, researchers at the US National Renewable Energy Laboratory have created a method to 3D print wind turbine blades using thermoplastics, increasing the efficiency of wind turbines in harnessing wind

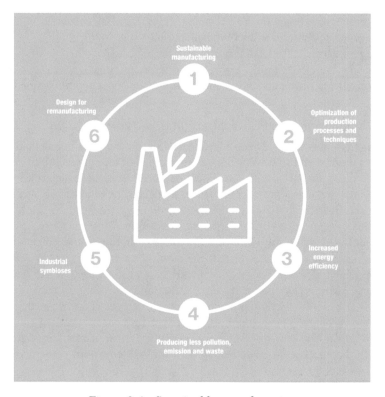

Figure 9.4   Sustainable manufacturing

energy. With this process, cutting-edge blades that are longer, lighter, more affordable, and more effective are produced.

Cost-effectiveness: Through the use of energy-efficient technology and resource optimisation, businesses may lower their operating expenses associated with utilities and raw materials. Furthermore, waste recycling techniques supported by sustainable manufacturing lower the costs associated with material procurement and disposal. This contributes to lower transportation costs and long-term cost savings, together with locally sourced resources and sustainable supply chain management.

Innovation opportunities: The development and processing of products can alter how vital resources are used if sustainability goals are met. This frequently results in the creation of novel, environmentally friendly designs, materials, and production processes. Businesses may look into using renewable resources, use energy-saving manufacturing techniques, or develop goods that are more robust and recyclable. These developments have the potential to improve consumer happiness and product performance in addition to having a negative influence on the environment.

9.2.2   Methods to improve sustainable manufacturing

Organisations have a number of options for improving the sustainability of their manufacturing processes.

Use digital technology and automation: Large volumes of data may be collected and analysed at any point during the production process thanks to digital technology. Manufacturers may obtain important insights into their operations, energy consumption, material utilisation, and waste creation by utilising data analytics. Making well-informed decisions to maximise resource use, pinpoint opportunities for development, and put more sustainable practices into effect is made possible by this data-driven approach. Digital technology combined with automation systems allows for real-time machinery and equipment monitoring. This makes predictive maintenance possible, which minimises energy loss and downtime by detecting any problems or inefficiencies early on [9]. Furthermore, digital technology helps with energy management by giving producers visibility into patterns of energy usage, which enables them to recognise and put energy-saving measures into place. Supply chain networks may be made intelligent and integrated through automation and digital technologies. Manufacturing, suppliers, and consumers may communicate and share data in real time, which enhances demand forecasting, production scheduling, and inventory management. This leads to a more sustainable and efficient supply chain by reducing waste from overproduction, excess inventory, and ineffective logistics (as shown in Figure 9.5).

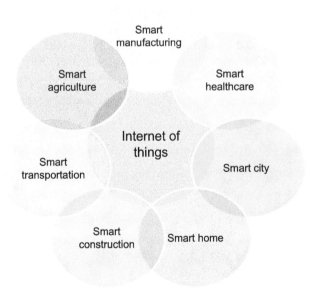

Figure 9.5   Digital technology and automation

Reduce subtractive manufacturing practice: In subtractive manufacturing, the material is taken out of a solid block or stock to form the product into the required shape. It uses equipment and methods including drilling, grinding, milling, turning, and cutting. 3D printing uses an additive method as opposed to traditional manufacturing procedures, which sometimes entail removing extra material from a bigger block. This implies that materials are added to things only where they are required, layer by layer. As the printer only utilises the material required to build the intended product, this approach minimises wasteful material usage. For example, 3D printing uses a lot less material to create an engine part prototype than machining it from a solid block of metal.

Reduces emission and pollution: The industrial sector is responsible for around 30% of greenhouse gas emissions in the United States, including indirect emissions from power usage. Manufacturers must implement strategies that drastically cut down on emissions and pollution to create a sustainable environment. These strategies include recycling materials, using renewable energy sources, converting to lower-emitting fuels like decarbonised petrol, and implementing additive manufacturing techniques. This is particularly crucial as industrial waste pollutes the environment and damages several significant ecosystems, making landfills a burden (as shown in Figure 9.6).

Prioritise local resources: The industrial sector is responsible for around 30% of greenhouse gas emissions in the United States, including indirect emissions from power usage. Long-distance procurement of resources and subsequent transportation back for on-site processing are common starting points for industrial supply chains. However, the process of moving things by land, air, or sea adds to energy waste and toxic emissions that are bad for the environment. Making local resources a priority entails obtaining manufacturing materials locally [10]. Manufacturers may lessen the carbon footprint associated with logistics and transportation by procuring materials and resources locally because shorter distances result in reduced emissions. More significantly, procuring materials locally gives businesses

Figure 9.6   Reduce emission and pollution

Reduce   Reuse   Recycle

Figure 9.7   Reduce, reuse, and recycle

the ability to manage their supply chain and manufacturing expenses. The capacity of the 'reduce, reuse, and recycle' strategy to save raw resources and lessen environmental effects makes it essential for sustainable production. Let us examine how these ideas especially relate to 3D printing: **Reduce:** Manufacturers may cut back on a number of production-related expenses by using 3D printing. Through waste reduction and design optimisation, it makes it possible to reduce the amount of materials required. Manufacturers may minimise surplus and lower total material consumption by precisely controlling the printing process so that only the essential quantity of material is used [3]. **Reuse** is the act of making use of materials that have previously been used once. Examples of this include reusing supports that have been used to print a component or printing new products using leftover filament. Reusing materials contributes to the decrease of trash sent to incinerators or landfills. **Recycling** is the process of disassembling items into their constituent elements and utilising those pieces to make new goods. Recycling may lessen pollution and help preserve natural resources. Reuse and recycling are two ways to cut waste in 3D printing. Nevertheless, the two techniques differ significantly in a few key ways. Generally speaking, reuse requires less energy and is easier than recycling. Reuse and recycling are two ways to cut waste in 3D printing. Nevertheless, the two techniques differ significantly in a few key ways. Generally speaking, reuse requires less energy and is easier than recycling. This is so that energy and resources may be saved since repurposed materials do not require processing or transportation. Reusing materials can result in new products that are distinct from the original. If the new product proves to be more valuable or useful than the original, then this might be advantageous. Reuse is typically the best choice when it comes to cutting waste in 3D printing. Recycling, however, can be a wise choice if the materials that can be reused cannot be used in any other way (as shown in Figure 9.7).

9.3   Conclusion

With the growing awareness of sustainable practices throughout the globe, 3D printing has a great chance to be a big part of manufacturing's future. The above information shows that 3D printing supports environmentally friendly production techniques. Not every 3D printer is thought to be environmentally friendly, even if it is a more sustainable alternative to traditional production To lessen their influence on the environment, sustainable 3D printers usually use environmentally acceptable materials. To use less energy and emit fewer harmful pollutants, they also have energy-saving features including sleep modes and low power consumption. Manufacturing sustainability may be attained through 3D printing by the following: *Cutting waste*: When compared to conventional production techniques, 3D printing may help minimise waste by up to 90% since it only utilises the material required to manufacture an item. *Resource conservation*: By using 3D printing to make parts out of recycled materials, natural resources may be preserved. *Encouraging local production*: By using 3D printing to make items locally, long-distance transportation and the emissions it produces may be minimised.

References

[1]   L. Agnusdei and A. Del Prete, "Additive manufacturing for sustainability: A systematic literature review," *Sustainable Futures*, 4, 100098, 2022, doi: 10.1016/j.sftr.2022.100098.

[2]   M. Hogan, "Sustainable manufacturing: How to use 3D printing for sustainable production," Nexa3D, 2023. [Online]. Available: https://nexa3d.com/blog/sustainable-manufacturing/.

[3]   "How sustainable is additive manufacturing?," *Roland Berger*, 2022. [Online]. Available: https://www.rolandberger.com/en/Insights/Publications/Sustainability-Is-Additive-Manufacturing-a-green-deal.html.

[4]   M. Javaid, A. Haleem, R. P. Singh, R. Suman, and S. Rab, "Role of additive manufacturing applications towards environmental sustainability," *Advanced Industrial and Engineering Polymer Research*, 4, 4, 312–322, 2021, doi:10.1016/j.aiepr.2021.07.005.

[5]   S. Ford and M. Despeisse, "Additive manufacturing and sustainability: An exploratory study of the advantages and challenges," *Journal of Cleaner Production*, 137, 1573–1587, 2016, doi:10.1016/j.jclepro.2016.04.150.

[6]   M. Mani, K. W. Lyons, and S. K. Gupta, "Sustainability characterization for additive manufacturing," *Journal of Research of the National Institute of Standards and Technology*, 119, 419, 2014, doi:10.6028/jres.119.016.

[7]   R. Sreenivasan, A. Goel, and D. L. Bourell, "Sustainability issues in laser-based additive manufacturing," *Physics Procedia*, 5, 81–90, 2010, doi:10.1016/j.phpro.2010.08.124.

[8]   T. Peng, K. Kellens, R. Tang, C. Chen, and G. Chen, "Sustainability of additive manufacturing: An overview on its energy demand and environmental impact," *Additive Manufacturing*, 21, 694–704, 2018, doi:10.1016/j.addma.2018.04.022.

[9]   G. N. Ganeshan, R. Kamalakannan, and S. Vijay, "Design, manufacture and sustainable analysis of biodegradable computer mouse using solid works and

3D printer," *Materials Today: Proceedings*, 21, 56–59, 2020, doi:10.1016/j. matpr.2019.05.360.

[10]   G. G. Natarajan, R. Kamalakannan, and R. Vijayakumar, "Sustainability analysis on polymers using life cycle assessment tool (OPENLCA)," 3rd International Conference on Frontiers in Automobile and Mechanical Engineering (Fame 2020), 2020, doi: 10.1063/5.0034820.

Chapter 10

AI Frontiers in medicine: engineering the future of patient care

Vaikunthan Rajaratnam[1] and Norana Abdul Rahman[2]

Integrating artificial intelligence (AI) into healthcare is a significant development in modern medicine. This chapter explores how AI is being used in healthcare. We focus on understanding how AI technologies, like machine learning and natural language processing, are applied in medical settings. These technologies are not just tools for automation; they enhance medical processes, aiding healthcare professionals and improving patient outcomes.

The chapter will explore various ways AI is currently used in healthcare, such as diagnostics, treatment planning, and patient management. This includes examining how AI contributes to tailored medical treatments and individual patient care. Alongside the benefits, the chapter also discusses the challenges and ethical considerations associated with AI in healthcare. These include issues related to data privacy, potential biases in AI algorithms, and the need for appropriate regulations.

Additionally, the chapter considers the global application of AI in healthcare. It looks at how AI is used in different parts of the world, both in well-resourced areas and places with fewer resources. This global view helps us understand how AI can help reduce healthcare quality and access gaps worldwide.

The literature documents how this AI works in the healthcare sector in various ways. From enhancing drug development, clinical trials, and patient care to improving overall healthcare management through efficient data manipulation. AI's current applications include diagnosis, treatment recommendations, patient engagement, and administrative tasks, often performing these tasks as effectively as humans [1]. AI is growing, particularly in early disease detection, diagnosis, treatment options, and predicting outcomes for significant diseases. It is constructive in the early detection of chronic diseases, which can reduce both the financial burden and severity of illnesses. AI-based services in healthcare show promise in improving patient outcomes, assisting caregivers, and cutting costs. Despite these advancements, ongoing discussions exist about how best to implement AI's potential in practical healthcare settings [2].

[1]Department of Orthopaedic Surgery, Khoo Teck Puat Hospital, Singapore
[2]Faculty of Science, Athena Institute, Vrije Universiteit, Amsterdam

In summary, this chapter is about the role of AI in enhancing healthcare. It presents a balanced view, recognising AI's potential and challenges in the medical field. The aim is to show how AI, when used effectively, can be a valuable tool for healthcare professionals, leading to better healthcare outcomes. This exploration is not only about technology but also involves understanding AI's ethical and practical implications in healthcare.

10.1 Understanding AI technologies in medicine

At its core, AI involves creating machines and software that can simulate human intelligence, encompassing a range of technologies and applications. This introductory section aims to define AI and explore its different types, each characterised by unique capabilities and functions.

AI can be broadly categorised into Narrow AI and General AI. Narrow AI, also known as Weak AI, is designed for specific tasks and is today's most commonly implemented form of AI. It operates within a limited context and does not possess consciousness or genuine intelligence. Examples include language translation apps, recommendation systems, and medical diagnostic tools. On the other hand, General AI, or Strong AI, refers to systems that exhibit human-like intelligence across a wide range of tasks. This type of AI can understand, learn, and apply its intelligence broadly, similar to a human being. Although a fascinating concept, General AI remains theoretical mainly at this stage.

Further exploring AI types, we delve into machine learning (ML) and deep learning (DL). ML, a subset of AI, involves algorithms that enable machines to improve tasks through experience. DL, a more advanced subset of ML, mimics the human brain's neural networks to process data and create patterns for decision-making. These technologies form the backbone of many AI applications in healthcare, from patient data analysis to advanced diagnostic procedures.

Understanding AI's definition and its types provides a foundational context for appreciating its applications and potential in healthcare. As we progress through this exploration, we will see how these AI types differ in their capabilities and how they are uniquely suited to various aspects of healthcare, from administrative tasks to complex clinical decision-making.

AI in healthcare utilises various technologies, each with unique applications. Key technologies include:

1. **Machine learning (ML)**: ML algorithms learn from data and improve over time. In healthcare, ML is used for tasks like analysing medical images for diagnostic purposes. For example, Google Health developed an ML model to detect diabetic retinopathy in eye scans.
2. **Natural language processing (NLP)**: NLP helps computers understand and interpret human language. A notable application is in the analysis of patient records and clinical notes. IBM Watson, for example, uses NLP to

analyse unstructured clinical notes and extract relevant information for patient care.

3. **Predictive analytics**: This involves analysing data to predict future events. In healthcare, predictive analytics can forecast patient admissions, potential disease outbreaks, or the likelihood of disease progression. An example is using predictive models in electronic health records (EHRs) to identify patients at risk of chronic diseases like diabetes.

10.2 AI in diagnostics and treatment

In diagnosis, AI involves sophisticated algorithms capable of analysing complex medical data, ranging from imaging scans to genetic information. These tools assist healthcare professionals in identifying diseases with greater precision and often at earlier stages than traditional methods. For example, AI systems can detect subtle patterns in medical images that might be overlooked by the human eye, leading to more accurate diagnoses of conditions like cancer, cardiovascular diseases, and neurological disorders [3].

Inpatient management: AI contributes to developing personalised treatment plans, monitoring patient progress, and predicting potential health risks. AI-driven applications can analyse vast patient data, including medical history, current health status, and lifestyle factors, to tailor treatment strategies to individual needs. This personalised approach enhances the effectiveness of treatments and improves patient engagement and adherence to care plans.

Moreover, AI technologies streamline administrative tasks in healthcare settings, such as scheduling appointments, managing patient records, and optimising hospital resource allocation. This efficiency not only saves time for healthcare providers but also enhances the overall patient experience.

The integration of AI in diagnosis and management is challenging. It raises important questions regarding data privacy, ethical considerations, and the need to balance technology and human oversight. Despite these challenges, the potential of AI to transform healthcare is immense, promising a future where medical practice is more informed, efficient, and patient-centred. As we delve deeper into this topic, we will explore specific examples, applications, and the evolving role of AI in shaping the future of healthcare diagnosis and management.

1. **Diagnostic imaging**: AI algorithms are increasingly used to interpret X-rays, CT scans, and MRIs. For instance, Zebra Medical Vision* offers AI solutions to identify various medical conditions from imaging data [4].
2. **Personalised medicine**: AI assists in developing tailored treatment plans based on a patient's genetic makeup, lifestyle, and other factors [5].

---

*Zebra Medical Vision's technology for detecting coronary artery calcium has been approved by the FDA. This technology aids clinicians in preventing heart events by analysing CT scans to categorize patients based on their cardiac calcium levels. This facilitates tailored patient care. The FDA's clearance of this technology is a significant step towards wider adoption of AI in healthcare.

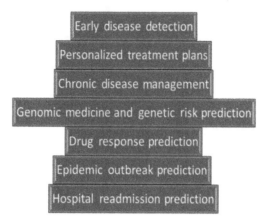

Figure 10.1   AI for personalised medicine

Oncology is a prime area where AI personalises cancer treatments based on genetic data (Figure 10.1).[†]

10.3   Patient management and engagement

AI's innovative applications redefine how healthcare providers interact with and manage patients, leading to more efficient care and enhanced patient experiences.

AI tools in patient management include systems for tracking patient health data, personalising treatment plans, and facilitating effective communication between patients and healthcare providers. These technologies allow for more precise monitoring of health conditions and quicker adjustments to treatment as needed.

Regarding patient engagement, AI creates more interactive and responsive healthcare experiences. This includes AI-powered chatbots for initial consultations, symptom checking, and providing health advice. These tools make healthcare more accessible and can lead to better patient understanding and involvement in their care.

This section explores how AI makes patient management more data-driven and patient engagement more personalised, ultimately leading to better health outcomes and more satisfied patients. We will examine various AI applications in this field, their benefits, and their challenges in the evolving healthcare landscape.

[†]The chapter highlights the transformative role of AI in healthcare, specifically in improving patient care and quality of life. It covers AI's impact on disease diagnosis, treatment plans, and patient engagement. The chapter also addresses ethical considerations and the necessity for human expertise in AI applications. Notably, AI contributes to increased accuracy, cost reduction, and time savings in healthcare services. It plays a significant role in personalising medicine and supporting mental health care. Furthermore, AI aids doctors in decision-making and fosters trust between patients and physicians.

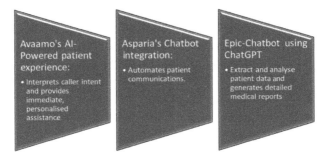

Figure 10.2   Examples of AI in patient engagement

AI tools are reshaping patient management and engagement in several ways. These examples provide an overview of three AI-powered applications designed to enhance patient engagement and streamline healthcare operations.

First, Avaamo's AI-powered patient experience platform is adept at discerning callers' intentions, enabling it to offer prompt and individualised assistance. This feature is precious in delivering a responsive and tailored patient experience. Second, Asparia's chatbot integration focuses on the automation of patient communication, which helps efficiently manage patient queries and reduce the workload on healthcare staff. Lastly, an application leveraging ChatGPT within the Epic-Chatbot framework is highlighted for its ability to meticulously extract and analyse patient data, thereby facilitating the creation of comprehensive medical reports. This integration is a powerful tool for healthcare providers, enhancing the accuracy and efficiency of patient data management and reporting (Figure 10.2) [6].

1. **Chatbots for patient interaction:**
   Chatbots for patient interaction utilise AI to communicate with patients, providing them with information and support. These AI chatbots can answer questions, offer health advice, and assist with scheduling appointments, making healthcare more accessible and efficient for patients.

   AI-powered chatbots, like Babylon Health's chatbot [7], provide initial patient consultations, symptom checks, and health advice.‡

2. **Remote monitoring:**
   With the increase of the ageing population, there has been a high demand for a wearable health monitoring device to assist the elderly in maintaining their physical health conditions. Such devices are highly required to monitor elderly people living alone with limited mobility. As healthcare costs are escalating, there has been a need for an affordable option to monitor elderly people from a remote location, provide them with vital medical data at home, and connect

---

‡Chatbots mimic human interactions through text or voice responses and are designed to deliver accurate information in response to user inquiries. The architecture of an AI-powered health chatbot incorporates elements of dialogue management, communication interfaces, and deep learning algorithms to facilitate this interaction.

them with the nearest professional healthcare providers. As a result, proposing an IoT-based wearable system that provides real-time health monitoring for elderly people becomes essential. It keeps track of the health status of elderly people, alerts healthcare professionals about any changes or abnormalities in real time, and reminds elderly people about the defined medicine time.

Furthermore, the use of a dynamic model that represents the changes in the elderly's behaviour is highly important. This work proposes a new approach for healthcare monitoring using the ThingSpeak IOT platform. This approach is modelled using multi-agent systems. It surveys vital signs such as heart rate, blood pressure, and body temperature and provides fall detection and emergency alert functions [8].§

Remote monitoring for the elderly is critical, in light of a growing ageing population and the need for cost-effective healthcare solutions. To this end, various technological approaches have been explored. IoT-based wearable systems are one such solution, providing continuous health monitoring by tracking vital signs and notifying medical staff of any irregularities.

Privacy-conscious methods have also been developed, using video surveillance and 3D modelling to observe unusual behaviours without compromising personal privacy. Moreover, telemedicine platforms can send automated alerts regarding health deterioration, focusing on risks common in old age. Additionally, there are systems employing microcontrollers and global system for mobile communications (GSM) technology to measure health parameters and send text message updates to caregivers and physicians.

Remote monitoring employs AI-integrated wearable devices to analyse patient data continuously, facilitating the surveillance of chronic health conditions from a distance. A notable instance is the Apple Watch, which tracks heart rate and identifies irregular patterns, exemplifying this technology's capability [9].¶

Technology is integral to managing daily routines and is increasingly important in addressing health and social care challenges among the elderly. Ambient assistive living (AAL) improves the quality of life by merging technology with

---

§The paper contributes to healthcare monitoring for the elderly by proposing an IoT-based wearable system that provides real-time health monitoring. This system tracks vital signs such as heart rate, blood pressure, and body temperature and includes fall detection and emergency alert functions. The system is mainly aimed at elderly people with limited mobility living alone, offering a way to monitor their health status remotely and connect them with healthcare providers as necessary. Additionally, the paper introduces a dynamic model using multi-agent systems to represent the behavioural changes of the elderly, which is a novel approach in this context. The ThingSpeak IoT platform is also a significant contribution, as it facilitates the collection and analysis of health data, enabling timely interventions by healthcare professionals. These contributions address the need for affordable, remote health monitoring solutions in the context of rising healthcare costs and an increasingly aging population.

¶Atrial fibrillation (AF), a prevalent heart rhythm disorder, heightens the risk of stroke among millions of people. Stroke risk can be mitigated in AF patients through oral anticoagulation therapy. However, many AF cases remain undiagnosed, posing serious health risks and economic costs. Continuous ECG monitoring has proven more effective in detecting AF compared to intermittent approaches. Smartwatches offer a potential solution for AF monitoring, provided they ensure accurate detection and are coupled with proper medical follow-up.

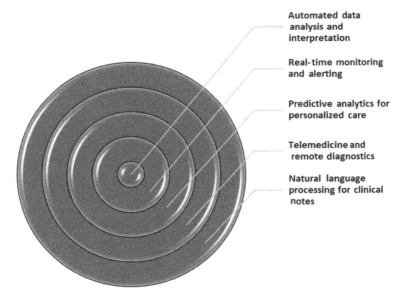

Automated data
analysis and
interpretation

Real-time monitoring
and alerting

Predictive analytics for
personalized care

Telemedicine and
remote diagnostics

Natural language
processing for clinical
notes

Figure 10.3   Patient diagnosis and data management

social settings and integrating different technologies to assist with daily activities
(Figure 10.3) [10].

10.4   Administrative applications of AI

AI applications are being harnessed to optimise the efficiency of healthcare
administration. AI's role in this domain includes automating routine tasks, enhancing data management, and facilitating decision-making processes, thus allowing
healthcare providers to focus more on patient care and less on administrative
burdens.

By leveraging AI, healthcare facilities are experiencing improvements in
operational workflows, patient scheduling, and resource allocation. For example,
AI-driven systems can analyse large volumes of data to predict patient admission
rates, which helps in staffing and bed management. Additionally, AI is instrumental
in managing EHRs, billing, coding, and insurance claims processing with greater
accuracy and speed.

1. **Automating administrative tasks**: AI applications like Nuance's Dragon
   Medical One-use voice recognition to automate clinical documentation, reducing doctors' time on paperwork. EHRs offer clinical benefits and are crucial
   in enhancing patient care. The adoption of EHRs has increased, driven by
   competition among vendors and ongoing innovations in the field. A critical
   aspect of using EHRs effectively is improving their usability, as it is essential
   for ensuring patient safety and delivering quality care [11].

Figure 10.4   AI for healthcare management

2. **Optimising hospital operations**: AI systems analyse hospital data to improve resource allocation and patient flow. For example, General Electric (GE) Healthcare's Command Centre software helps hospitals manage bed availability and patient care coordination (Figure 10.4) [12].[||]

10.5   Challenges and considerations

All stakeholders need to address the complexities and ethical dilemmas that arise with the integration of AI in medical settings. While AI promises significant advancements in patient care, diagnostics, and administrative efficiency, its implementation is challenging [13].

This section explores various issues, from technical challenges such as data security and algorithm accuracy to ethical concerns including patient privacy and bias in AI decision-making. The reliability and interpretability of AI systems, especially in life-critical healthcare applications, demand rigorous scrutiny. Moreover, integrating AI into existing healthcare frameworks poses logistical and infrastructural challenges that require careful planning and resource allocation.

Another critical aspect is the regulatory landscape governing AI in healthcare. This includes understanding and complying with evolving standards and guidelines

---

[||]Health systems frequently struggle with managing their capacity effectively, highlighting a need for improved governance and management. Critical elements of successful capacity management include dedicated leadership, central analytics, alignment of incentives, and an engaged workforce. Various case studies demonstrate the use of tools that have enhanced patient flow and operations in hospitals operating at high capacity.

that ensure AI technologies' safe and ethical use. Collaboration between technologists, healthcare providers, patients, and policymakers is paramount to navigate these challenges effectively.

One of the most critical challenges in implementing AI in healthcare is the protection of data privacy and security. The sanctity of patient data, which often includes highly sensitive and personal health information, is of utmost importance. This section expands on safeguarding this data in an AI-enhanced healthcare environment [14].

10.5.1   Critical concerns in data privacy and security

1. **Vulnerability to data breaches**: AI systems, which often process and store large amounts of patient data, can be susceptible to cyberattacks and data breaches. This risks patient confidentiality and undermines public trust in healthcare institutions.
2. **Compliance with regulations**: Healthcare providers must navigate a complex web of regulations like Health Insurance Portability and Accountability Act (HIPAA) in the United States and General Data Protection Regulation (GDPR) in the European Union. These laws dictate stringent guidelines for handling, storing, and sharing patient data.
3. **Ensuring anonymity in data usage**: When patient data is used to train AI models, it is crucial to anonymise this information to prevent any potential identification of individuals, which could lead to privacy violations [15].

10.5.2   Strategies for enhancing data privacy and security

1. **Robust encryption methods**: Employing advanced encryption for storing and transmitting patient data can significantly reduce the risk of unauthorised access.
2. **Regular security audits and updates**: Monitoring and updating AI systems to guard against new cyber threats is essential for maintaining data security.
3. **Training and awareness**: Healthcare staff should be trained in best practices for data privacy, including understanding the risks involved in handling sensitive patient information.
4. **Developing secure AI models**: Designing AI algorithms that inherently prioritise data security and include safeguards against potential breaches is a crucial consideration for developers.
5. **Patient consent and transparency**: Maintaining transparency with patients regarding how their data is used and obtaining their consent, mainly when their information is utilised for AI model training or similar purposes, is vital.

In summary, while AI offers immense potential to enhance healthcare delivery and outcomes, protecting patient data is paramount. A multifaceted approach involving technological solutions, regulatory compliance, staff training, and transparent patient communication is essential to address the challenges of data privacy and security in AI-driven healthcare. By prioritising these aspects, healthcare providers can harness the benefits of AI while upholding the trust and confidentiality that patients expect.

10.5.3   Bias and ethical concerns in AI-driven healthcare

Implementing AI in healthcare brings to light significant concerns regarding bias and ethics. A critical issue is AI algorithms' potential to inherit biases in the training data. This section will explore how such biases can impact healthcare delivery and the ethical implications thereof [16].

10.5.4   Understanding the origin of bias in AI

Bias in training data: AI algorithms learn from historical data, and if this data contains biases – whether related to race, gender, socioeconomic status, or other factors – the AI is likely to replicate these biases in its outputs. For example, an AI model trained predominantly on data from a specific demographic might need to be more accurate and effective for underrepresented groups.

Implications of biased algorithms: Using biased AI tools in healthcare can lead to unequal treatment recommendations, diagnostic accuracy, and patient outcomes. This not only raises ethical concerns but also undermines the goal of equitable healthcare provision [17].

10.5.5   Addressing bias and ethical challenges

Diverse and representative data sets: Ensuring that training data encompasses a broad spectrum of patient demographics can help reduce the risk of bias in AI models. This includes data from diverse racial, ethnic, and socioeconomic backgrounds.

Algorithmic transparency: Healthcare providers and AI developers should strive for transparency in how AI models are developed and how they make decisions. This involves explaining the data sources, the learning process, and the rationale behind AI-generated recommendations.

Regular audits for bias: Implementing routine checks and audits of AI systems to identify and correct any biases is crucial. This ongoing process helps ensure that AI tools remain fair and effective over time.

Interdisciplinary collaboration: Combining the expertise of healthcare professionals, ethicists, data scientists, and patient advocates can provide a more comprehensive approach to identifying and addressing potential biases in AI systems.

Ethical guidelines and standards: Developing and adhering to robust ethical guidelines specific to AI in healthcare can guide the responsible development and use of these technologies.

AI's integration into healthcare is a multifaceted process with profound implications for patient care, diagnosis, treatment, and healthcare management. While AI holds great promise, it is crucial to navigate its implementation thoughtfully, considering ethical, privacy, and bias-related challenges [18].

10.5.6   The foundation of generative AI

Generative AI marks a significant technological leap with its ability to read and comprehend human-written text. At the heart of this capability is sophisticated NLP technology, which allows machines to parse and understand language in a way that mirrors human comprehension. This foundational skill enables AI to process vast amounts of information, learn from context, and recognise nuances in text.

Generative AI is a branch of AI that generates new data that resembles the training examples it learns from, employing techniques such as generative adversarial networks (GANs) and variational autoencoders (VAEs).**

10.5.7   Example of generative AI

Some examples of generative AI platforms:

1. **Google**: Google has two large language models; Palm, a multimodal model, and Bard, a pure language model.
2. **Microsoft and OpenAI** are embedding generative AI technology CoPilot into their products.
3. **ChatGPT by OpenAI**: This chatbot is capable of very human-seeming interactions.
4. **DALL·E 2 by OpenAI**: This tool generates images from text.
5. **Ada**: A doctor-developed symptom assessment app that offers medical guidance in multiple languages.
6. **Skinvision**: An app for early detection of skin cancer.
7. **Virtual volunteer/Be My Eyes**: An AI app designed for visually impaired individuals.
8. **Hyro**: A conversational AI explicitly designed for health systems to enhance patient engagement

A prominent example is ChatGPT, a state-of-the-art language model by OpenAI that uses Transformer architecture to create text closely mimicking human writing from given prompts. These models are extensively used in tasks requiring understanding natural language, including operating chatbots and generating content.

Understanding language allows AI to interact meaningfully with user inputs and is a powerful tool for information retrieval and knowledge dissemination [19][††].

**Generative adversarial networks (GANs) and variational autoencoders (VAEs) are used for generative modelling but differ in their approaches. GANs use a competitive framework between a generator and a discriminator, excelling at creating high-fidelity and diverse data, but their training can be unstable and resource-intensive. On the other hand, VAEs use an encoder and a decoder to model complex data distributions and generate diverse outputs, offering interpretable latent representations. However, they may struggle with generating highly detailed data. The choice between GANs and VAEs depends on the specific application and desired data characteristics.

[††]Generative AI models like OpenAI's GPT series are poised to transform healthcare through advanced natural language processing capabilities. These models support clinical decision-making, enhancing patient care and service quality. They offer valuable assistance in early disease diagnosis and developing personalised treatment plans. Additionally, GPT models improve the accuracy and speed of medical image interpretation in radiology. In pharmaceuticals, they expedite drug discovery by efficiently predicting effective and safe drug candidates, potentially bringing new treatments to market more quickly. Beyond clinical applications, GPT models also interact directly with patients, providing educational materials and addressing health inquiries, thus improving patient engagement and self-management of health.

10.5.8   *Generating text: the creative power of AI*

Beyond comprehension, generative AI possesses the remarkable ability to produce human-like text. From crafting responses to user queries to generating creative content, AI algorithms can mimic the style and tone of human authors. This text generation is not just a regurgitation of memorised phrases; it is a complex synthesis of language patterns learned from training data. As a result, AI can write informative articles, compose poetry, or generate reports, making it an invaluable asset in fields like journalism, marketing, and education.

10.5.9   *Conversation: interacting with AI*

The conversational aspect of generative AI is perhaps its most engaging feature. These AI systems are designed to engage in text-based conversations with users, providing a responsive and interactive experience. Whether it is a customer service chatbot, a personal assistant, or an educational tutor, AI can converse with users, answer questions, and provide assistance. This interactive capability transforms customer service and support, making it more accessible and efficient.

10.5.10   *Prompt engineering*

This refers to the skilful and strategic formulation of prompts or queries to effectively communicate with and extract desired responses from AI models, particularly those based on language. This practice is vital in fields that leverage AI for generating text, solving problems, or providing information. It involves understanding the capabilities and limitations of the AI model and crafting prompts that are clear, specific, and aligned with the intended goal [20].

Prompt generation is crucial in interacting with AI models, involving formulating input to elicit specific responses. It comprises several steps to enhance clarity and effectiveness:

1. **Defining the objective**: Identifying the specific information or assistance needed.
2. **Clarity and precision**: Using clear language to minimise ambiguity and include essential details without over-complication.
3. **Context consideration**: Providing relevant background or context to inform and guide the response.
4. **Tone and style specification**: If relevant, specify the desired tone (e.g. formal, casual) and style (such as a summary or detailed explanation) of the response.
5. **Direct questions**: Formulating prompts as direct questions when seeking specific information.
6. **Neutral crafting to avoid bias**: Ensuring the prompt is crafted neutrally to avoid leading questions or biased responses.
7. **Testing and refinement**: Experiment with different phrasings, observing the impact on responses and refining the prompt accordingly.
8. **Ethical and privacy considerations**: Adhering to ethical guidelines and avoiding requests or revelations of sensitive or private information.

Each step is integral to creating effective prompts that produce targeted and accurate responses from AI models.

Practical prompt engineering is crucial in leveraging AI for diverse applications, including research, content creation, problem-solving, and interactive dialogue. It requires a blend of technical understanding, linguistic skills, and creativity (see Table 10.1).

Healthcare prompt engineering checklist: Creating a checklist for prompt engineering in healthcare will ensure that prompts are effectively tailored to elicit accurate and helpful information from AI systems. This checklist would include considerations specific to the healthcare domain, such as medical accuracy, ethical considerations, and patient privacy (Figure 10.5).

Table 10.1   Examples of good and bad prompts in healthcare

Bad prompts	Comments	Good prompts	Comments
Tell me about heart problems.	It is too vague and lacks focus and context.	Summarise the diagnostic criteria for Congestive Heart Failure according to the latest ACC/AHA guidelines.	Specific, focused, and references a reputable source.
What drugs are suitable for high BP, diabetes, and heart issues?	Overly complex, risks dangerous oversimplification.	List the first-line antihypertensive medications according to the latest guidelines.	Focused on a single condition, it asks for evidence-based treatment.
What is the best treatment for a 45-year-old male named John Smith with these symptoms?	Contains potentially identifiable information, risking patient confidentiality.	What are the treatment options for a 45-year-old male presenting with these generic symptoms?	Generalised and anonymised, preserving patient confidentiality.

Figure 10.5   Tips for healthcare prompt engineering

10.5.11 Response validation

Validation is where human critical thinking – natural intelligence (RI) meets with AI. This can be summarised as follows:

1. **Reviewing the response**: Assessing if the AI-generated response aligns with the initial requirements and objectives outlined in the prompt.
2. **Limitation acknowledgement**: Recognising that AI does not have access to real-time data can affect the relevance and accuracy of its responses, especially in dynamic fields like healthcare.
3. **Continuous validation**: Emphasising the importance of rigorously validating the AI's responses. This involves ensuring the information is accurate, relevant, and applicable to the specific context.
4. **Iterative process**: Implementing a cycle of prompting, receiving a response, and refining the in-essence, response validation, particularly in healthcare, involves a careful and iterative approach to ensure that the information provided by AI is reliable, accurate, and fitting to the specific needs of the healthcare domain.

10.5.12 Applications: the versatility of generative AI

The applications of generative AI are diverse and continually expanding. Virtual assistants powered by AI are becoming commonplace in smart homes and smart-phones, helping users perform tasks and access information. In education, AI aids in creating learning materials and providing personalised tutoring. Content creation, too, has been revolutionised by AI's ability to generate original articles, stories, and even code. The versatility of generative AI makes it a pivotal tool across various industries, enhancing productivity and fostering innovation.

10.5.13 Not a human: the machine behind the words

Despite its advanced capabilities, it is essential to recognise that generative AI is not human. It generates text through complex algorithms without feelings, desires, or consciousness. The content produced by AI lacks the emotional depth and subjective experiences that human authors bring to their writing. While AI can simulate human-like writing, distinguishing between machine-generated and human-created content is essential, particularly in understanding AI technology's limitations and appropriate applications. Recognising this boundary ensures that while we leverage AI's strengths, we continue to value the irreplaceable qualities of human touch in communication and creativity.

10.5.14 Integration of AI in medical education and training

Integrating AI in medical education and training reshapes future healthcare professionals' preparation. This section explores the various facets of AI's role in medical education. It highlights how it enhances learning outcomes and equips medical students and professionals with advanced skills suited to the evolving healthcare environment (Figure 10.6) [21].

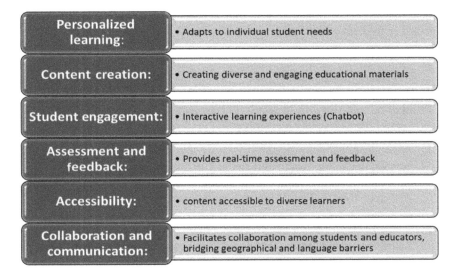

Figure 10.6   Role of AI in medical education

10.5.15   AI-enhanced curriculum development

At the heart of this endeavour is the creation of learning outcomes that are not only aligned with academic standards and professional competencies but also adaptive to the changing demands of the job market and technological advancements. AI-enhanced curriculum development transcends traditional boundaries, offering insights based on extensive data analysis and predictive modelling. This enables educators and curriculum designers to tailor learning outcomes that are precisely calibrated to bridge gaps in knowledge, skills, and competencies.

Furthermore, using AI in writing learning outcomes empowers a more personalised educational experience. By analysing learner data, AI can help identify individual learning patterns, preferences, and challenges, creating outcomes that cater to diverse learning needs and promote inclusivity [22].

10.5.15.1   Example of prompts for developing curriculum

To develop a comprehensive and practical curriculum for teaching nutrition in diabetes management to medical students, the following prompt could be utilised for ChatGPT-4‡‡:

This prompt is designed to guide ChatGPT-4 in creating a comprehensive, clinically relevant curriculum tailored to medical students' educational needs. It balances theoretical knowledge with practical application, considering the multidisciplinary and multicultural aspects of diabetes management. The focus on

---

‡‡This prompt was generated by ChatGPT 4 using this prompt – 'Write a prompt for ChatGPT4 to develop a curriculum for teaching nutrition for diabetes management for medical students'.

evidence-based practice aligns with the academic and professional standards expected in medical education.

Prompt for ChatGPT-4:

Design a detailed and evidence-based curriculum for medical students focused on nutrition in diabetes management. The curriculum should encompass the following key components:

1. **Foundational knowledge**: Provide a module on basic nutritional science, including macronutrients and micronutrients and their roles in the human body. Emphasise the pathophysiology of diabetes, types of diabetes (Type 1, Type 2, gestational diabetes), and the impact of nutrition on blood glucose levels.
2. **Dietary management in diabetes**: Offer in-depth content on the dietary management of diabetes. This should include nutritional recommendations for diabetic patients, understanding the glycaemic index, meal planning strategies, portion control, and the importance of dietary fibre.
3. **Case studies and practical applications**: Integrate case studies that allow students to apply nutritional knowledge in clinical scenarios involving diabetic patients. This should encourage critical thinking and clinical decision-making skills.
4. **Cultural and individual considerations**: Address the cultural diversity in dietary habits and its implications in diabetes management. Include strategies to tailor nutritional advice to individual patient needs, preferences, and socioeconomic backgrounds.
5. **Interdisciplinary approach**: Highlight the importance of an interdisciplinary approach in managing diabetes, involving dietitians, endocrinologists, primary care physicians, and diabetes educators.
6. **Current research and advances**: Provide insights into the latest research and advances in nutrition and diabetes. This could include emerging dietary strategies, novel therapeutic foods, and future directions in diabetes management.
7. **Ethical and patient education aspects**: Cover ethical considerations in patient management and effective patient education and counselling strategies in dietary practices.
8. **Assessment and evaluation**: Outline methods for assessing and evaluating the student's understanding and competency in applying nutritional knowledge in diabetes management.

The curriculum should be structured to facilitate blended learning, incorporating lectures, interactive sessions, workshops, and online resources. Ensure all content is up-to-date, evidence-based, and referenced from credible sources. Additionally, provide guidelines for continual professional development in this area.

This is the response from ChatGPT in executing the prompt it generated, as shown above.

The curriculum you have outlined for teaching nutrition in diabetes management to medical students is comprehensive and well structured. It encapsulates the essential elements to equip future medical professionals with the knowledge and skills to effectively manage diabetes through nutritional interventions. Let us explore how each component contributes to the overall educational objectives:

1. **Foundational knowledge**: This module sets the stage for understanding the intricate relationship between nutrition and diabetes. By covering basic nutritional science and the pathophysiology of different types of diabetes, students gain a solid foundation. Understanding how food impacts blood glucose levels is crucial for managing diabetes effectively.

2. **Dietary management in diabetes**: This section is central to the curriculum, providing practical knowledge on managing diabetes through diet. Teaching about the glycaemic index, meal planning, and dietary fibre is essential for students to advise patients on controlling their blood glucose levels through diet.

3. **Case studies and practical applications**: Applying theoretical knowledge to real-life scenarios enhances learning and prepares students for clinical practice. Through case studies, students develop critical thinking and decision-making skills essential for patient care.

4. **Cultural and individual considerations**: Recognising and respecting cultural diversity in dietary practices is vital in personalised care. This module ensures that future medical professionals are sensitive to their patient's needs and backgrounds, tailoring dietary advice accordingly.

5. **Interdisciplinary approach**: Diabetes management is most effective when it is collaborative among various healthcare professionals. Students learn the value of teamwork in patient care by understanding the roles of dietitians, endocrinologists, and others.

6. **Current research and advances**: Staying abreast of the latest research and advances ensures that medical professionals know the most current and effective practices. This component keeps the curriculum dynamic and forward thinking.

7. **Ethical and patient education aspects**: Addressing ethical considerations and patient education strategies is crucial. This part of the curriculum ensures that students are prepared to handle the complexities of patient interactions and education responsibly and effectively.

8. **Assessment and evaluation**: Finally, assessing and evaluating student competency is critical to ensuring that the educational objectives of the curriculum are met. This component ensures that students are knowledgeable and capable of applying their knowledge in clinical settings.

In summary, your curriculum is meticulously designed to cover all critical aspects of nutrition in diabetes management. It balances theoretical knowledge with practical application, considering cultural diversity and the importance of an interdisciplinary approach. This curriculum is well-positioned to prepare medical students for the challenges of managing diabetes in diverse patient populations. When prompted to 'expand on Item 2', the following was the response.

Expanding on the 'Dietary Management in Diabetes' component of the curriculum, we can delve deeper into several key areas. This module is crucial as it directly addresses how medical students can apply nutritional knowledge to manage and advise patients with diabetes. Here is a more detailed breakdown:

1. **Overview of dietary management in diabetes**: Begin with an introductory lecture that outlines the importance of diet in diabetes management. This should include a historical perspective, current standards, and an overview of how dietary management has evolved.

2. **Nutritional recommendations for diabetic patients**: Provide detailed guidance on nutritional needs for diabetic patients. This includes optimal intake of carbohydrates, proteins, and fats, considering the unique metabolic challenges faced by individuals with diabetes. Discuss the role of diet in maintaining glycaemic control, preventing complications, and improving overall health outcomes.

3. **Understanding the glycaemic index and glycaemic load**: Offer in-depth information on foods' glycaemic index (GI) and glycaemic load (GL). Explain how different foods affect blood sugar levels and how to use this knowledge to make better dietary choices. Include practical examples and tools for estimating the GI and GL of various meals and food items.

4. **Meal planning strategies**: Teach students to develop effective meal plans for diabetic patients. This should cover balancing macronutrients, portion sizes, and meal timing to stabilise blood glucose levels. Include the use of technology and apps in meal planning.

5. **Importance of dietary fibre**: Emphasise the role of dietary fibre in diabetes management. Discuss how fibre impacts glucose absorption and blood sugar levels and the health benefits of soluble and insoluble fibre. Guide how to incorporate high-fibre foods into a diabetic diet.

6. **Managing hypoglycaemia and hyperglycaemia through diet**: Address the dietary management of common diabetes-related conditions like hypoglycaemia and hyperglycaemia. Teach students about the appropriate nutritional adjustments needed in these situations.

7. **Nutrition label reading and food choices**: Teach students how to read and interpret nutrition labels effectively. This skill is essential for

making informed food choices and guiding patients in selecting appropriate foods.

8. **Workshops and simulations**: Conduct interactive workshops and simulations where students can practice developing meal plans and dietary advice for hypothetical patients. Use case studies to simulate real-life scenarios and encourage group discussions for peer learning.
9. **Guest lectures and collaborations**: Invite dietitians, nutritionists, and endocrinologists to deliver guest lectures or participate in panel discussions. This can give students diverse perspectives and insights into practical dietary management in diabetes.
10. **Resource compilation and access**: Compile a comprehensive list of resources, including guidelines, research papers, websites, and apps, which students can use for reference and further learning.

Expanding on these areas (a concept known as Skelton of Thought),[§§] the curriculum equips medical students with theoretical knowledge and the practical skills necessary for effective dietary management in diabetes. This comprehensive approach ensures that future healthcare professionals are well prepared to advise and support their diabetic patients nutritionally. *(The content generated has been validated by a Diabetic subject matter expert and is evidence based).*

Here is a sample assessment generated by ChatGPT with appropriate prompts (Figure 10.7).

10.5.16 AI-driven simulations and virtual patients

1. **Realistic clinical scenarios**: AI-driven simulations provide medical students with realistic, interactive clinical scenarios. These simulations can mimic

---

[§§]Skeleton of thought, or SoT, is a prompting technique to improve the efficiency and quality of outputs from large language models (LLMs). It works by guiding the LLM to generate an answer in two stages:

1. Skeleton stage:
   - The LLM is provided with a 'skeleton prompt' that focuses on identifying the answer's key points, structure, and overall flow. This prompt often takes the form of bullet points or short phrases.
   - The LLM generates a concise outline of the answer, covering the main ideas without getting bogged down in details.
2. Point expanding stage:
   - For each point in the skeleton, the LLM is presented with a separate 'point-expanding prompt'. This prompt delves deeper into the specific subtopics and supporting information.
   - The LLM expands on each point independently, filling in the details and specificities.

1.Question: Which of the following statements best describes the role of HbA1c in managing diabetes?

1. A) HbA1c is primarily used to diagnose acute diabetic complications.
2. B) HbA1c levels reflect the average blood glucose levels over the past 2-3 months.
3. C) HbA1c is a short-term marker for blood glucose fluctuation.
4. D) HbA1c measures the immediate postprandial blood glucose level.

Answer: B) HbA1c levels reflect the average blood glucose levels over the past 2-3 months.

1.Question: A 54-year-old patient with Type 2 diabetes presents with an HbA1c of 8.0%. What is the most appropriate initial approach?

1. A) Immediate hospitalisation for insulin therapy.
2. B) Evaluation and optimisation of the current treatment regimen.
3. C) Disregarding the HbA1c level as it is within the normal range.
4. D) Starting an antihypertensive medication.hospitalisation

---

Figure 10.7   Example of assessment generated by ChatGPT

various medical conditions and patient responses, offering a safe environment for students to practice and hone their diagnostic and treatment skills [23].

2. **Virtual patients**: AI-powered virtual patients are increasingly used as a teaching tool. These sophisticated models can simulate patient interactions, allowing students to practice communication skills, clinical reasoning, and decision-making in a risk-free setting (Figure 10.8).

10.5.17   Enhancing clinical training with data analysis

1. **Data-driven decision-making**: Training in data analysis is becoming essential in medical education. AI tools can analyse vast medical data, helping students understand patterns and correlations in patient care, disease progression, and treatment outcomes.

2. **Personalised learning paths**: AI can tailor educational content to each student's individual learning style and pace. By analysing students' performance and learning behaviours, AI systems can suggest areas for improvement and additional resources.

10.5.18   AI for content creation in healthcare education

AI's role in content creation is not just about automating tasks; it is about augmenting the human creativity process with advanced analytical capabilities and insights, thereby expanding the horizons of what can be achieved in this field.

At the core of AI-driven content creation is the ability to analyse vast amounts of data, recognise patterns, and generate relevant and engaging content for diverse

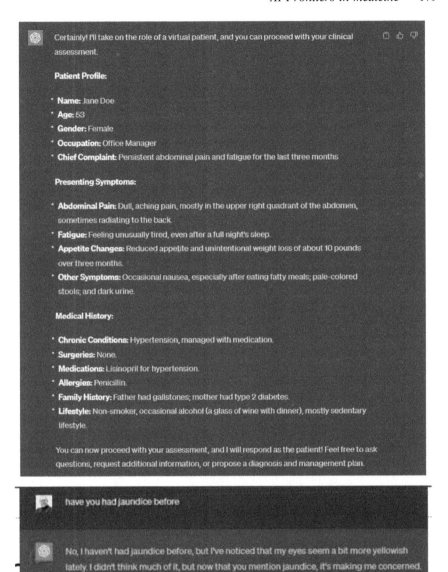

Figure 10.8   A virtual patient generated by ChatGPT

audiences. This involves utilising ML algorithms and NLP tools to create content that ranges from written articles to visual media, all tailored to specific contexts and audience preferences. The implications of this are profound, as it enables creators to produce highly targeted and adaptive content at a scale and speed that was previously unattainable [24].

Furthermore, AI in content creation is not merely a tool for efficiency; it is a catalyst for creativity. It opens up new avenues for storytelling, design, and multimedia expression, allowing creators to push the boundaries of conventional formats and explore innovative narratives. This technology also democratises content creation, making it more accessible to a broader range of individuals and organisations, irrespective of their size or resources.

1. **Objective setting**: Define the purpose and goals of the content creation process. This could include identifying the target audience, the type of content required (e.g. text, images, videos), and the intended impact or outcome of the content.
2. **Data collection and analysis**: Gather relevant data to inform content creation. This can include audience demographics, content preferences, historical engagement data, and other relevant information. AI can be used to analyse this data to identify trends, patterns, and insights that will guide the content creation.
3. **Selection of AI tools**: Choose the appropriate AI tools and technologies based on the content needs. This could include NLP tools for text generation, ML algorithms for content personalisation, or AI-driven design tools for visual content.
4. **Training the AI model**: If using ML, this step involves training the AI model with relevant data sets to ensure it understands the context and can generate appropriate content. This might include feeding the AI examples of similar content, style guides, or other relevant materials.
5. **Content generation**: Utilise the AI tool to generate initial content drafts. This could involve AI writing assistants for textual content, AI-based graphic design tools for visual content, or automated video creation tools.
6. **Review and refinement**: Human intervention is crucial to review the AI-generated content at this stage. This involves editing, refining, and ensuring the content meets quality standards and aligns with the set objectives.
7. **Optimisation based on feedback**: Gather feedback and engagement data after publishing or using the content. Use this information to train further and refine the AI models, optimising them for future content creation tasks.
8. **Ethical and compliance considerations**: Throughout the process, it is essential to consider ethical implications, such as bias in AI-generated content, and ensure compliance with relevant laws and regulations (e.g. copyright laws and data privacy regulations).
9. **Continuous learning and adaptation**: AI in content creation is evolving. Stay informed about new technologies and methods, and continuously adapt strategies to leverage these advancements effectively.

By following these steps, content creators can effectively harness AI technologies to produce high-quality, relevant, and engaging content while ensuring ethical and regulatory compliance.

10.5.19   *Creating engaging video content with an AI platform*

AI is revolutionising video content creation for medical education by simplifying and enhancing the production process. AI-driven platforms offer an array of customisable templates suitable for various educational purposes. Users can quickly adapt these templates to their educational content needs, modifying text, colours, and design elements with intuitive editing tools [25].

Content creators can further personalise their videos by adding their media or choosing from a vast collection of stock images, footage, and audio tracks on the platform. This flexibility allows for the assembly of an informative and engaging narrative catering specifically to medical education audiences.

Key to the creation process is the ability to overlay text, insert captions, and incorporate educational graphics, which are essential for explaining complex medical concepts and data. Selecting the right audio can significantly enhance the learning experience, providing a sensory backdrop that supports the educational material.

AI tools also enable real-time previews, allowing creators to adjust and maintain narrative coherence. Once the video content meets the desired educational standards and objectives, it can be exported in the preferred format and resolution for use in various educational settings.

The feedback on the completed video becomes a valuable asset, driving continuous improvement in content delivery and presentation. In medical education, where accurate and effective communication is critical, the blend of AI technology and human creativity sets new standards for educational video content (Figure 10.9).

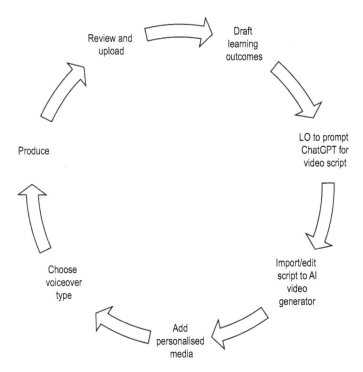

Figure 10.9   Steps for video content creation in medical education with AI (LO – learning outcome)

10.5.20   Ethical and practical considerations

1. **Ethical training**: It is important to incorporate training on the ethical use of AI in healthcare. Students should know the ethical dilemmas and privacy concerns associated with AI technologies.
2. **Critical thinking and human oversight**: While AI provides valuable tools and insights, medical education must emphasise critical thinking, human judgement, and empathy in patient care.

In summary, integrating AI in medical education and training is an enhancement of traditional methods and a fundamental shift towards a more interactive, data-driven, and patient-centred approach to medical education. This shift promises to equip future healthcare professionals with the skills and knowledge necessary to excel in an increasingly AI-integrated healthcare landscape.

10.5.21   AI in healthcare research

AI's ability to process vast amounts of data, recognise complex patterns, and provide predictive insights offers unparalleled opportunities for advancements in diagnosis, treatment, and disease prevention [26].

In healthcare research, AI tools and algorithms are transforming how researchers approach complex medical challenges. From analysing genetic information to identifying potential new drugs, AI enables faster, more accurate, and more efficient research processes. This technological revolution is crucial in global health crises, where rapid responses and solutions are essential.

The potential of AI in healthcare research extends beyond conventional methodologies, offering a new lens through which the intricacies of human health can be understood and addressed. AI's role in this field is multifaceted, involving the enhancement of research capabilities, the improvement of patient outcomes, and the optimisation of healthcare systems. The following discussion explores the various dimensions of using AI in healthcare research, examining its impact, challenges, and prospects in transforming the landscape of medical science and patient care.

The advancement of AI tools in academic research has opened new avenues for scholars and researchers, particularly in handling and analysing research papers. Platforms like Elicit.org and SCISPACE have emerged as valuable assets in this domain [27].

10.5.21.1   Elicit.org: enhancing research paper analysis

Elicit.org is a cutting-edge tool designed to assist researchers in navigating through the vast ocean of academic literature. Its capabilities extend to various aspects of research paper handling:

1. **Advanced search functionality**: Elicit.org uses AI algorithms to help researchers find relevant papers quickly. Unlike traditional databases, it can understand the context of your search query, providing more targeted results.

2. **Summarisation and analysis**: The platform can summarise research papers, highlighting key findings and methodologies. This feature is handy for researchers who need to process large volumes of literature efficiently [28].

3. **Trend identification**: Elicit.org can identify trends and emerging research areas within specific fields. This is invaluable for staying up-to-date with the latest developments.

4. **Collaborative features**: The tool also offers features for collaboration, allowing research teams to work together seamlessly on literature reviews and data analysis.

10.5.21.2 SCISPACE: leveraging generative AI for document analysis

SCISPACE offers a suite of tools that utilise generative AI to analyse and enhance academic documents:

1. **Document analysis with generative AI**: SCISPACE's core feature involves using AI to analyse academic papers. This includes understanding the structure, extracting critical information, and offering insights into the content.

2. **Integration with SciSpace copilot**: By adding SciSpace Copilot to your browser, you can access SCISPACE's features directly while working on your documents online. This integration simplifies the research process, making it more efficient and user-friendly.

3. **Real-time assistance**: The platform provides real-time suggestions and improvements to your writing based on the context and content of your document.

4. **Data visualisation and interpretation**: SCISPACE can help visualise complex data and interpret statistical results, a crucial aspect of research papers involving substantial quantitative analysis.

AI tools like Elicit.org and SCISPACE represent the forefront of technological integration into academic research. They streamline the research process and enhance the quality and depth of academic work. By leveraging these tools, researchers can navigate the complexities of academic literature more effectively, leading to more robust and insightful research outcomes. These platforms symbolise the synergy between AI and academic research, paving the way for a more informed and efficient scholarly environment.

10.5.22 *Preparing for the future healthcare environment*

Tomorrow's healthcare environment promises to be drastically different from what we know today, driven by rapid technological advancements, evolving patient needs, and changing global health dynamics. This imminent evolution calls for a proactive and strategic approach to embrace these changes and leverage them for improved health outcomes and healthcare delivery.

Preparing for the future healthcare environment involves understanding and adapting to critical trends such as integrating AI and digital health technologies, personalised medicine, the increasing importance of mental health, and the shift towards patient-centred care models. It also requires anticipating and addressing challenges such as healthcare accessibility, ethical considerations in medical technology, and the need for continuous professional development among healthcare providers.

10.5.23   Keeping pace with technological advancements

The rapid evolution of AI presents both a challenge and an opportunity for medical education systems worldwide. To ensure that future healthcare professionals are not only proficient but also innovative in their use of AI tools, several key aspects need to be addressed:

1. **Curriculum integration**: Medical education curricula must evolve to include comprehensive training in AI and its applications in healthcare. This integration should extend beyond theoretical knowledge, encompassing practical, hands-on training with AI tools. Topics like ML algorithms, data analytics, and their ethical use in clinical settings should become standard components of medical education.

2. **Interdisciplinary learning**: AI in healthcare is inherently interdisciplinary, blending aspects of computer science, data science, and clinical practice. Medical education should reflect this interdisciplinarity, encouraging collaboration and learning across these fields. This approach will equip future healthcare professionals with a holistic understanding of how AI can be leveraged in clinical settings.

3. **Clinical application**: Understanding the practical implications of AI in clinical practice is crucial. This includes training on integrating AI tools into clinical decision-making, patient management, and personalised medicine. Real-world case studies and simulations can effectively illustrate AI's practical uses and limitations in healthcare.

4. **Ethical and legal considerations**: As AI becomes more prevalent in healthcare, so do concerns about ethics and privacy. Medical education must, therefore, include robust training on the ethical use of AI, including patient data privacy, informed consent, and the potential biases inherent in AI algorithms.

5. **Lifelong learning**: AI is dynamic and rapidly evolving. Thus, medical education should foster a culture of lifelong learning and adaptability. Continuous education programmes, workshops, and seminars on the latest AI advancements should be readily available to healthcare professionals throughout their careers.

6. **Research and innovation**: Encouraging research and innovation in AI within medical education can spur the development of new AI tools tailored to healthcare needs. This involves not only using existing AI technologies but also contributing to the creation and refinement of these tools.

7. **Collaboration with the tech industry**: Partnerships between medical institutions and the technology sector can facilitate the sharing of knowledge and

resources. Such collaborations can lead to developing cutting-edge AI applications in healthcare and ensure that medical training stays current with technological advancements.

10.6   Conclusion

In summary, keeping pace with technological advancements in AI necessitates a multifaceted approach to medical education. By integrating AI into curricula, promoting interdisciplinary learning, focusing on practical applications, and emphasising ethical considerations, the future healthcare workforce can be adequately prepared to utilise AI effectively and responsibly in clinical practice.

As AI and related technologies advance, medical education must keep pace. This ensures that future healthcare professionals are adept at using AI tools and understanding their implications in clinical practice.

In conclusion, the rapidly evolving landscape of healthcare, underscored by the integration of AI and advanced technologies, calls for a paradigm shift in how we approach medical innovation. The future of patient care hinges not just on the advancements in technology alone but critically on the synergistic collaboration between engineers, AI developers, and healthcare professionals. This multidisciplinary alliance is the cornerstone for developing relevant, authentic, and patient-centred innovations that will define the future of healthcare.

The collaboration between these diverse fields offers a unique confluence of expertise – engineers and AI developers bring technical prowess and innovative thinking. At the same time, healthcare professionals contribute clinical insight and patient-centric perspectives. This partnership is essential in ensuring that the technologies developed are advanced and align seamlessly with the practical realities of patient care. Through such collaborative efforts, AI and technology can be harnessed to their fullest potential, resulting in solutions that are both technically sound and deeply attuned to the nuances of human health and wellness.

Looking ahead, we encourage continued and deepened collaboration across these disciplines. We can collectively address modern healthcare's complex challenges through open dialogues, shared projects, and joint ventures. Institutions, industries, and individuals in these fields are urged to seek partnerships, foster interdisciplinary teams, and create platforms for exchange and innovation. Doing so will pave the way for a future where healthcare is more advanced technologically, compassionate, personalised, and effective.

The patients of the future deserve a healthcare system that is the best amalgamation of human ingenuity and technological advancement. As we close this chapter, we look forward to a future where the combined efforts of engineers, AI developers, and healthcare professionals create a healthcare landscape that is transformative, sustainable, and, above all, focused on the well-being of every patient. Let us embrace this collaborative spirit and work together towards a future where healthcare innovation is not just about creating new technologies but enriching lives and fostering healthier communities.

References

[1] Shaheen, M. Y. (2021). *Applications of artificial intelligence (AI) in healthcare: A review. ScienceOpen Preprints*, doi:10.14293/S2199-1006.1. SOR-.PPVRY8K.v1.

[2] Maddox, T. M., Rumsfeld, J. S., and Payne, P. R. O. (2019). Questions for artificial intelligence in health care. *JAMA, 321*(1), 31.

[3] Maity, N. G., and Das, S. (2017). Machine learning for improved diagnosis and prognosis in healthcare. In *2017 IEEE Aerospace Conference* (pp. 1–9). https://doi.org/10.1109/AERO.2017.7943950

[4] Alexander, A., Jiang, A., Ferreira, C., and Zurkiya, D. (2020). An intelligent future for medical imaging: A market outlook on artificial intelligence for medical imaging. *Journal of the American College of Radiology, 17*(1, Part B), 165–170. https://doi.org/10.1016/j.jacr.2019.07.019

[5] Alowais, S. A., Alghamdi, S. S., Alsuhebany, N., *et al.* (2023). Revolutionizing healthcare: The role of artificial intelligence in clinical practice. *BMC Medical Education, 23*(1), 689. https://doi.org/10.1186/s12909-023-04698-z

[6] Scalia, P., Ahmad, F., Schubbe, D., *et al.* (2021). Integrating option grid patient decision aids in the epic electronic health record: Case study at 5 health systems. *Journal of Medical Internet Research, 23*(5), e22766. https://doi.org/10.2196/22766

[7] Khadija, A., Zahra, F. F., and Naceur, A. (2021). AI-powered health Chatbots: Toward a general architecture. *Procedia Computer Science, 191*, 355–360. https://doi.org/10.1016/j.procs.2021.07.048

[8] Aldousari, A., Alotaibi, M., Khajah, F., Jaafar, A., Alshebli, M., and Kanj, H. (2023). A wearable IOT-based healthcare monitoring system for elderly people. In *2023 5th International Conference on Bio-engineering for Smart Technologies (BioSMART)* (pp. 1–4). https://doi.org/10.1109/BioSMART58455.2023. 10162041

[9] Turakhia, M. P., Desai, M., Hedlin, H., *et al.* (2019). Rationale and design of a large-scale, app-based study to identify cardiac arrhythmias using a smartwatch: The Apple Heart Study. *American Heart Journal, 207*, 66–75. https://doi.org/10.1016/j.ahj.2018.09.002

[10] Al-khafajiy, M., Baker, T., Chalmers, C., *et al.* (2019). Remote health monitoring of elderly through wearable sensors. *Multimedia Tools and Applications, 78*(17), 24681–24706. https://doi.org/10.1007/s11042-018-7134-7

[11] Kumar, A., and Gond, A. (2023). Natural language processing: Healthcare achieving benefits via NLP. *ScienceOpen Preprints.* https://doi.org/10. 14293/PR2199.000280.v1

[12] Kane, E., Kobayashi, K., Dunn, P. F., and Scheulen, J. J. (2018). Transforming hospital capacity management: Experience from two academic medical centres. *Management in Healthcare, 3*(4), 339–348.

[13] Rahman, N., Thamotharampillai, T., and Rajaratnam, V. (2023). Ethics, guidelines, and policy for technology in healthcare. In *Medical equipment*

engineering: Design, manufacture and applications (pp. 119–147). IET Digital Library. https://doi.org/10.1049/PBHE054E_ch9

[14]   Murdoch, B. (2021). Privacy and artificial intelligence: Challenges for protecting health information in a new era. *BMC Medical Ethics*, *22*(1), 122. https://doi.org/10.1186/s12910-021-00687-3

[15]   Duraku, L. S., Hoogendam, L., Hundepool, C. A., *et al.* (2022). Collaborative hand surgery clinical research without sharing individual patient data; Proof of principle study. *Journal of Plastic, Reconstructive & Aesthetic Surgery*, *75*(7), 2242–2250. https://doi.org/10.1016/j.bjps.2022.02.065

[16]   Reddy, S., Allan, S., Coghlan, S., and Cooper, P. (2020). A governance model for the application of AI in health care. *Journal of the American Medical Informatics Association*, *27*(3), 491–497. https://doi.org/10.1093/jamia/ocz192

[17]   Varona, D., and Suárez, J. L. (2022). Discrimination, bias, fairness, and trustworthy AI. *Applied Sciences*, *12*(12), Article 12. https://doi.org/10.3390/app12125826

[18]   Goirand, M., Austin, E., and Clay-Williams, R. (2021). Implementing ethics in healthcare AI-based applications: A scoping review. *Science and Engineering Ethics*, *27*(5), 61. https://doi.org/10.1007/s11948-021-00336-3

[19]   Zhang, P., and Kamel Boulos, M. N. (2023). Generative AI in medicine and healthcare: Promises, opportunities and challenges. *Future Internet*, *15*(9), Article 9. https://doi.org/10.3390/fi15090286

[20]   Meskó, B. (2023). Prompt engineering as an important emerging skill for medical professionals: Tutorial. *Journal of Medical Internet Research*, *25*(1), e50638. https://doi.org/10.2196/50638

[21]   Wood, E. A., Ange, B. L., and Miller, D. D. (2021). Are we ready to integrate artificial intelligence literacy into medical school curriculum: Students and faculty survey. *Journal of Medical Education and Curricular Development*, *8*, 23821205211024078. https://doi.org/10.1177/23821205211024078

[22]   Grunhut, J., Marques, O., and Wyatt, A. T. M. (2022). Needs, challenges, and applications of artificial intelligence in medical education curriculum. *JMIR Medical Education*, *8*(2), e35587. https://doi.org/10.2196/35587

[23]   Harder, N. (2023). Advancing healthcare simulation through artificial intelligence and machine learning: Exploring innovations. *Clinical Simulation In Nursing*, *83*. https://doi.org/10.1016/j.ecns.2023.101456

[24]   Dickey, E., and Bejarano, A. (2023). A model for integrating generative AI into course content development. https://doi.org/10.48550/ARXIV.2308.12276

[25]   Son, J. W., Han, M. H., and Kim, S. J. (2019). Artificial intelligence-based video content generation. *Electronics and Telecommunications Trends*, *34*(3), 34–42. https://doi.org/10.22648/ETRI.2019.J.340304

[26]   Stokel-Walker, C., and Van Noorden, R. (2023). What ChatGPT and generative AI mean for science. *Nature*, *614*(7947), 214–216. https://doi.org/10.1038/d41586-023-00340-6

[27] Freer, J., Dolan, N. T., and Wiersma, G. (2023). The fast and the curious: Accelerating literature reviews with AI. *Presentations and other scholarship.* https://scholarworks.rit.edu/other/1026

[28] Salvagno, M., Taccone, F. S., and Gerli, A. G. (2023). Can artificial intelligence help for scientific writing? *Critical Care, 27*(1), 75. https://doi.org/10.1186/s13054-023-04380-2

Chapter 11

Green sustainable technology in physical education and sports

Viswanath Sundar¹ and Vinodhkumar Ramalingam²

11.1 Introduction

Physical education and sports play a pivotal role in promoting overall health and wellness. Engaging in physical activity and participating in sports have numerous health benefits [1]. These activities help in maintaining a healthy weight, improving cardiovascular health, strengthening muscles and bones, and enhancing flexibility and coordination. Physical Education, as a formal educational programme, focuses on teaching students about the importance of physical activity and developing their motor skills. It also plays a crucial role in promoting the holistic development of students by incorporating social, emotional, and cognitive activities [2]. In addition, participating in sports teaches valuable life skills such as discipline, perseverance, teamwork, and goal setting.

Furthermore, having access to outdoor spaces like fields or tracks for outdoor sports activities is crucial. These facilities and infrastructure not only provide a conducive environment for physical education and sports but also ensure the safety and well-being of individuals participating in these activities. By having the necessary infrastructure, schools, and communities can effectively implement physical education programmes and provide ample opportunities for individuals to engage in sports [3]. Moreover, integrating sustainability practices in sports and physical education ensures that these activities have a minimal negative impact on the environment and promote long-term ecological balance. By incorporating principles of sustainability, such as using eco-friendly materials for sports equipment, promoting energy-efficient practices in facilities, and encouraging the use of renewable energy sources, sports, and physical education can contribute to a greener and more sustainable future [4]. In the present situation, there is a global shift towards green sports technology and the introduction of innovative and sustainable solutions that improve performance while reducing adverse environmental impacts.

¹Department of Physical Education and Sports Science, Visva-Bharati (Central University), India
²Saveetha College of Physiotherapy, SIMATS, India

11.2   Understanding green sports technology

Green sports technology refers to the use of innovative and sustainable practices, materials, and technologies in physical education and sports. In the current scenario, the adoption of eco-friendly practices in sports activities has become crucial. The primary goal is to minimize the environmental impact of sports activities while simultaneously enhancing performance and sustainability. One of the key strategies is integrating eco-friendly measures into various aspects of sports activities, infrastructure, and operations. This includes recycling initiatives, implementing energy-efficient lighting systems, adopting sustainable water management practices, and harnessing renewable energy sources. The aim is not only to provide top-notch facilities and equipment for athletes and fans but also to minimize harm to the environment. As highlighted, it is not about creating an eco-friendly sporting experience but about actively participating in the broader mission of preserving our planet [5]. Through green sports technology, sports organizations can reduce the carbon footprint, mitigate pollution, preserve natural habitats, and contribute to broader environmental education within the sports community.

11.3   Need of green sports technology

The understanding of health and engagement in regular fitness activities, as well as participation in competitions, is rapidly growing [5]. As a result, there is an increasing demand for sports facilities and equipment. Different schemes were introduced all over the world to promote physical fitness, such as Fit India, Fit Malaysia, the National Fitness Program in China [6,7], and various sports initiatives in different countries. These initiatives highlight the importance of physical activity and the need for proper infrastructure and equipment to support it. However, sports can also have a negative impact. For example, the construction of golf courses or motor racing speedways often requires clearing large areas of land, which in many cases leads to the loss of fertile farmlands or thriving forests [4]. In addition, organizing large-scale sports events often results in excessive energy consumption, waste generation, and transportation-related emissions [5,8]. However, the traditional approach to sports and physical education often neglects the environmental impact of these activities [9]. This not only contributes to the depletion of natural resources but also results in pollution and the emission of greenhouse gases. To address these concerns, the adoption of green sports technology has become imperative.

11.4   Sustainable sports infrastructure and equipment

Sustainable sports infrastructure and equipment are a major contribution towards green development in the sports industry. Many international sports federations have taken the initiative for the sustainability of sports. In terms of infrastructure, the International Hockey Federation took a challenge to change the playing field from water-based to dry turf worldwide (Figure 11.1(a)). According to the Sports

Figure 11.1   (a) Water based field hockey turf and (b) synthetic feather shuttlecocks

Turf Research Institute, one modern hockey field requires 6000 L of water for one session. This becomes a major threat to the growing environmental issues and increased water scarcity. Similarly, regarding equipment, the Badminton World Federation adopted synthetic feather shuttlecocks in international tournaments. Historically, the feathers of ducks or geese are used to produce high-quality feathers, which poses a challenge related to sustainability and ethical concerns in the future. Currently, all international competitions are played with synthetic feathers (Figure 11.1(b)).

11.5   IOC case study

The International Olympic Committee's (IOC) headquarters, also known as the Olympic House, is a symbol of athletic excellence; it reflects a deep commitment to health, sports, and environmental sustainability. The design of this building, brought to life by the Danish architecture firm 3XN, uses eco-friendly methods in construction, making it a model of green architecture around the world.

The Olympic House journey starts by looking upwards; the Solar panels cover over 1150 m^2 of rooftop. It creates about 20% of the electricity the building needs. Additionally, Lake Geneva's innovative lake water heat pumps offer natural heating and cooling, reducing reliance on fossil fuels. This seamless interplay between solar energy and lake resources highlights the building's commitment to renewable energy sources, serving as a compelling model for future architectural projects.

The entire structure serves as insulation, which acts as a barrier, preserving warmth during winter and cool air during summer, reducing the overall use of energy. The Olympic House carefully manages lighting, heating, and ventilation to ensure efficient energy utilization. Switching to energy-efficient LED lighting contributes even more to the eco-friendly approach, minimizing the building's impact on the environment.

Water, the essential source of life, in Olympic House, the rainwater was carefully gathered and utilized for various purposes such as toilet flushing, irrigation, and car washing. Installing water-saving fixtures throughout the building established an important example for a world grappling with growing water shortages. Furthermore, Olympic House showcases a state-of-the-art waste management system that promotes recycling and minimizes waste generation. This waste management system reduces the building's environmental footprint and sets an example for effective waste reduction and recycling practices in the broader community.

The designers of Olympic House recognized that genuine sustainability goes beyond physical construction. They emphasized integration with the local environment by using predominantly Swiss-sourced materials for 80% of the building's construction. This minimizes transportation emissions and contributes to the local economy, promoting a sense of interconnectedness between ecology and economics. Moreover, more than 60% of the site is preserved as open space, creating a sanctuary for biodiversity and removing the barrier between the building and the lively city surroundings.

Olympic House prioritizes the well-being of its occupants by providing natural ventilation whenever possible, allowing fresh air to refresh the workspace. Low-volatile organic compound (VOC) materials minimize indoor air pollution, ensuring a healthy and comfortable environment for those who dedicate their lives to the Olympic movement. Large windows clean the interior in natural light, reducing reliance on artificial lights and lowering the building's energy consumption (Figure 11.2).

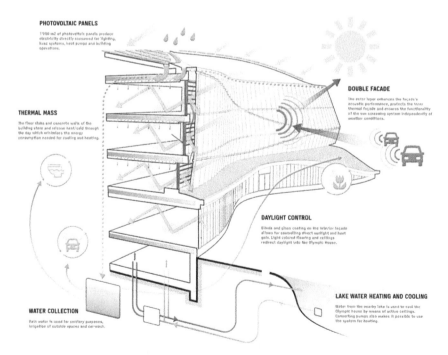

Figure 11.2   Blue print of sustainable building of Olympic house

The IOC building represents more than just a structure; it stands as evidence of human creativity and a symbol of optimism for a more sustainable future. It illustrates that even in the pursuit of athletic greatness, environmental stewardship can and should be our guiding principle. As this extraordinary building's legacy unfolds, let us draw inspiration from its values to create a world where every step towards sporting success is also a move towards an environmentally friendly tomorrow. Though the focus is on sustainable sports environments, sportsmen play a vital role in standardizing by framing the guidelines for adaptation, so the following section of this chapter will address the sportsmen's contribution towards green sustainability [10].

11.6   Green sustainability contribution from sportsman

Sports professionals, such as athletes, coaches, and other stakeholders, can make important contributions to environmental sustainability through their activities and decisions [11]. Here are a few ways that athletes can support the sustainability of the environment.

11.6.1   Environmentally friendly behaviours

Sporty people who commute to training facilities or competitions should choose eco-friendly modes of transportation like electric automobiles or bicycles to reduce their carbon emissions [12]. In order to lessen the carbon imprint of single-use plastics, practising using reusable water bottles or containers in training and competitions is the best choice for maintaining environmental sustainability.

11.6.2   Energy conservation

On the other hand, in order to preserve energy resources, sports facilities encourage athletes, coaches, and sports-supporting team members to use energy-efficient lighting and equipment in training areas. To conserve energy, athletes should turn off lights and equipment when not in use, create energy-efficient travel and home routines, and incorporate these energy-saving habits into their personal spaces and surroundings [13]. This healthy practice may reduce the utility of energy resources that help maintain environmental sustainability.

11.6.3   Healthy diet and adequate hydration

A balanced diet plays a vital role in athletic performance. However, while considering environmental sustainability, athletes have to consider incorporating plant-based diets into their daily routines rather than meat [14,15]. Reducing meat intake can aid in greenhouse gas reduction and promote sustainable agriculture. Besides a plan to preserve water and maintain proper hydration, which is essential for sportspeople, they also have to monitor their daily water intake quantity during exercise and for their daily activities [16].

11.6.4   Encouragement of environmental initiatives

As such, it is a very important concern regarding environmental initiatives. In order to initiate and enhance sustainability, individual athletes utilize their social media platforms to promote and advocate for the environment within the sports sector [17]. Encourage teammates and other athletes to adopt eco-friendly practices and make sustainable judgements. In addition, they need to collaborate with environmental programmes or groups to raise awareness and encourage long-term change. Additionally, they have to engage in dialogue with surrounding communities to promote environmental awareness and sustainability [18].

11.6.5   Participation in green events

Emphasize the importance of green events; even the public has a responsibility to attend athletic events where sustainability and environmental friendliness are prioritized. Also, they have to promote and support the organization of athletic events that employ litter reduction, environmentally friendly venue management, and eco-friendly transportation techniques [19]. On the other hand, athletes and public supporters should make sure their sporting wear is made of environmentally friendly materials and support companies that prioritize using environmentally friendly production materials [20]. Furthermore, they have to use energy-efficient workout equipment, wearable technology that is friendly to the environment, and other sustainable advances in their training.

11.6.6   Offsetting carbon footprint

Athletes must also think about taking part in carbon offset schemes when they compete in environmentally friendly events in order to offset the carbon emissions that are inevitably generated by travel and other activities [21]. This participation in sportsmanship will serve as a sustainable model for other athletes, supporters, and the community at large. This can be accomplished by performing honourably both on and off the field. Sportsmen can use their position to positively move towards environmental protection by actively participating in sustainable activities and promoting green projects. Their dedication to sustainability has the potential to encourage positive changes in the sports sector and inspire a community of supporters to lead healthier and more sustainable lives [22].

11.7   Protecting the physical education environment

In order to reduce their negative effects on the environment, athletes and teams should encourage the proper usage and upkeep of sporting facilities. Motivate athletes, coaches and spectators to properly dispose of their trash and take part in post-event cleanup efforts of their venues [23]. Encourage and support the use of sporting goods manufactured using environmentally friendly manufacturing techniques and sustainable materials. Promote the use of energy- and water-saving techniques in competition and training. Sports organizations, coaches, and athletes all have the opportunity to set an example for others by modelling sustainability and good sportsmanship in their choices and actions [24].

11.7.1 Social responsibility

Sports teams and athletes can work with nearby communities to promote sustainability through programmes like tree planting, environmental education, or cleanup activities. Volunteer organizations in the sports industry that encourage sustainability by leveraging competitions to generate money or awareness for environmental issues [25].

11.7.2 Equity and fair play

By ensuring that athletes from all backgrounds have the opportunity to participate in sports, we can promote inclusivity and diversity in the sport and promote social sustainability. Besides, encourage gender equality in sports by fostering an environment where everyone, regardless of gender, has equal opportunities [26].

11.7.3 Minimizing the effects of travel and promoting green projects

Considering the green environment is a vital part of sustainability, it is recommended that even participants consider sustainable transportation options when attending events and opt for eco-friendly travel and public transportation. Encourage and engage in sports initiatives that prioritize sustainability, such as reducing waste, conserving energy, and practising responsible resource management [27]. Conversely, athletes and sports groups can also choose sponsors who uphold moral and environmentally friendly business practices.

11.7.4 Initiatives for education

Incorporate teaching about environmental sustainability with initiatives promoting good sportsmanship, and encourage athletes to act as green brand ambassadors. Engage in conversation with young athletes and teach them the value of sustainability and sportsmanship at a young age. In addition, the physical education course has to introduce the importance of environmental suitability in the student curriculum, which has an impact on sports participants globally [28].

11.7.5 Green certification

Seek out and promote green certifications for athletic facilities to ensure that these establishments are managed sustainably. Make use of media outlets to spread the word about the importance of good sportsmanship and sustainability [29]. Prominent athletes can serve as ambassadors for eco-friendly products, lifestyle choices, and practices [30]. By merging sustainability with the principles of sportsmanship, athletes and sports organizations may contribute to the betterment of society and the environment. These programmes may inspire athletes, fans, and the general public to embrace moral values that promote justice, fair play, and a sustainable future.

11.8 Conclusion

To promote sustainability in physical education and sports, it is crucial to adopt sustainable approaches. These approaches include integrating environmental education into sports curriculum, promoting active transportation such as walking or cycling to sports venues, and encouraging the use of eco-friendly sports equipment and materials. Furthermore, sports organizations can also prioritize the use of environmentally friendly transportation methods for athletes and spectators, such as public transportation or carpooling. By implementing these sustainable approaches, physical education and sports can play a significant role in raising awareness about environmental issues and inspiring behaviour change among students and athletes.

References

[1]   Leyton-Román M, Guíu-Carrera M, Coto-Cañamero A, and Jiménez-Castuera R. Motivational variables to predict autotelic experience and enjoyment of students. Analysis in function of environment and sports practice. *Sustainability*. 2020;12(6):2352.

[2]   Eime RM, Young JA, Harvey JT, Charity MJ, and Payne WR. A systematic review of the psychological and social benefits of participation in sport for children and adolescents: informing development of a conceptual model of health through sport. *International Journal of Behavioral Nutrition and Physical Activity*. 2013;10(1):1–21.

[3]   Sibarani HEI. Survey of physical education facilities and teaching and learning process of physical education in state high schools in the Middle City of Tebing Tinggi Academik Year 2018/2019. 2020. https://doi.org/10.2991/ahsr.k.200305.052

[4]   Chen X, Niu J, Nakagami KC, Zhang Q, Qian X, and Nakajima J. Green sports supporting a low-carbon society: inspiration from Japan. *International Journal of Global Warming*. 2018;14(1):61–80.

[5]   Zhang L. Research on the evaluation of sports events based on the concept of green environmental protection. In *IOP Conference Series: Earth and Environmental Science* 2021 (Vol. 651, No. 4, p. 042028).

[6]   Ayub SH, Hassim N, Yahya AH, Hamzah M, and Abu Bakar MZ. Exploring the characteristics of healthy lifestyle campaign on social media: A case study on FIT Malaysia. *Malaysian Journal of Communication*. 2019;35(4):322–336. https://doi. org/10.17576/JKMJC-2019-3504-20.

[7]   Xiang C, Zhao J, Tengku Kamalden TF, Dong W, Luo H, and Ismail N. The effectiveness of child and adolescent sports engagement in China: an analysis of China's results for the 2016–2022 global matrix report cards on physical activity. *Humanities and Social Sciences Communications*. 2023;10(1):1–2.

[8]   Meng Y. Research on sports system based on the concept of green environmental protection. In *IOP Conference Series: Earth and Environmental Science* 2021 (Vol. 714, No. 2, p. 022037).

[9] Hanna RK and Subic A. Towards sustainable design in the sports and leisure industry. *International Journal of Sustainable Design*. 2008;1(1):60–74.

[10] Boykoff J and Mascarenhas G. The Olympics, sustainability, and green-washing: The Rio 2016 summer games. *Capitalism Nature Socialism*. 2016; 27(2):1–1.

[11] Sotiriadou P and Hill B. Raising environmental responsibility and sustainability for sport events: a systematic review. *International Journal of Event Management Research*. 2015;10(1):1–1.

[12] Yin G, Huang Z, Fu C, Ren S, Bao Y, and Ma X. Examining active travel behavior through explainable machine learning: Insights from Beijing, China. *Transportation Research Part D: Transport and Environment*. 2024;127:104038.

[13] Mamajonova N, Oydin M, Usmonali T, Olimjon A, Madina A, and Marg'uba M. The role of green spaces in urban planning enhancing sustainability and quality of life. *Holders of Reason*. 2024;2(1):346–58.

[14] Messina M, Duncan AM, Glenn AJ, and Mariotti F. Plant-based meat alternatives can help facilitate and maintain a lower animal to plant protein intake ratio. *Advances in Nutrition*. 2023;14(3):392–405.

[15] Thakur S, Pandey AK, Verma K, Shrivastava A, and Singh N. Plant-based protein as an alternative to animal proteins: a review of sources, extraction methods and applications. *International Journal of Food Science & Technology*. 2024;59(1):488–97.

[16] Pérez-Castillo ÍM, Williams JA, López-Chicharro J, *et al.* Compositional aspects of beverages designed to promote hydration before, during, and after exercise: concepts revisited. *Nutrients*. 2023;16(1):17.

[17] Kim YD, Nam C, and LaPlaca AM. Marketing and communicating sustainability through college athletics: the effects of pro-environmental initiatives on the belief-attitude-intention hierarchy. *Journal of Marketing for Higher Education*. 2023;33(1):58–78.

[18] Le Duc A. *Interreligious dialogue to promote environmental flourishing: an ongoing imperative*. Available at SSRN 4490844. 2023 Jun 25.

[19] Marrucci L, Daddi T, and Iraldo F. *Sustainable Football: Environmental Management in Practice*. Taylor & Francis; 2023 May 10.

[20] Xu X and Wang L. Assessing the role of sports economics and green supply chain management for the coordinative and coupling development of china's green economic growth. *Environmental Science and Pollution Research*. 2023:1–3.

[21] Vienažindienė M, Perkumienė D, Atalay A, and Švagždienė B. The last quarter for sustainable environment in basketball: the carbon footprint of basketball teams in Türkiye and Lithuania. *Frontiers in Environmental Science*. 2023;11(1197798):1–1.

[22] Gionfriddo G, Rizzi F, Daddi T, and Iraldo F. The impact of green marketing on collective behaviour: experimental evidence from the sports industry. *Business Strategy and the Environment*. 2023.

[23] Al Mohannadi F. *Critical analysis of the environmental component of the sustainable stadiums built for the 2022 FIFA World Cup in Qatar*. Doctoral dissertation, Hamad Bin Khalifa University, Qatar.

[24] Xu L, Huang D, He Z, and Cao J. Evolutionary game analysis of the innovation and diffusion of water-saving technology. *Water Economics & Policy.* 2023;9(4).

[25] Choi W, Chung MR, Lee W, Jones GJ, and Svensson PG. A resource-based view of organizational sustainability in sport for development. *Journal of Sport Management.* 2023;37(6):429–39.

[26] Camarasa J, Boned-Gomez S, García-Taibo O, and Baena-Morales S. *Game on for gender equality: an evaluation of ultimate team in primary physical education.*

[27] McCullough BP, Hardie A, Kellison T, and Dixon M. Environmental perspectives of external stakeholders in sport. *Managing Sport and Leisure.* 2023;28(6):670–83.

[28] Larneby M. School sport education and sustainability: towards ecological and inclusive student-athletes? In *Sport, Performance and Sustainability* 2023 May 19 (pp. 130–147). Routledge.

[29] Azadi A, Rahimi G, and Nazari R. Presenting a model for the role of sport on Iran's sustainable development: an approach to the role of sport in GDP. *Sports Business Journal.* 2023;3(1):37–52.

[30] Stålstrøm J, Iskhakova M, and Pedersen ZP. Role models and athlete expression at the Youth Olympic Games as impactful sport communication practices. *International Journal of Sport Communication.* 2023;1:1–5.

Chapter 12

Medical equipment engineering in ESG and green sustainable technology

Wendy Wai Yeng Yeo¹ and Chia Chao Kang²

12.1 Introduction

The medical equipment that is extensively utilized in the healthcare market ranges from simple devices like nebulizers to sophisticated imaging systems such as computed tomography (CT) scans and magnetic resonance imaging (MRI) scans. With the current healthcare landscape transitioning toward a sustainable and resilient future, the incorporation of environmental, social, and governance (ESG) and adoption of green sustainable technology in medical equipment engineering is getting more attention. This entails the imperative of mitigating environmental impact, embedment elements of social responsibility when it comes to their employees, customers, and suppliers as well as ensuring governance transparency. This chapter explores medical equipment engineering that is motivated by ESG principles and green technology, which offer insights into the current advancement of a sustainable healthcare system.

12.2 Environmental impact on society

The healthcare industry plays an important role in our society, and it also has a significant environmental footprint. The manufacturing utilization and disposal of medical equipment have the potential to contribute to the release of greenhouse gas emissions, pollution, and resource depletion [1]. However, cutting-edge manufacturing methods are being developed to enhance the environmental friendliness and sustainability of medical equipment engineering.

First, the poorly considered designs and unsustainable materials used in medical equipment innovation and design can contribute to environmental burdens. The selection of materials used during the manufacturing of medical equipment significantly has an impact on the environment. For instance, the usage of unsustainable polymeric materials that are derived from fossil sources or improper handling of the release of toxic chemicals such as cyanide derivatives during the

¹School of Pharmacy, Monash University Malaysia, Malaysia
²School of Electrical Engineering and Artificial Intelligence, Xiamen University Malaysia, Malaysia

manufacturing of medical equipment is harmful to the environment and human health [2]. Furthermore, the environmental effects of medical device development throughout their lifecycles are exacerbated by greenhouse gas emissions in addition to the increasing of single-use medical devices that are primarily disposed of by incineration or landfilling [3,4].

Hence, the environmental effect of medical equipment can be substantially achieved through a transition to more sustainable materials such as bioplastics and recycled metals, aligning with ESG principles and green technology practices. This shift helps in reducing the usage of non-renewable resources and minimizing the carbon footprint associated with production. For example, biodegradable bioplastics include poly(lactic acid) (PLA) and poly(ε-caprolactone) (PCL) that exhibit chemical and solvent resistance, making them suitable for use in the manufacturing of medical devices [5]. Herein, the exploration of design processes and biodegradable options for non-critical components are promising alternatives to minimize environmental impact and foster an eco-friendly approach.

On the other hand, reuse and refurbishment of medical equipment such as medical imaging equipment, operating room equipment, and patient monitoring devices do help with the conservation of natural resources and minimize the environmental impact of the waste [6,7]. However, it is essential to consider aspects such as product performance and quality as well as legal responsibility when using refurbished medical devices [8]. This is to ensure that the reprocessed medical equipment does not pose any risk of health damage particularly in relation to infections, pyrogen-induced reactions, toxic or even allergic reactions due to altered technical or functional properties [6].

Moreover, the energy required for medical equipment to operate and the energy consumed in the manufacturing process both have an influence on the environment. Implementing energy-efficient practices throughout the manufacturing process also can significantly reduce greenhouse gas emissions and energy consumption. This includes using renewable energy sources like solar or wind power as shown in Figure 12.1, optimizing production lines to minimize waste, and employing heat recovery systems to capture and reuse waste heat. Thus, devices that are energy-efficient and designed to consume less power during use can help mitigate this impact within the ESG and green technology framework.

Other than that, traditional packaging practices in medical equipment often rely heavily on virgin plastics [4], which require fossil fuels for production. A substantial amount of medical waste ends up in landfills each year and improper disposal can lead to environmental contamination (Figure 12.2). This waste takes centuries to decompose, polluting our environment and releasing harmful toxins. This unsustainable practice adversely affects ecosystems and biodiversity. Therefore, advocating and facilitating the reuse of packaging materials wherever feasible can help to reduce waste but also align with ESG goals and green technology principles, promoting a more sustainable and environmentally responsible approach. Besides that, designing equipment with less complex packaging, eliminating unnecessary layers, and using smaller container sizes also will greatly reduce waste.

Figure 12.1   Windmills near solar panel boards. Source: https://www.pickpik. com/electricity-sun-wind-forces-of-nature-resources-solar-energy-127376.

Figure 12.2   Medical waste. Source: https://www.smchealth.org/solidwaste.

12.3   Social commitment

The current advancement in telemedicine platforms using different sophisticated medical equipment which enables live tracking of patient status and remote patient monitoring is bridging the geographical barriers and improving access to the healthcare system in rural areas [9–11]. Technological innovation has brought about a key aspect of the social impact of medical equipment engineering, particularly in terms of healthcare accessibility and affordability (Figure 12.3). This driving innovation in responsible and sustainable medical equipment engineering exemplifies the transformative potential of technology in facing societal challenges and ensuring equitable healthcare access for underserved communities.

Notably, the innovation in medical equipment engineering not only serves primary and specialty care in the community but also brings medical care to remote and underdeveloped areas all over the world. Therefore, the ESG framework is vital in ensuring the companies that are involved in the medical equipment engineering field are working toward equitable healthcare access for the underserved in local communities. Meanwhile, the booming of the medical equipment industry creates employment opportunities for engineers, researchers, technicians, and other professionals. Hence, by integrating ESG principles, these companies also emphasize providing fair employment opportunities as well as fostering a diverse and inclusive workplace. Herein, the social impacts encompass the organization's relationships not only with customers and communities where it operates but also with its employees and suppliers (Figure 12.4).

Features and devices

1. Tablet – Dell Touch-Display

2. PTZ (Pan, Tilt and Zoom) camera

3. Conference Speaker

4. RCS-100 Medical Camera System

5. ri-sonic digital stethoscope

6. Blood pressure management cuffs

7. ri-thermo sensioPRO+ thermometer

8. ECG

9. Ultrasound probe

10. Vital Signs Monitor

Figure 12.3   Patient assessments can be performed in designated remote locations worldwide and involve capturing and transferring data from multiple medical devices to enable healthcare providers to deliver on-site patient treatment using the Riester Telemedicine Case, a mobile diagnostic case. Source: https://riester.de/products/telemedicine/ telemedicine-case.

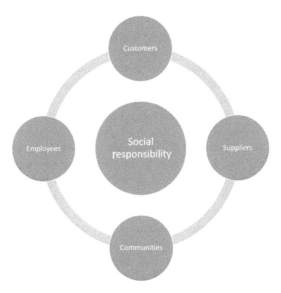

Figure 12.4   This figure illustrates the multifaceted dimensions of social impact, emphasizing the organization's interconnected relationships among customers, suppliers, communities, and employees

12.4   Governance principles

The ESG and green sustainable technology play crucial roles in shaping the medical equipment engineering industry today [12]. The impact of governance on this sector is multifaceted and extends to various aspects of product development, manufacturing, and overall corporate responsibility. Here are some key aspects.

The first aspect related to the regulatory compliance of governance. Noncompliance can lead to legal consequences, damage to reputation and thus impact on financial losses. Therefore, governance structures can establish clear lines of accountability and reporting mechanisms for ESG performance. This ensures stakeholders hold decision-makers responsible for sustainable practices and have access to information. In addition to reducing legal and reputational risks, strong governance also can guarantee adherence to pertinent social and environmental requirements [13].

Besides that, in the medical equipment industry, violations of ethical standards, such as unauthorized use of patient data or conflicts of interest, can impact the trust in healthcare institutions and technologies [14]. Basically, ethical decision-making is intrinsically tied to governance principles, by taking into account patient safety, privacy, and informed consent during the design and utilization of medical devices [15]. Thus, integrating ethical considerations into the broader ESG framework alongside the existing governance structure will foster a socially responsible and environmentally sustainable approach in the development and deployment of medical equipment industries.

Inadequate quality control procedures can result in product recalls, equipment malfunctions, and compromised patient safety in the healthcare system. Thus, establishing a robust ESG principles governance framework is important to ensure that effective systems are in a position to maintain the highest standards of product safety, quality, and reliability. For example, it is necessary to have a governance structure that emphasizes systematic documentation such as document access limitations, systematic document review and approval procedures, lastly ongoing training for staff members to keep them up to date on industry standards [16].

In addition, the lack of accountability and transparency in governance can create uncertainty among all parties involved, including patients, regulatory agencies, and healthcare providers [17]. Meanwhile, the evolving technology in the healthcare sector may lead to resistance to the adoption of new medical engineering technology [18]. Thus, adopting clear decision-making protocols, documenting decision criteria, and communicating decisions openly to relevant stakeholders are essential steps to enhance transparency in accordance with ESG principles within the medical equipment engineering industries.

On the other hand, the accountability of leadership for strategic choices and organizational performance is determined by governance systems [19]. Defined roles and responsibilities within the leadership team, establishing performance metrics, and conducting regular performance evaluations will lead to achieving organizational goals. By effectively integrating ESG principles into governance models, medical equipment engineering can contribute significantly to a greener and more sustainable future for healthcare.

12.5   Challenges

In today's increasingly conscious business environment, the healthcare sector is no exception. The embracement of the social pillar of ESG efforts and green sustainable technology in medical equipment engineering has become crucial for promoting equitable healthcare access and addressing environmental concerns. Millions of dollars have been invested in ESG funds, demonstrating the growing recognition of its business value as shown in Figure 12.5.

Nevertheless, deploying ESG strategies within the medical equipment engineering field is inevitably complex as well as poses several challenges. First, there will be increasing pressure from all stakeholders in terms of sustainability expectations with more data quality control which is needed during medical equipment manufacturing. Data challenges during collection and analysis including reliability, accessibility, and timeliness are the major issues faced by these companies for the assessment of ESG efforts [20]. In addition, the lack of standardized guidelines for ESG reporting and ESG disclosure by companies also contributes to the challenge for the medical equipment industry [21]. The companies are required to stay up to date and comply with the evolving ESG requirements while optimizing scoring, reporting, planning, and performance.

The inequitable distribution of healthcare access or utilization, resource allocation, prioritization, and quality of care or health outcomes are often key barriers that occur between and within countries around the world [22]. Similarly, medical equipment engineering companies also often encounter obstacles to healthcare

Figure 12.5   Rising investment in ESG

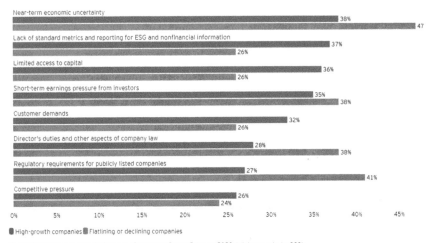

Figure 12.6   External factors that impact the ability to generate long-term value through a strong ESG proposition

access due to financial constraints, geographical limitations, and cultural disparities in order to deliver life-saving technology to diverse communities. Thus, tackling these challenges requires a collective effort, and herein, ESG plays a pivotal role as a transformative force shaping the future of our healthcare system in ensuring equitable healthcare access. However, this may lead to short-term financial pressures on the companies in order to balance short-term financial objectives with long-term ESG commitments. This was reported in a recent survey, which revealed that that near-term economic uncertainty and short-term earnings pressure from investors are the major roadblocks hindering the ability to generate long-term value through strong ESG commitments as shown in Figure 12.6.

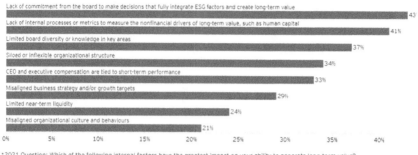

*2021 Question: Which of the following internal factors have the greatest impact on your ability to generate long-term value?
Source: 1. EY Long-Term Value and Corporate Governance Survey February 2022 (total respondents: 200) 2. EY Long-Term Value and Corporate Governance Survey March 2021 (total respondents: 100).

Figure 12.7   Internal factors that impact the ability to generate long-term value through a strong ESG proposition

For the companies to deliver the ESG agenda, it is essential for the boards to incorporate ESG factors into strategic decision-making as part of their sustainability mandate. Yet, another major challenge lies in how to generate long-term value when the board's role in ESG is still evolving as shown in Figure 12.7 based on the recent survey. The initial substantial costs of adopting ESG practices that require upfront investment may hinder these companies from committing themselves fully to integrating ESG values and green technology into their business strategy and structure for long-term value in return. These costs entail changes in business processes, adoption of new technologies, or providing additional training for employees [23].

12.6   Future directions

In the long run, embracing ESG and green technology enables the medical equipment engineering industry to gain benefits in different aspects. The implementation of ESG strategies aligns with the environmental initiatives to reduce carbon footprint through the engineering and design of eco-friendly and environmentally sustainable manufacturing medical equipment [24,25]. This serves as a catalyst for stimulating research and development of innovative as well as sustainable solutions and driving technological advancements within the medical equipment engineering industry (Table 12.1). For instance, medical equipment can be designed to allow for the reuse, remanufacturing, or recycling of specific components. In addition, selecting safe and sustainable materials for medical devices should be a key consideration for responsible medical device development. In addition, it is imperative to improve the end stages of medical equipment, which is important to help reduce medical waste, protect the environment, and build a more resilient and environmentally responsible healthcare system [26].

Prioritizing social responsibility is one of the main pillars of ESG integration in medical equipment engineering, which will contribute to a more stable and

Table 12.1 *Examples of green medical devices (adapted from https://www.medicaldevice-network.com/features/feature128184/?cf-view&cf-closed)*

Medical devices	Function	Diagram	Source
Welch Allyn Green Series Exam Light IV	Among the first medical exam lights available in the United States to feature energy-efficient, light-emitting diodes (LEDs) that provide 3× the typical output of halogen and consume less energy for improved facility efficiencies. The Welch Allyn Green Series Lights were designed to improve patient exams while reducing environmental impact.		https://www.hillrom.com/en/products/green-series-exam-light-iv/
Philips Mammo-Diagnost Digital mammography	The Mammo Diagnost DR digital radiography system is designed to support high-volume screening and meet the workflow challenges of hospitals and health-care centers, as well as mobile screening programs in rural communities.	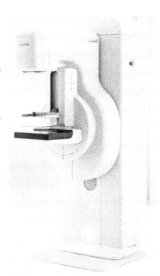	https://clinicalimaging-systems.com/product/philips-mammodiagnost-and-mammodiagnost-fd-eleva-mammography/

(Continues)

Table 12.1   (Continued)

Medical devices	Function	Diagram	Source
Syreen pre-filled syringe, Cambridge Consultants	A novel pre-filled syringe concept that has been designed with sustainability, as well as patient safety and support, is a highly effective, safe, and easy-to-use drug delivery device for self-administration use, designed to ease resource intensity and material wastage.		http://dev.sergeroux.com/portfolio/syreen/
Neptune 2 waste management system, Stryker	A mobile unit collects surgical waste from operating rooms while reducing exposure to surgical fluids. It is constructed of environmentally preferred Polypropylene #5 plastic, Neptune's 1.6oz disposable manifold reduces the volume of waste; it takes 77 manifolds to equal one full 3-1 canister in a landfill.		https://techweb.stryker.com/Field_Service/Neptune/702_N2/0702-002-619C.pdf

resilient global supply chain. The ESG is not only a framework for tackling a series of global challenges but nurturing fair labor practices by offering an equal, inclusive, and dignified environment in which every employee can grow and develop. It also provides a golden opportunity to amplify its social impact, opening the door for skill development and upskilling with more investment in training programs for employees and communities, empowering them to participate in medical equipment manufacturing and maintenance [27]. Moreover, the incorporation of ESG principles in medical equipment engineering is vital to pave the way for greater equity and inclusivity in healthcare delivery by operating with more socially responsible practices and procedures.

The lack of standardized ESG frameworks and reporting practices has resulted in varying levels of readiness among businesses and investors, potentially hindering the implementation of ESG and green technology in the medical equipment engineering industry. Hence, in moving forward, there is a need for a clear ESG framework to facilitate the companies to act responsibly for the interest of their customers and society. The development of ESG strategies and green technology positions in the medical equipment engineering industry has provided an opportunity to share key insights around sustainability. Consequently, this leads to long-term sustainability and viability transformation within the healthcare system.

Hence, the adoption of a transparent and actionable ESG framework is crucial as it helps to guide them on how to outline what will be reported and how it will be measured. Besides that, embracing ESG fosters ethical business practices, and thereby enhances the industry's integrity and mitigates legal issues.

12.7 Conclusion

In summary, various sectors including medical equipment engineering, are increasingly prioritizing ESG framework and green technology. This paradigm shift aims to foster long-term value creation and contribute positively to environmental and social outcomes. This industry relies heavily on global manufacturing, and supply chain networks will be impacted by evolving expectations around carbon emissions and environmental impact. Besides, adopting the ESG framework builds a diverse and inclusive workforce and promotes good governance, therefore enhancing business resilience. Thus, it is important for these companies to begin revisiting business models and developing new ESG strategies and green technology for a greener and healthier future.

References

[1] D. H. S. Good, "What methods can the health care industry implement to reduce energy consumption and waste production from patient care to limit greenhouse gas emissions," 2020.

[2] A. E. Ongaro, Z. Ndlovu, E. Sollier, *et al.*, "Engineering a sustainable future for point-of-care diagnostics and single-use microfluidic devices," *Lab on a Chip*, vol. 22, no. 17, pp. 3122–3137, 2022, doi:10.1039/d2lc00380e.

[3] P. Arun Kumar, "Regulating environmental impact of medical devices in the United Kingdom—A scoping review," *Prosthesis*, vol. 3, no. 4, pp. 370–387, 2021, doi:10.3390/prosthesis3040033.

[4] B. Joseph, J. James, N. Kalarikkal, and S. Thomas, "Recycling of medical plastics," *Advanced Industrial and Engineering Polymer Research*, vol. 4, no. 3, pp. 199–208, 2021, doi:10.1016/j.aiepr.2021.06.003.

[5] U. Kong, N. F. Mohammad Rawi, and G. S. Tay, "The potential applications of reinforced bioplastics in various industries: A review," *Polymers (Basel)*, vol. 15, no. 10, 2023, doi:10.3390/polym15102399.

[6]   S. Shukla, V. Kalaiselvan, and R. S. Raghuvanshi, "How to improve regulatory practices for refurbished medical devices," *Bull World Health Organ*, vol. 101, no. 6, pp. 412–417, 2023, doi:10.2471/BLT.22.289416.

[7]   K. Oturu, W. L. Ijomah, A. Broeksmit, *et al.*, "Investigation of remanufacturing technologies for medical equipment in the UK and context in which technology can be exported in the developing world," *Journal of Remanufacturing*, vol. 11, no. 3, pp. 227–242, 2021, doi:10.1007/s13243-021-00102-5.

[8]   V. Pabalkar, R. Chanda, and J. Sachin, "Refurbished medical imaging equipment through technology," *Presented at the 2022 Interdisciplinary Research in Technology and Management (IRTM)*, 2022.

[9]   D. Whitehead and J. Conley, "The next frontier of remote patient monitoring: Hospital at home," *Journal of Medical Internet Research*, vol. 25, p. e42335, 2023, doi:10.2196/42335.

[10]   M. Volterrani and B. Sposato, "Remote monitoring and telemedicine," *European Heart Journal Supplements*, vol. 21, pp. M54–M56, 2019, doi:10.1093/eurheartj/suz266.

[11]   C. C. Kang, Y. H. Teh, J. D. Tan, M. M. Ariannejad, and S. S. Balqis, "Review of 5G wireless cellular network on Covid-19 pandemic: Digital healthcare & challenges," *Jurnal Kejuruteraan*, vol. 35, no. 3, pp. 551–556, 2023, doi:10.17576/jkukm-2023-35(3)-02.

[12]   A. S. Kang and S. Arikrishnan, "Sustainability reporting and total quality management post-pandemic: The role of environmental, social, governance (ESG), and smart technology adoption," *Journal of Asia Business Studies*, 2024, doi:10.1108/jabs-03-2022-0080.

[13]   A. V. Wirba, "Corporate social responsibility (CSR): The role of government in promoting CSR," *Journal of the Knowledge Economy*, 2023, doi:10.1007/s13132-023-01185-0.

[14]   D. Dhagarra, M. Goswami, and G. Kumar, "Impact of trust and privacy concerns on technology acceptance in healthcare: An Indian perspective," *International Journal of Medical Informatics*, vol. 141, p. 104164, 2020, doi:10.1016/j.ijmedinf.2020.104164.

[15]   P. Citron, "Ethics considerations for medical device R&D," *Program in Cardiovascular Disease*, vol. 55, no. 3, pp. 307–15, 2012, doi:10.1016/j.pcad.2012.08.004.

[16]   S. Reddy, S. Allan, S. Coghlan, and P. Cooper, "A governance model for the application of AI in health care," *Journal of the American Medical Informatics Association*, vol. 27, no. 3, pp. 491–497, 2020, doi:10.1093/jamia/ocz192.

[17]   E. B. S. Çubuk, B. Demirdöven, and M. Janssen, "Policies for enhancing public trust and avoiding distrust in digital government during pandemics: Insights from a systematic literature review," in *Pandemic, Lockdown, and Digital Transformation*, Public Administration and Information Technology, 2021, Chapter 1, pp. 1–23.

[18]   W. W. Y. Yeo and C. C. Kang, "Wireless communication technologies for medical equipment engineering," 2023, doi:10.1049/pbhe054e_ch12.

[19] D. Tipurić, "The rise of strategic leadership," in *The Enactment of Strategic Leadership*, 2022, Chapter 3, pp. 55–92.

[20] M. Shapsugova, "ESG principles and social responsibility," *E3S Web of Conferences*, vol. 420, 2023, doi:10.1051/e3sconf/202342006040.

[21] W. M. W. Mohammad and S. Wasiuzzaman, "Environmental, social and governance (ESG) disclosure, competitive advantage and performance of firms in Malaysia," *Cleaner Environmental Systems*, vol. 2, 2021, doi:10.1016/j.cesys.2021.100015.

[22] J. Love-Koh, S. Griffin, E. Kataika, P. Revill, S. Sibandze, and S. Walker, "Methods to promote equity in health resource allocation in low- and middle-income countries: an overview," *Global Health*, vol. 16, no. 1, p. 6, 2020, doi:10.1186/s12992-019-0537-z.

[23] M. Shahzad, Y. Qu, S. U. Rehman, and A. U. Zafar, "Adoption of green innovation technology to accelerate sustainable development among manu-facturing industry," *Journal of Innovation & Knowledge*, vol. 7, no. 4, 2022, doi:10.1016/j.jik.2022.100231.

[24] S. Hinrichs-Krapels, J. C. Diehl, N. Hunfeld, and E. van Raaij, "Towards sustainability for medical devices and consumables: The radical and incre-mental challenges in the technology ecosystem," *Journal of Health Services Research and Policy*, vol. 27, no. 4, pp. 253–254, 2022, doi:10.1177/13558196221110416.

[25] P. Reynolds, "Designing eco-friendly medical devices," *IEEE Pulse*, vol. 13, no. 4, pp. 24–26, 2022, doi:10.1109/MPULS.2022.3191817.

[26] B. Mang, Y. Oh, C. Bonilla, and J. Orth, "A medical equipment lifecycle framework to improve healthcare policy and sustainability," *Challenges*, vol. 14, no. 2, 2023, doi:10.3390/challe14020021.

[27] L. Straub, K. Hartley, I. Dyakonov, H. Gupta, D. van Vuuren, and J. Kirchherr, "Employee skills for circular business model implementation: A taxonomy," *Journal of Cleaner Production*, vol. 410, 2023, doi:10.1016/j.jclepro.2023.137027.

Chapter 13

Decreasing the risks of smart supply chain systems based on the Internet of Things under the ESG concept

Ali Ala[1,2]

This study investigates the integration of smart supply chain systems, leveraging the Internet of Things (IoT) within the framework of environmental, social, and governance (ESG) principles. In the research, develop a nuanced understanding of these systems' complex interactions and uncertainties by employing fuzzy non-linear mathematical modeling. By incorporating fuzzy logic, the study addresses the imprecision and ambiguity associated with ESG considerations, allowing for a more comprehensive assessment of risks, responsibility, and comparisons of various factors such as supply, production, and distribution practices within IoT-driven smart supply chains. The proposed modeling approach seeks to enhance decision-making processes, optimize resource utilization, and contribute to developing resilient and sustainable supply chain strategies under the ESG concept. The outcomes indicate that privacy in interaction with suppliers and customers is the most essential and practical benchmark that should be taken to address these risks.

13.1 Introduction

The intersection of technology, environmental responsibility, social impact, and robust governance has given rise to innovative approaches in various sectors. One such groundbreaking advancement is the implementation of a smart supply chain system based on the IoT within the overarching framework of ESG principles. This integration represents a transformative paradigm, marrying technological efficiency with a conscientious commitment to environmental sustainability, social welfare, and governance excellence. As businesses globally seek more responsible and sustainable practices, the convergence of IoT and ESG principles in supply chain management emerges as a potent solution, promising operational efficiency and a positive impact on the broader socio-environmental landscape. This

[1]Department of Industrial Engineering and Management, School of Mechanical Engineering, Shanghai Jiao Tong University, China
[2]Faculty of Engineering and Quantity Surveying, INTI International University, Malaysia

introduction sets the stage for a deeper exploration into the intricacies and benefits of this symbiotic relationship between IoT technology and ESG considerations in shaping the future of supply chain systems. The Internet of Things, or IoT, is a network that connects items via Internet connectivity and applications. It increases contact and communication, which raises the viability of the system. IoT-based applications simplify services and make them less complicated for users. IoT helps increase manufacturing productivity for consumer goods in smart supply chains. Reduced product costs and pollution are the main goals of an IoT-enabled smart supply chain. IoT gives smart supply chain products access to a wide range of services and capabilities by giving each product a unique code that contains vital information. Through wireless sensors, IoT monitors and provides users with the best datasets. It connects to multiple devices and collects data, producing relevant information for product development. Through improved overall performance and effectiveness, IoT increases user demand for products. Guidelines on developing supply chain management approaches from the ground up are provided by the ESG framework. The organization's priorities determine how well an ESG supply chain performs. ESG performance aspects are taken into account by financial institutions during the investment decision-making process, resulting in a surge in investments made in businesses that demonstrate a sustained dedication to ESG responsibility. It is important to remember that although green finance is primarily concerned with environmental issues, social and economic issues may receive little attention. A wide range of funding activities that support sustainable development are included in sustainable finance. The study presented here offers a novel method of evaluating supply chain economies through blockchain technology in an ESG framework. ESG concepts and blockchain technology are used in this plan to improve supply chain transparency, auditability, and responsibility globally.

This concept is significant because it can lead to beneficial changes in the social and environmental spheres, foster moral business practices, and create a more optimistic outlook for intelligent supply chains. By using it, supply chain management may be improved, ESG goals can be supported, and the potential applications of blockchain technology for sustainable development can be further explored. This study's main contribution is developing an IoT-based supply chain assessment model to identify and mitigate risk factors related to functional growth. The experimental results show that the suggested model improves suggestion and evaluation rates while effectively minimizing risk factors within a regulated evaluation timeframe. The model determines economic needs and supply distributions for different types of transactions.

13.2   Literature review

Research on ESG elements focuses mainly on risk, sustainable development, firm performance, and associated fields. Interestingly, this article argues that corporate social responsibility (CSR) and the notion of ESG can be used interchangeably despite specific philosophical differences. Pratap *et al.* [1] introduced the integration of ESG within the industry. This enabled monitoring components' journeys throughout the chain of custody, thereby enhancing overall sustainability. Using ratio or long-term profitability to evaluate the state-owned enterprises' financial performance, Asif *et al.* [2] found no significant relationship between environmental choices and shareholder value. Most academics continue to agree that there is a positive correlation between corporate financial success and ESG policies despite the continuous discussion about how ESG affects corporate economic performance. Many theories, including smart enterprise and, consequently, smart supply chains, have evolved with the advent of the IoTs and digital technology. Liu *et al.* [3] examined a study on how numerous businesses and organizations depend on data and telecom technology to boost productivity, cut expenses, and improve the quality of their output. The IoT is closely related to analyzing vast amounts of data and powerfully penetrates many fields to optimize energy efficiency and reduce harmful environmental effects. Zhang *et al.* [4] proposed an idea that is mainly due to the effective services offered including utilization of natural resources, astute facility and infrastructure leadership, and promotion. Environmental protection is also a factor. New ideas like innovative supply chains have emerged due to the advent of titles like "smart supply chains" and "business intelligence," which describe how technology is used to generate massive amounts of information and organize, store, procedure, and preserve it. Liao *et al.* [5] introduced a novel circular supply chain (CFC) model to mitigate waste within a management system. The existing CFC model generates significant waste, contributing to an elevated environmental pollution rate. The newly proposed method primarily focuses on recycling, aiming to decrease waste within the CFC by Nagarajan *et al.* [9].

Park *et al.* [6] proposed a method that effectively minimizes the overall food waste rate within the CFC system by maximizing efficiency and reliability. Nozari *et al.* [7] demonstrated an IoT-enhanced supply chain management solution. This approach's main goal is to identify the key components and principles of a management system. Through the use of various analysis and prediction techniques, data extraction serves the purposes of detection and identification, producing an appropriate dataset for the management procedure [14]. Azizi *et al.* [8] introduced a new approach that helps the management system function more dependably and efficiently. The volume of data generated within the smart supply chains is on the rise. Abundant information spanning different facets of the supply chain, including raw material procurement, supplier communication, material transportation to production sites, manufacturing processes, program production planning, and the subsequent systematic distribution, is readily available promptly under the ESG concept by De Vass *et al.* [12] and Shao *et al.* [13]. The intersection of the IoT and artificial intelligence (AI) presents various prospects within the healthcare industry. Ala *et al.* (2023) presented a novel model for enhancing the standard of treatment in smart healthcare systems (SHSs) based on the convergence of AI and IoT. This wealth of data holds significant value, and through collaboration between supply chain planners, decision-makers, and information and communication technology specialists, it can be harnessed for the advancement of environmental sustainability and financial gains [15]. Al-Talib *et al.* [10] noted that companies focus primarily on meeting customer demands, which lengthens and complicates the supply chain (SC). This complexity makes SC management more challenging and increases the risk of disruptions, such as losing a key supplier or experiencing a fire at a production facility. Integrating IoT into the SC enhances flexibility, improves product quality analysis, and helps companies bolster their supply chain resilience (SCRes). Ala *et al.* [11] developed a smart deals system (SDS) that utilizes advanced machine learning algorithms for enhanced performance and developed secure, user-friendly consumer applications to make the system operational. The advancement of intelligent commerce in these large-scale retail platforms is propelled by effective supply chain management.

13.2.1   Research gaps

Research gaps in smart supply chain systems based on the IoT under the ESG concept include a need for more sophisticated models incorporating fuzzy logic and nonlinear techniques for a more nuanced evaluation of ESG principles. Considering the evolving nature of both operational processes and sustainability challenges, there is a need to explore dynamic and real-time ESG risk management strategies within smart supply chains. Quantifying social impact and human rights within IoT-enabled supply chains still needs to be explored, requiring in-depth studies to measure the effectiveness of these systems in promoting fair labor practices across the supply chain. Additional research gaps that need attention include cross-sector collaboration for standardized ESG frameworks, understanding behavioral aspects of ESG adoption, and developing comprehensive long-term impact assessment models. Additionally, addressing ethical considerations and privacy concerns in IoT data governance, along with studying the influence of consumer perceptions on ESG-focused smart supply chain systems, represents critical areas for future research.

13.3   Methodology

This research used a fuzzy ranking method based on hierarchical analysis. In this method, first, the integration matrix of fuzzy pairwise comparisons is obtained based on triangular numbers and linguistic criteria in Eq. (13.1) below. Then, these obtained values are used to rank the criteria using a fuzzy nonlinear mathematical modeling method.

$$
\begin{aligned}
&\max \lambda \\
&s.t. \\
&(m_{ij} - l_{ij})\lambda w_j - w_i + l_{ij}w_j \leq 0 \\
&(u_{ij} - m_{ij})\lambda w_j + w_i - u_{ij}w_j \leq 0 \\
&\sum_{k=1}^{n} w_k = 1 \\
&w_k \geq 0, k = 1, 2, ..., n; i = 1, 2, ..., n-1; j = 2, 3, ..., n. \\
&j \geq i
\end{aligned}
\tag{13.1}
$$

The values were obtained from the fuzzy numbers from the pairwise comparisons matrix. Since the created model is non-linear, it cannot be solved using simple mathematical programming methods. Therefore, LINGO is used to solve the created model.

To identify shortcomings and evaluate probability in the risk factor assessment, the classification process is based on ESG performance standards. Risk analysis is used to determine the assessment rate, and the random forest classifier is used to forecast the financial assessment length based on the available parameters. The first category depends on the maximum ESG concept request (ESG$_r$), and F as financial assessment is calculated as follows:

$$F(C_{gr}, ESG_r) = \left[Y_d - \left(\frac{\alpha_{ESG}}{\alpha_{Pa}} \right) \times \frac{1}{n} \right] - Pa(n) + 1 \tag{13.2}$$

In (13.2), the ESG concept recommendation depends on the risk factor analysis in the supply chain for distribution and economic management, as in ESG and peril aspects (n). Here, the chances of functional growth through fewer perilous aspects, achieving sequential supply distribution is computed in (13.3):

$$\alpha_{ESG} \left(\frac{T}{\alpha_{Pa}} \right) = \frac{1}{\sqrt{2n}} \exp \left[\frac{S_d - \alpha_{ESG} \times E_d}{n} \right] \tag{13.3}$$

13.4   Proposed research framework

Considering that the IoT is one of the most significant sources of massive data, it is imperative to ascertain the data-generating foundation in this chain to conduct a risk analysis. Every input inside the supply chain has data collected utilizing various IoT tools, including computers, smartphones, tablets, and other sensors. The smart supply chain framework receives and analyzes these data through reliable systems and extensive evaluations to aid in decision-making. Figure 13.1 shows the most essential

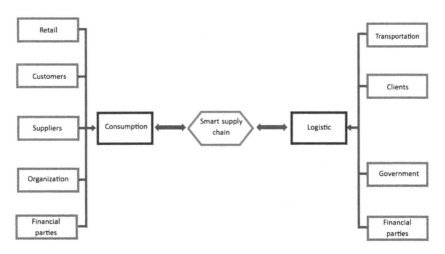

Figure 13.1   The framework of significant data sources in the smart supply chain

Table 13.1   Various risks in three main areas for IoT-based smart supply chain

Values	Risks definition	Supply chain sectors
W1	Lack of security and privacy program in interaction with suppliers	Supply
W2	Disruption in transportation and delivery	Supply
W3	Lack of monitoring of devices and systems to detect security incidents	Production
W4	Failure to include security in product and ecosystem design	Production
W5	Impaired identification and treatment of product hazards	Production
W6	Lack of awareness and sufficient security training for engineers	Production
W7	Implementation of security and privacy risk management in interaction with customers	Distribution
W8	Disruption in checking product inventory	Distribution
W9	Disruption in environment detection	Distribution
W10	Interference in schedules for distribution	Distribution

sources of ample data supply for the supply chain in the IoTs by selecting production, supply, and distribution sectors.

The supply chain, methods for integrating large amounts of data, an analysis of the relevant literature, and the insights of working specialists were used to determine the risks associated with the supply chain in all three domains of distribution, production, and supply. According to the data in Table 13.1, protecting one's privacy is one of the most important considerations while communicating with clients and providers of products and services. Electronic devices getting details from suppliers and conducting business extensively through smartphones and tablets can provide significant private and economic details. As a result, one of the top obstacles and issues can be ensuring security for these kinds of industries. The lack of security programs for interactive programs is always one of the most critical threats that can endanger the business. For industries that consider smart supply chains, timely delivery of raw materials and no disruption in this sector can be essential.

Receiving traffic information and the status of regular distribution, interference in schedules, and accurate information on the ground of raw material inventory and products ready for delivery are all information that can cause irreparable damage to business systems if there is a risk. Therefore, a proper understanding of the risks in businesses that have become intelligent through the IoT can be a factor that creates organizational peace and, as a result, increases efficiency.

13.4.1   Definition of proposed steps

This study employed a survey-style research methodology, which is helpful for its objective since it attempts to utilize a decision-making approach to rate the risks of the smart supply chain based on IoTs while accounting for ESG. Most of the information in this research has been collected by sending and completing questionnaires by specialists and experts in the field of study. The stages and steps of

*Table 13.2   Qualitative standards for
imprecise pairwise comparisons*

Linguistic terms	Fuzzy equivalent
Very low (VL)	(1,2,3)
Low (L)	(4,2,3)
Average (A)	(5,4,3)
High (H)	(6,5,4)
Very high (VH)	(7,6,5)

the proposed methodology for the ESG performance evaluation of listed companies are considered.

Step 1: To identify risks in the smart supply chain based on the IoTs achieve this goal, library studies and literature on the subject and risks and threats resulting from using the IoTs in the supply chain were investigated to make it brighter. These risks were refined and validated using experts' opinions.

Step 2: Design a structure that is hierarchical. IoT-based smart supply chain risks were organized into a hierarchical structure employing objectives, criteria, and choice levels. This structure is shown in Table 13.1. This framework aims to analyze the risks associated with the smart supply chain. This analysis has been conducted throughout the supply, production, and distribution sectors.

Step 3: The expert who was invited furnishes pairwise comparison matrices for the assessed criteria within the E, S, and G aspects of the evaluation. Subsequently, these linguistic matrices are converted into model comparison matrices, as depicted in (13.2).

Step 4: The expert who has been invited provides linguistic assessments for the criteria of each alternative, creating an initial linguistic evaluation, as outlined in (13.3).

Step 5: Creation of fuzzy judgment matrices. In the smart supply chain and analytic hierarchy process (AHP) technique, risks are analyzed and ranked using fuzzy matrixes based on expert judgments.

Step 6: Table 13.2 displays the linguistic criteria utilized in this study for fuzzy pairwise analyses. Fuzzy triangular numbers provide the foundation of these scales.

Step 7: Developing and resolving fuzzy mathematical nonlinear models. A fuzzy and nonlinear computational ranking method is applied to assess the risks in the smart supply chain according to the IoT. This approach is predicated on the fuzzy hierarchical analysis technique.

13.5   Calculation steps and results

There are two primary components to the smart supply chain risk assessment and rating system. Using fuzzy surveys, opinions from experts are first gathered and incorporated. The risks are then ranked utilizing the non-linear computational framework that the research presents.

A thorough process, involving the sequential steps of choosing preferred companies, developing the indicator system, gathering data, and constructing the decision matrix, is carried out to determine the ESG performance evaluation matrix. Initially, alternatives such as supply (A1), production (A2), and distribution (A3) were selected and employed for the empirical validation of the proposed method described earlier in Table 13.3.

Choose a group of supply chain-related professionals with various backgrounds and expertise. These specialists ought to be able to illuminate the different facets of every field you are assessing, as well as the aggregated findings, fuzzy pairwise comparison matrices, and the logic underlying the judgments. Share the results with the relevant parties transparently to promote comprehension. In terms of the results in Table 13.4, for the two W1 aspects, the W2 aspects related to the production sector carry the most weight, making up 40%, while criterion W1 has the least weight.

Fuzzy pairwise comparisons were used to combine expert viewpoints and determine the significance of each criterion. Weighted criteria were produced by aggregating fuzzy matrices, which helped in decision-making. A similar strategy was used in the supply chain, resulting in fuzzy weighted matrices representing experts' agreement on the essential criteria for well-informed supply chain decisions.

The results obtained in Tables 13.5 and 13.6 by utilizing expert opinions, fuzzy pairwise comparisons were employed to assess criteria relevance in both the production and distribution sectors. Among the two W6, W10 aspects, with distribution and production sector has the largest weight, accounting for 45%, while criterion W3, W7 have the smallest weight. By placing the data from the paired comparison tables obtained from integrating experts' opinions, we have formed fuzzy nonlinear mathematical models and solved the resulting models using

Table 13.3   Fuzzy pairwise comparisons integrate expert opinions for different areas of the supply chain

Value symbols	A1			A2			A3		
A1	–	–	–	–	–	–	–	–	–
A2	3.2	3	7	–	–	–	–	–	–
A3	3	3.1	5.6	2	2.6	2.6	–	–	–

Table 13.4   Utilizing expert opinions to perform fuzzy pairwise comparisons for the supply sector

Value symbols	W1			W		
W1	–	–	–	–	–	–
W2	3.2	4.3	6.2	–	–	–

LINGO software. The weight and rank of each risk in supply, production, and distribution were obtained.

In the dynamic landscape of the production, supply, and distribution sectors, understanding and managing risks are crucial. The ESG framework emphasizes sustainable and responsible business practices, making it integral to the risk assessment process. Simultaneously, the rating of each risk provides a qualitative measure of its potential impact and likelihood. Risks associated with environmental factors, such as resource depletion or pollution, social factors like labor practices, and governance issues such as regulatory compliance are assessed with a nuanced understanding of their potential consequences. Incorporating ESG considerations ensures that the weight and rating assigned to each risk align with the broader goals of sustainability, ethical business conduct, and long-term resilience. Based on results obtained in Tables 13.7–13.9 the results of each of the three alternatives, the risk assessment in the supply sector has the highest weight for ranks 1 and 2, while the risk assessment in the production and distribution sector is the lowest for rank 4 in both sectors.

Table 13.5   Utilizing expert opinions to perform fuzzy pairwise comparisons for the production sector

Value symbols	W3				W4			W5			W6		
W3	–	–	–	–	–	–	–	–	–	–	–	–	–
W4	5.2	4.3	6	–	–	–	–	–	–	–	–	–	–
W5	2.3	3.2	4.3	2.2	2.2	3.2	–	–	–	–	–	–	–
W6	2.6	4.1	4.4	3	6.1	3	2	2.2	3.3	–	–	–	–

Table 13.6   Utilizing expert opinions to perform fuzzy pairwise comparisons for the distribution sector

Value symbols	W7				W8			W9			W10		
W7	–	–	–	–	–	–	–	–	–	–	–	–	–
W8	5.2	4.3	6	–	–	–	–	–	–	–	–	–	–
W9	2.3	3.2	4.3	2.2	2.2	3.2	–	–	–	–	–	–	–
W10	2.6	4.1	4.4	3	6.1	3	2	2.2	3.3	–	–	–	–

Table 13.7   Evaluating and ranking individual risks within the supplier sector

Ranking	Weighting	Symbols	Risk definition
1	0.620	W1	Lack of security and privacy program in interaction with suppliers
2	0.40	W2	Disruption in transportation and delivery

Table 13.8   Evaluating and ranking individual risks within the production sector

Ranking	Weighting	Symbols	Risk definition
2	0.28	W3	Lack of monitoring of devices and systems to detect security incidents
1	0.310	W4	Failure to include security in product and ecosystem design
3	0.240	W5	Impaired identification and treatment of product hazards
4	0.177	W6	Lack of awareness and sufficient security training for engineers

Table 13.9   Evaluating and ranking individual risks within the distribution sector

Ranking	Weighting	Symbols	Risk definition
1	0.310	W7	Implementation of security and privacy risk management in interaction with customers
4	0.210	W8	Disruption in checking product inventory
3	0.230	W9	Disruption in environment detection
2	0.250	W10	Interference in schedules for distribution

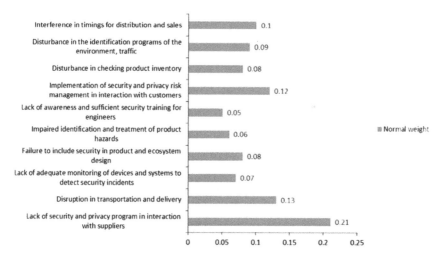

Figure 13.2   Standard weight diagram of hazards in the smart supply chain network

The ranking depicted in Figure 13.2 underscores the significance of maintaining privacy in interactions with both suppliers and customers as the foremost concern among smart supply chain risks. This implies that safeguarding sensitive information and data confidentiality within the supply chain ecosystem is paramount for ensuring the integrity and security of the overall operation. Setting privacy first in consumer interactions demonstrates that the sensitive nature of consumer data and information is acknowledged. In a time when data privacy laws are getting stricter, ensuring customer-related data is handled securely is not only required by law but also essential to upholding credibility and confidence. The effectiveness and longevity of a smart supply chain can be significantly impacted by improper treatment of consumer data, leading to reputational harm, legal issues, and a loss of customer confidence.

13.6 Discussion

The paper addressed how IoT analytics and data may lower risks by increasing decision-making, streamlining processes, and increasing visibility. One noteworthy discussion point is giving privacy first priority when interacting with suppliers and consumers. The focus on protecting sensitive data recognizes how crucial data security is becoming in the digital era. By tackling this risk within the context of the ESG framework, the conversation draws attention to the moral issues and conscientious corporate practices needed to maintain privacy standards, which adds to the ESG's social responsibility component. Under the ESG framework, it is crucial to recognize some restrictions related to risk mitigation in IoT-based smart supply chains. First, firms may need help to successfully understand and comply with new ESG standards because technology may evolve more quickly than legal frameworks. Furthermore, the present ESG rules may not eliminate the additional risks brought about by the reliance on networked systems, such as the danger of online threats and data attacks.

Even though the talk emphasizes the progress made in lowering risks in IoT-based smart supply chains under the ESG concept, it is important to remember that technology is constantly changing and that implementing ESG may cause operational difficulties. Continual endeavors to adjust to changing regulatory environments, strengthen cybersecurity protocols, and guarantee a scalable and comprehensive implementation of ESG principles will augment the robustness and sustainability of intelligent supply chain networks.

13.6.1 Managerial implications

In the context of the ESG theory, the managerial implications of reducing the risks associated with IoT-based smart supply chain systems are complex and call for a proactive and deliberate approach. Managers must carefully incorporate ESG principles into the overarching company plan. This entails matching ESG objectives with the business's purpose, values, and operational procedures.

Establishing a well-defined ESG methodology is the basis for recognizing and managing the risks connected to IoT-based smart supply chains. Given the subject's emphasis on privacy, managers must set up strong governance structures for privacy and data security. Safeguarding sensitive data inside intelligent supply chain systems entails implementing encryption technologies, access restrictions, and frequent audits.

By considering these managerial implication factors, organizations can effectively decrease the risks associated with IoT-based smart supply chain systems under the ESG concept, contributing to long-term sustainability and resilience.

13.7   Conclusion and future studies

Given that the supply chain encompasses a significant portion of organizational operations, ranging from raw material procurement to distribution, a thorough grasp of technology's applications can aid in optimizing many organizational processes. One of the most important sources of data may be IoTs. If appropriately examined and handled, these data can greatly aid the company in making timely decisions. However, because this technology is intertwined with the Internet, there will always be issues and worries. By selecting the major dozen production concepts in the smart supply chain as an area of study, the current research has attempted to identify and comprehend the dangers an IoT-based supply chain can handle. To assess the relative importance of each of these threats, nonlinear computational modeling has been combined with a fuzzy ranking technique. According to the research findings, one of the most critical hazards associated with the smart supply chain for these firms is maintaining privacy in their interactions with suppliers and consumers. While the suggested multi-criteria decision making (MCDM) framework can comprehensively and successfully handle the ESG sustainable performance evaluation, there are still issues with the techniques and thorough treatment. The ambiguous expert opinions may lead to departure because of the subjective variations in articulating the qualitative linguistic assessments. The qualitative conclusions should be supported by more quantitative evidence.

Future studies in smart supply chain systems based on the IoT under the ESG can explore several emerging trends and potential research areas, such as investigating how IoT technologies can optimize resource use, minimize waste, and support sustainable practices across the entire product lifecycle. Moreover, blockchain technology can contribute to ESG objectives by creating immutable records of ESG practices throughout the supply chain.

References

[1]   Pratap, S., Jauhar, S. K., Gunasekaran, A., and Kamble, S. S. (2023). Optimizing the IoT and big data embedded smart supply chains for sustainable performance. *Computers & Industrial Engineering*, 109828.
[2]   Asif, M., Searcy, C., and Castka, P. (2023). ESG and Industry 5.0: The role of technologies in enhancing ESG disclosure. *Technological Forecasting and Social Change*, 195, 122806.
[3]   Liu, X., Wu, H., Wu, W., Fu, Y., and Huang, G. Q. (2021). Blockchain-enabled ESG reporting framework for sustainable supply chain. In *Sustainable design and manufacturing 2020: Proceedings of the 7th international*

conference on sustainable design and manufacturing (KES-SDM 2020) (pp. 403–413). Springer Singapore.

[4] Zhang, M., Yang, W., Zhao, Z., Pratap, S., Wu, W., and Huang, G. Q. (2023). Is digital twin a better solution to improve ESG evaluation for vaccine logistics supply chain: An evolutionary game analysis. *Operations Management Research*, 1–23.

[5] Liao, H. T., and Pan, C. L. (2021). The role of resilience and human rights in the green and digital transformation of supply chain. In *2021 IEEE 2nd international conference on technology, engineering, management for societal impact using marketing, entrepreneurship and talent (TEMSMET)* (pp. 1–7). IEEE.

[6] Park, A., and Li, H. (2021). The effect of blockchain technology on supply chain sustainability performances. *Sustainability*, 13(4), 1726.

[7] Nozari, H., Fallah, M., and Szmelter-Jarosz, A. (2021). A conceptual framework of green smart IoT-based supply chain management. *International Journal of Research in Industrial Engineering*, 10(1), 22–34.

[8] Azizi, N., Malekzadeh, H., Akhavan, P., Haass, O., Saremi, S., and Mirjalili, S. (2021). IoT–blockchain: harnessing the power of internet of thing and blockchain for smart supply chain. *Sensors*, 21(18), 6048.

[9] Nagarajan, S. M., Deverajan, G. G., Chatterjee, P., Alnumay, W., and Muthukumaran, V. (2022). Integration of IoT based routing process for food supply chain management in sustainable smart cities. *Sustainable Cities and Society*, 76, 103448.

[10] Al-Talib, M., Melhem, W. Y., Anosike, A. I., Reyes, J. A. G., and Nadeem, S. P. (2020). Achieving resilience in the supply chain by applying IoT technology. *Procedia Cirp*, 91, 752–757.

[11] Ala, A., Sadeghi, A. H., Deveci, M., and Pamucar, D. (2024). Improving smart deals system to secure human-centric consumer applications: Internet of Things and Markov logic network approaches. *Electronic Commerce Research*, 24(2), 771–797.

[12] De Vass, T., Shee, H., and Miah, S. J. (2021). Iot in supply chain management: A narrative on retail sector sustainability. *International Journal of Logistics Research and Applications*, 24(6), 605–624.

[13] Shao, X. F., Liu, W., Li, Y., Chaudhry, H. R., and Yue, X. G. (2021). Multistage implementation framework for smart supply chain management under industry 4.0. *Technological Forecasting and Social Change*, 162, 120354.

[14] Lee, K. L., Wong, S. Y., Alzoubi, H. M., Al Kurdi, B., Alshurideh, M. T., and El Khatib, M. (2023). Adopting smart supply chain and smart technologies to improve operational performance in manufacturing industry. *International Journal of Engineering Business Management*, 15, 18479790231200614.

[15] Rejeb, A., Simske, S., Rejeb, K., Treiblmaier, H., and Zailani, S. (2020). Internet of Things research in supply chain management and logistics: A bibliometric analysis. *Internet of Things*, 12, 100318.

Chapter 14

ESG green corrosion inhibition of mild steel in 0.1 M HCl, NaOH and H₂O using aloe vera gel

Olayide R. Adetunji[1], Wai Yie Leong[2], Emmanuel A. Asesanya[1] and Waliat O. Adetunji-Popoola[3]

Mild steel returns back to iron oxide (rust) in the presence of water, oxygen, and ions leading to loss of desired structural properties and functional integrity which poses expensive problems for industry, including unsafe structural damage. This study investigated the impact of aloe vera gel inhibitors on the corrosion behavior of mild steel in 0.1 M HCl, NaOH, and water. Mild steel coupons in the media using the inhibitor (aloe vera gel) were studied in duplicates over a period of 10 days during which average weight loss measurements were taken to obtain the inhibition efficiencies. The electrochemical behavior of the mild steel in the inhibited and uninhibited solution was investigated by recording cathodic and anodic potentiodynamic polarization readings, and the morphology of the mild steel coupon surface was recorded using the scanning electron microscope machine. From the gravimetric experiment, the corrosion parameters showed a relatively substantial inhibition efficiency in water. The potentiodynamic polarization results indicated a high potential of the aloe vera gel as a corrosion inhibitor in the 0.1 M NaOH solution and water with efficiency of 33.71% and 57.62%, respectively. A negative efficiency of 7.82% was obtained for the inhibition in 0.1 M HCl.

14.1 Introduction

Corrosion of metals and alloys resulting from their reactions with the environment is a major problem in the industry thereby making its prevention and protection inevitable for the safety and economic progress of the environment. Different classes of corrosion inhibitors exist providing protection against corrosion for

[1]Department of Mechanical Engineering, College of Engineering, Federal University of Agriculture, Nigeria
[2]Faculty of Engineering and Quantity Surveying, INTI International University, Malaysia
[3]Department of Biosciences and Biotechnology, University of Medical Sciences, Nigeria

different metals by decreasing the corrosion rate of the metal or alloy. However, commercially available corrosion inhibitors in the market are mostly toxic, giving room to biodegradable, non-toxic, inexpensive, and reusable green corrosion inhibitors [1,2]. In addition to being acceptable with respect to ecology, it is also inexpensive and environmentally friendly as reported by Sorkhabi and Asghari [3] and Trindade and Goncalves [3,4].

Recently, many researchers acknowledged the use of plant parts, including their products like organic compounds extracted from their seeds, roots leaves, flowers, and fruits as green corrosion inhibitors [5].

Aloe vera L. is a member of the Asphodelaceae family, cultivated for its agricultural and medicinal use. Aloe vera leaf flesh has been reportedly employed in treating ailments [6]. However, a simple literature survey indicates that little study has been carried out on the inhibitive effects of the Aloe vera leaf gel on the corrosion of mild steel. Aloe vera grows about 160 cm tall, without a stem or a short stem about 30 cm long, forming dense groups. It is an important medicinal plant belonging to the Liliacea family. The gel, making up the bulk of the substance, serves the purpose of the organ that stores water for the plant. This gel, which can be separated as a semisolid "fillet" prior to being processed, constitutes over 200 different substances. The major ones included glycoproteins, vitamins, polysaccharides, minerals, and enzymes. Aloe vera gel is organic in nature and can be employed in making green corrosion inhibitors. It is a colorless mucilaginous gel obtained from the parenchymatous cell in fresh leaves of aloe vera and constitutes various active compounds including salicylates, magnesium lactate, acemannan, lupeol, campesterol, sterol, linolenic, aloctin, and anthraquinones [7].

Mild steel is the most widely used, and also one of the cheapest metals in the world. Carbon content played a high role in shaping the properties of mild steel [8]. In constituent, among other elements, carbon is up to 0.25% in mild steel. This makes it readily useful in making a wide range of products, including structural beams, car bodies, and kitchen appliances [9].

Over the years, the inhibitive effects of several chemicals have been successfully exploited to combat corrosion. However, the use of these chemicals as corrosion inhibitors has been limited by their inherent toxicity, availability, and cost. In addition, as environmental awareness increases, there is a need for green inhibitors with zero environmental effects instead of the traditional chemicals employed as corrosion inhibitors. Natural products were found to have excellent inhibitive effects and to be much eco-friendlier. This trend of employing green inhibitors (plant extracts) as alternatives to using chemicals has gained wide adoption among scholars and researchers on the merit of its availability, non-toxicity, low cost, and environmental friendliness. This research addressed four sustainable development goals: good health and well-being (SDG3), clean water and sanitation (SDG6), industry, innovation, and infrastructure (SDG9), and sustainable cities and communities (SDG11). This research therefore investigated the impact of aloe vera gel inhibitor on the corrosion behavior of mild steel in 0.1 M HCl, NaOH, and water.

14.2 Materials and methods

14.2.1 Materials

The materials used for the experiments included aloe vera gel as an inhibitor, mild steel, acidic and basic compounds (HCl, NaOH), and water (Plate 14.1).

Plate 14.1 Aloe vera leaves

14.2.2 Methods

14.2.2.1 Materials preparation

The chemical composition of the mild steel was confirmed using X-ray fluorescence spectroscopy (XRF). Aloe vera gel was obtained by peeling the leaves and scooping them into a blender using a small spoon. The gel is blended to get fluid.

14.2.2.2 Potentiodynamic polarization

The electrochemical behavior of mild steel samples in the inhibited and uninhibited solution was investigated by recording cathodic and anodic potentiodynamic polarization curves [10].

A platinum electrode was used as an auxiliary electrode and a saturated calomel electrode as the reference electrode. The measurements were carried out in the 0.1 M HCl, NaOH solution, and water-containing concentrations of the inhibitor (aloe vera gel extract) by altering the electrode potential automatically from -250 mV to $+250$ mV *vs* corrosion potential at a scan rate of 1 mV/s. The values for corrosion current density (I_{corr}), corrosion potential (E_{corr}), anodic Tafel slope (β_a), and cathodic Tafel slope (β_c) were obtained from the Tafel fit routine.

The inhibition efficiency was calculated as follows [11]:

$$\text{IE}(\%) = \left\{ \left[i_{corr} - i_{corr(inh)} \right] / i_{corr} \right\} \times 100$$

14.2.2.3   Scanning electron microscope analysis

The morphology of the mild steel coupon surface was recorded using the SEM machine (and microscope). To obtain the required images, samples of the mild steel coupons were immersed in 0.1 M HCl, NaOH, and water for 3 hours, and the surface morphology was compared with the surface images obtained in the presence of the concentration of the inhibitor (aloe vera gel extract) (Plate 14.2).

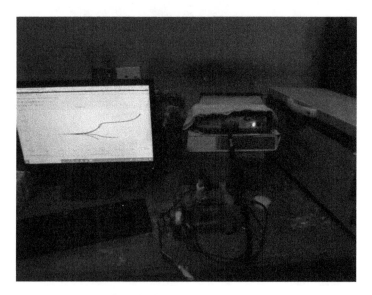

Plate 14.2   Potentiostat set-up

14.3   Results and discussion

14.3.1   Results

The results are contained in Tables 14.1–14.4 and illustrated in Figures 14.1–14.3.

14.3.2   Scanning electron microscope analysis

The SEM images of the mild steel surfaces exposed to 0.1 M each of HCL, H_2O, and NaOH in the absence and presence of the inhibitor concentration (1:50 ml v/v) after 200 hours are presented in Figures 14.4–14.9.

Table 14.1   Composition of mild steel coupons

Elements	Fe	C	Si	Mn	Cr	Ni	Cu	Al	V	Co	Nb	Others
Comp. in %	99.19	0.19	0.01	0.13	0.03	0.03	0.02	0.05	0.01	0.01	0.03	0.41

Table 14.2   Polarization measurement for mild steel in solution with inhibitor

Solution	E_{corr} (V vs S_{CE})	I_{corr} (A cm^{-2})	β_a (V dec^{-1})	β_c (V dec^{-1})	R_p (Ω cm^2)
H_2O	−0.404	9.58×10^{-5}	1.6869	16.5901	6940.23
HCl	−0.469	1.15×10^{-4}	0.0972	0.1119	196.40
NaOH	−0.451	3.53×10^{-5}	5.2965	0.8800	9282.41

Table 14.3   Polarization measurement for mild steel in solution without inhibitor

Solution	E_{corr} (V vs S_{CE})	I_{corr} (A cm^{-2})	β_a (V dec^{-1})	β_c (V dec^{-1})	R_p (Ω cm^2)
H_2O	−0.313	1.51×10^{-4}	5.1025	10.0297	9725.19
HCl	−0.475	1.06×10^{-4}	0.0881	0.1328	216.96
NaOH	−0.372	2.64×10^{-5}	1.2404	0.5614	6356.67

Table 14.4   Aloe vera inhibition efficiency

Solution	Aloe vera inhibition efficiency (%)
H_2O	57.62
HCl	−7.82
NaOH	33.71

14.3.3   Discussion

The polarization resistance values from Tables 14.2 and 14.3 increased with the application of the inhibitor in NaOH and H_2O, indicating the potential of the inhibitive effect of the inhibitor in both, the polarization resistance in the case of HCl being higher than when the inhibitor is present is due to the aggressiveness of HCl and the low concentration of the inhibitor. The results obtained corroborated the earlier findings from past researchers as reported by Vashi and Chaudhari [12].

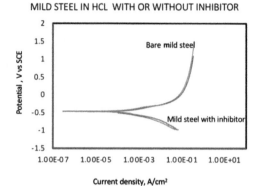

Figure 14.1   Polarization curves for mild steel in 0.1 M HCl with and without inhibitor

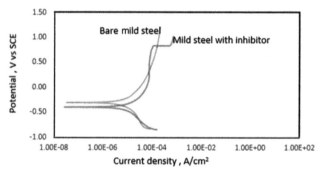

Figure 14.2   Polarization curves for mild steel in H₂O with and without inhibitor

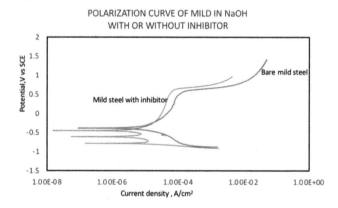

Figure 14.3   Polarization curves for mild steel in 0.1 M NaOH with and without inhibitor

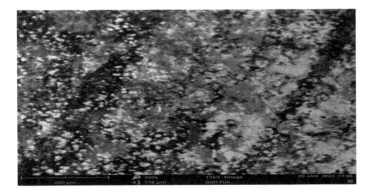

Figure 14.4   SEM image of mild steel in water with inhibitor

Figure 14.5   SEM image of mild steel in water without inhibitor

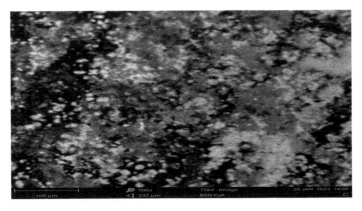

Figure 14.6   SEM image of mild steel in NaOH without inhibitor

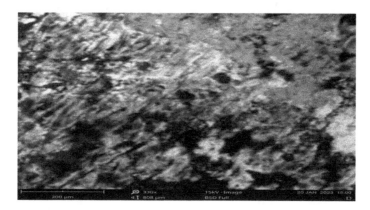

Figure 14.7   SEM image of mild steel in NaOH with inhibitor

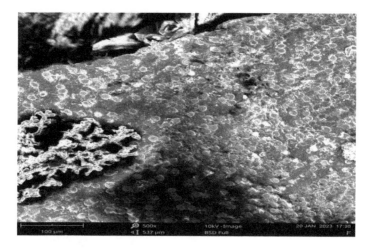

Figure 14.8   SEM image of mild steel in HCl with inhibitor

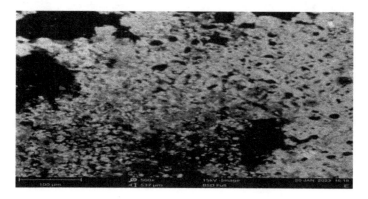

Figure 14.9   SEM image of mild steel in HCL without inhibitor

14.4 Conclusions

The potentiodynamic polarization study was carried out to evaluate the corrosion behavior of mild steel in 0.1 M HCl, NaOH, and H_2O using aloe vera gel as an inhibitor and the following are the findings:

- The inhibition efficiencies of 33.71% and 57.62% were obtained for mild steel in NaOH and H_2O, respectively, while negative inhibition of 7.82% was obtained for HCl.
- The SEM results only give an observable difference for the effect of the inhibitor in water only. The SEM images for the mild steel in HCl and NaOH with and without inhibitor showed little difference.

References

[1] Quraishi, M.A., and Singh, A.K. (2010). Green approach to corrosion inhibition of mild steel in hydrochloric acid and sulphuric acid solutions by the extract of Murraya Koenigii Leaves. *Materials Chemistry and Physics*, 122(1): 114–122.

[2] Chauhan, L.R., and Gunasekaran, G. (2007). Corrosion inhibition of mild steel by plant extract in dilute HCl medium. *Corrosion Science*, 49(3): 1143–1161.

[3] Ashassi-Sorkhabi, H., and Asghari, E. (2008). Effect of hydrodynamic conditions on the inhibition performance of L-methionine as a green inhibitor. *Electrochimica Acta*, 54(2): 162–167.

[4] Olawale, O., Bello J.O., Ogunsemi, B.T., Uchella, U.C., Oluyori, A.P., and Oladejo, N.K. (2019). Optimization of chicken nail extracts as corrosion inhibitor on mild steel in 2 M H_2SO_4. *Heliyon*, 5(11): 1–9.

[5] El-Etre, A. (2006). Khillah extract as inhibitor for acid corrosion of SX 316 steel. *Applied Surface Science*, 252(24): 8521–8525.

[6] Abiola, O.K., and James, A.O. (2010). The effects of Aloe vera extract on corrosion and kinetics of corrosion process of zinc in HCl solution. *Corrosion Science*, 52: 661–664.

[7] Pal, S., Lgaz, H., Tiwari, P., Chung, I.-M., Ji, G., and Prakash, R. (2018). Experimental and theoretical investigation of aqueous and methanolic extracts of Prunus dulcis peels as green corrosion inhibitors of mild steel in aggressive chloride media. *Journal of Molecular Liquids*, 11: 347–361.

[8] Haldhar, R., Prasad, D., and Saxena, A. (2018). Myristica fragrans extract as an eco-friendly corrosion inhibitor for mild steel in 0.5 M H_2SO_4 solution. *Journal of Environmental Chemical Engineering*, 6(2): 2290–2301.

[9] Jiang, L., Volovitch, P., Wolpers, M., and Ogle, K. (2012). Activation and Inhibition of Zn-Al and Zn-Al-Mg coatings on steel by nitrate in phosphoric acid solution. *Journal of Science Corrosion*, 60: 256–264.

[10] Adetunji, O.R., Oyelowo, M.A., Adekunle, N.O., Adeogun, A.I., and Okediran, I.K. (2022) Corrosion of some metallic pipes in water using

electrochemical methods, *Corrosion Protection*, 65(2): 43–50 www.ochro-naprzedkorozja.pl.

[11] Newman, D.J., Cragg, G.M., and Snader, K.M. (2000). The influence of natural products upon drug discovery. *Natural Product Reports*, 17: 215–234.

[12] Vashi, R.T., and Chaudhari, H.G. (2017). The study of aloe-vera gel extract as green corrosion inhibitor for mild steel in acetic acid. *International Journal of Innovative Research in Science, Engineering and Technology*, 6(11): 22081–22091.

Chapter 15

Augmented reality and ESG

Nassirah Laloo[1] and Mohammad Sameer Sunhaloo[1]

Organisations and businesses increasingly use emerging technologies to sustain environmental, social, and governance (ESG) principles and standards to boost their sustainability goals. One such emerging technology is augmented reality (AR). AR 'represents a system where a view of a live real physical environment is supplemented by computer-generated elements, such as sound, video, graphic or location data' [1]. AR refers to the dynamic integration of textual, graphical, auditory, and virtual enhancements with real-world objects in real time. AR enhances and enriches the user's interaction with the physical environment, distinguishing it from mere simulation. AR is becoming a game changer in numerous fields with its ability to align with sustainability objectives. AR is contributing to inclusive experiences. With its capacity to augment the physical world with digital content, it provides its viewers with interactive and immersive experiences while also allowing consideration for usability and accessibility (Figure 15.1).

This chapter explores the combination of ESG and AR, outlining how the amalgamation and integration of these two spheres can bring a more significant impact, awareness, action, and accountability in support of environmental issues while ensuring social equity and enhancing responsible governance practices. By exploiting the powers and potentials of AR, organisations, and businesses can be driven to innovative solutions to pressing and urgent sustainability issues.

Throughout this chapter, the potential applications of AR in advancing ESG initiatives are highlighted. Areas of application that provide a better insight into the real-world impact of AR-enabled ESG initiatives are considered. While providing a better vision of AR to advance ESG principles, this chapter offers direction for organisations and practitioners seeking to address sustainable development through this transformative technology.

[1]School of Innovative Technologies and Engineering, University of Technology, Mauritius

Figure 15.1   AR and ESG

15.1   The environmental factor

The environmental factor in ESG represents a crucial aspect of sustainable and responsible business practices [2]. Environmental considerations include various factors related to how organisations and businesses interact with the natural world and impact ecosystems. Resource scarcity, environmental degradation, and social disparities have encouraged governments and businesses to acknowledge the urgency of addressing sustainability challenges. This interest has also brought the 'E'-sector of ESG into the limelight. While continuous research into how technologies may help achieve this endeavour is under consideration, AR technologies can contribute significantly to the environmental factor in ESG by facilitating sustainable practices, promoting environmental awareness, and enabling businesses to reduce their ecological footprint [3].

AR offers many benefits for manufacturers and businesses, enabling them to reduce their carbon footprint while strengthening competitiveness against less environmentally friendly counterparts. These advantages include practising responsible consumption and production, minimising waste during early product design, and encouraging and enlightening recycling approaches [4].

AR is emerging as a crucial solution provider in various sectors, such as healthcare, telecommunication, manufacturing, and beyond. Its capability to address challenges is now evident, empowering responsible organisations to

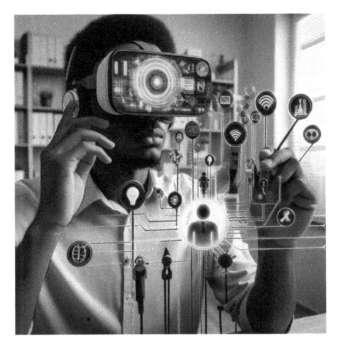

Figure 15.2   Using AR for technical intervention

contribute to sustainability efforts. In the face of ongoing global climate change, executive-level managers in businesses and organisations face increasing demands from stakeholders – customers, employees, and investors – to prioritise ESG initiatives within their operations [5].

Two instances where AR can help organisations and businesses achieve their ESG goals are described below. As reported in [5], AR can help organisations and businesses achieve their ESG objectives by minimising the need for field service visits. In traditional models, particularly in information technology or telecommunication services, customers would request assistance, prompting dispatching field technicians to their location. This way of doing things may prove inefficient and significantly impact the environment due to the associated CO_2 emissions. AR can allow service providers to swiftly and effectively address customer issues. Instead of potentially travelling long distances, service technicians can virtually connect with customers quickly to assess and troubleshoot problems (Figure 15.2).

AR can enhance energy efficiency within organisations and businesses. By integrating digital information into the physical environment, AR can provide real-time data regarding energy consumption, thus enabling organisations and businesses to pinpoint inefficiencies and provide targeted solutions [6]. For instance, AR can help follow consumption patterns within manufacturing companies and identify wasteful areas. In the sequel, ways of optimising energy usage can be recommended, thus enabling these companies to lower their carbon footprint and energy expenses (Figure 15.3).

Figure 15.3   Using AR to monitor energy usage

15.2   The social element

Fundamentally, the ESG social criteria concern human rights and equity. This dimension examines an organisation's interactions with people, policies, and practices that influence individuals, communities, and society. Below are some examples illustrating how AR reshapes organisational processes, functions, and society.

AR is changing how training works in different industries [7]. One significant benefit is that AR makes learning more immersive. It incorporates digital information into the real world so trainees can interact with it, thus enabling the trainees to understand complex ideas better (Figure 15.4). Also, AR lets the trainees practice in a safe place. They can do things like healthcare or manufacturing tasks without any real danger. This hands-on practice helps them learn faster and feel more confident. It is essential to acknowledge significant hindrances in creating AR technology for customised training: the expenses related to acquiring devices and setting up infrastructure. Additionally, technical difficulties and employee resistance to the new approach pose additional challenges.

Researchers have been exploring the relationship between AR technology and consumer engagement, particularly examining its influence on purchasing behaviour [8–10]. Given the fast technological progress and widespread adoption of AR across industries, it is essential to understand how AR affects consumer engagement and its potential to improve purchasing behaviour. AR boosts consumer engagement by

Figure 15.4   Using AR for training

offering immersive experiences that enable personalised product visualisation [8]. The connection between consumers and products is strengthened, increasing interest and purchase motivation. Moreover, AR positively influences consumer perceptions, as real-time product interaction enhances perceived value and reliability. It was also observed in [8] that AR helps decision-making by providing detailed information and virtual try-on experiences, reducing uncertainty and increasing confidence in purchase choices, ultimately leading to higher conversion rates (Figure 15.5).

AR is also transforming the healthcare industry and is considered the future of medical technology [11]. In [12], Mesko outlined some exemplary applications of AR in medicine and healthcare. These include using AR to locate nearby defibrillators to save lives, employing Google Glass to help new mothers facing challenges with breastfeeding, enabling patients to articulate their symptoms more effectively through AR, facilitating nurses in locating veins with greater ease, and supporting surgeons in the operating room with AR assistance. Moreover, AR presents noteworthy opportunities to boost wellness. AR enables personalised engagement and data-driven insights [13]. Furthermore, in recent years, there has been a global rise in the adoption of fitness mobile apps, especially among individuals seeking to lose weight and improve their health. Many smartphone users use fitness apps. AR has improved the use of mobile devices and apps, with particular benefits observed for users, including those with declining cognitive abilities, such as older adults (Figure 15.6) [14].

Figure 15.5   Using AR for try ons

Figure 15.6   AR for health

15.3   The governance aspect

Governance in ESG covers vital decision-making principles spanning governmental policy to stakeholder rights allocation. Key corporate governance areas include defining objectives, board structure, and executive oversight for accountability and sustainability [15]. As per standard practice, businesses rely on experienced managers and human input for decision-making. The emergence of AR and related technologies has transformed decision-making, helping organisations achieve strategic advantages and improve business operations [16]. For example, AR can help companies by showing real-time data on-site. The company operators and employees can thus check how the systems are doing and make changes for the systems to work better. AR also helps organisations by showing data visually and improving processes quickly. It connects business goals with data analysis for better understanding (Figure 15.7) [17].

Governments are using AR more to improve processes and make things better for people [18]. The military, law enforcement, and national security were among the first to try it. AR adds context and immersion, changing training and improving communication and public services. AR can offer improved productivity, transparency, and public engagement regarding government services [19]. For example, the public can easily access information like safety guidelines at a construction site using AR smartphone applications. AR can also help engage the public by

Figure 15.7   AR can provide on-site data

Figure 15.8   AR facilitating paperwork

providing interactive and immersive experiences at government buildings, whereby the public can revisit historical events and access information on administrative procedures and forms. Additionally, AR makes administrative tasks easier for government officers by streamlining paperwork and offering pertinent information during inspections with tools such as AR glasses (Figure 15.8).

15.4   Insights into AR: key statistics

In this section, we present selected AR statistics sourced from the following [20].

- The AR market exceeds $32 billion in value.
- AR revenue is forecasted to surpass $50 billion by 2027.
- The number of AR user devices is around 1.4 billion.
- Roughly 75% of adults under 44 are familiar with AR.
- Almost half of consumers report spending more due to AR-based shopping experiences.
- The AR market is expected to almost double from 2024 to 2028.
- Mobile AR market revenue may reach over $39 billion by 2027.
- Almost half of consumers are inclined to spend more through AR-based shopping.
- Up to 75% of individuals aged 16–44 are familiar with AR.

- The value of the AR agriculture market exceeds half a million dollars.
- The value of the AR manufacturing sector is expected to double from 2025 to 2030.
- The value of AR in offices could potentially increase by a factor of 10 within a span of five years.
- Projected for 2030, the value of AR within hospitals could potentially reach $90 billion.
- The value of the AR vehicle sector is expected to double between 2025 and 2030.
- From 2023 to 2026, revenue from AR glasses is projected to increase more than fivefold.
- The AR headset market is expected to deliver 50 million units by 2026.

The global mobile AR market revenue forecast is depicted in Figure 15.9. The most predominant use of AR by Americans is listed in Figure 15.10.

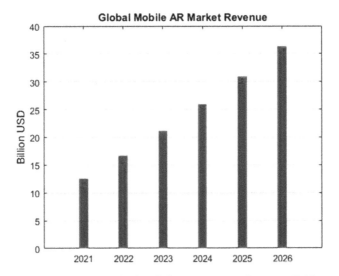

Figure 15.9   Global mobile AR revenue forecast [21]

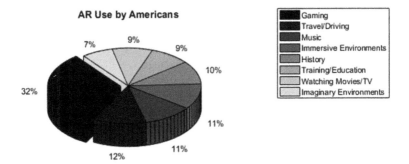

Figure 15.10   AR use by Americans as of October 2022 [22]

15.5   Conclusion

To conclude, it is evident that AR is a powerful tool for achieving ESG objectives. AR is transforming activities in various fields by facilitating business operations, promoting community engagement, and improving decision-making processes. Integrating AR into ESG initiatives holds great promise for building a more equitable and environmentally responsible future.

References

[1]   Taqvi, Z. (2013). Reality and perception: Utilization of many facets of augmented reality. In *2013 23rd International Conference on Artificial Reality and Telexistence (ICAT)*, Tokyo, Japan, 2013, pp. 11–12, doi:10.1109/ICAT. 2013.6728899.

[2]   Petruse, R. E., Grecu, V., and Chiliban, B. M. (2016). Augmented reality applications in the transition towards the sustainable organization. In: Gervasi, O., *et al.* (eds.). *Computational Science and Its Applications – ICCSA 2016. ICCSA 2016. Lecture Notes in Computer Science*, vol. 9788. Springer, Cham. https://doi.org/10.1007/978-3-319-42111-7_33.

[3]   Garcia, C. (2023). *Green Manufacturers Are Discovering the Merits of Augmented Reality, Blue and Green Tomorrow.* Available at: https://blue-andgreentomorrow.com/features/green-manufacturers-are-discovering-mer-its-of-augmented-reality/ (Accessed: March 2024).

[4]   How augmented reality is leading the sustainability charge. (2022). *Sherwen.* Available at: https://www.sherwen.com/insights/how-augmented-reality-is-leading-the-sustainability-charge (Accessed: March 2024).

[5]   Waicberg, S. (2022). *Augmented Reality Advances Organisation ESG Initiatives, Sustainability Magazine.* Available at: https://sustainabilitymag. com/esg/augmented-reality-advances-organisation-esg-initiatives   (Accessed: March 2024).

[6]   Fogarty, N. (2023). *Sustainability Through AR: Pioneering Eco-Friendly Business Innovations, nexposai.* Available at: https://nexposai.com/sustain-ability-through-ar-pioneering-eco-friendly-business-innovations/ (Accessed: March 2024).

[7]   Jimmy, H. (2024). *Augmented Reality in Training and Development, LinkedIn.* Available at: https://www.linkedin.com/pulse/augmented-reality-training-devel-opment-jimmy-ho-l2ryc?trk=public_post_feed-article-content (Accessed: March 2024).

[8]   Tran, N. N. B (2024). Reality and perception: Utilization of many facets of augmented reality, *Journal of Business Leadership and Management* 2(1). doi:10.59762/jblm845920462120240205151921.

[9]   Yang, J., and Lin, Z. (2024). *From Screen to Reality: How AR Drives Customer Engagement and Purchase Intention.* Available at SSRN: http://dx.doi. org/10.2139/ssrn.4726408.

[10] Jessen, A., Hilkena, T., Chylinski, M., *et al.* (2020). The playground effect: How augmented reality drives creative customer engagement, *Journal of Business Research* 116, pp. 85–98. doi:10.1016/j.jbusres.2020.05.002.

[11] Orji, J. U., Chan, G., and Orji, R. (2023). Augmented reality and machine learning in health: A systematic review. In *2023 the 7th International Conference on Virtual and Augmented Reality Simulations (ICVARS 2023), March 3–5, 2023, Sydney, Australia.* ACM, New York, NY, 9 pages. https://doi.org/10.1145/3603421.3603430.

[12] Mesko, B. (November, 2021). Augmented reality in healthcare: 9 examples. *The Medical Futurist.* Available at: https://medicalfuturist.com/augmented-reality-in-healthcare-will-be-revolutionary/ (Accessed: March 2024).

[13] Augmented Reality in Wellness Training: Opportunities and Challenges. (2023). *RSS.* Available at: https://www.corporatewellnessmagazine.com/article/augmented-reality-in-wellness-training-opportunities-and-challenges (Accessed: March 2024).

[14] Alturki, R., and Gay, V. (2019). Augmented and virtual reality in mobile fitness applications: A survey. In Khan, F., Jan, M., and Alam, M. (eds) *Applications of Intelligent Technologies in Healthcare. EAI/Springer Innovations in Communication and Computing.* Springer, Cham. https://doi.org/10.1007/978-3-319-96139-2_7.

[15] Campbell, M. (2022). What is the G in ESG: Governance explained, vation ventures research. Available at: https://www.vationventures.com/research-article/the-g-in-esg (Accessed: March 2024).

[16] Dehimi, Y. (2021). How can AR contribute to business decision making, ClickZ. Available at: https://www.clickz.com/how-can-ar-contribute-to-business-decision-making/264689/ (Accessed: March 2024).

[17] Shenoy, V. (2023). Augmented reality in business, fingent. Available at: https://www.fingent.com/blog/augmented-reality-in-business/ (Accessed: March 2024).

[18] United Nations E-Government Survey. (2018). Available at: https://publicadministration.un.org/egovkb/en-us/Reports/UN-E-Government-Survey-2018 (Accessed: March 2024).

[19] Government augmented reality revolutionizing government services: The role of augmented reality in entrepreneurship. (2024). *FasterCapital.* Available at: https://fastercapital.com/content/Government-Augmented-Reality-Revolutionizing-Government-Services–The-Role-of-Augmented-Reality-in-Entrepreneurship.html (Accessed: March 2024).

[20] Howarth, J. (2024). 24+ augmented reality stats (2024-2028). *Exploding Topics.* Available at: https://explodingtopics.com/blog/augmented-reality-stats (Accessed: 7 April 2024).

[21] Alsop, T. (2022). *Mobile Augmented Reality (AR) Market Size 2026, Statista.* Available at: https://www.statista.com/statistics/282453/mobile-augmented-reality-market-size/ (Accessed: 8 April 2024).

[22] Alsop, T. (2024). *US: Frequency of AR Use 2022, Statista.* Available at: https://www.statista.com/statistics/1310579/augmented-reality-use-frequency-us/ (Accessed: 8 April 2024).

Chapter 16

The role of ESG in the REE industry

Ain Nadirah Binti Romainor[1] and Wan Nada Marhamah Binti Wan Abdul Razak[2]

The rare earth element (REE) industry is crucial in developing modern technologies. It plays a significant interest in consumer electronics, electric vehicles as well as military equipment. However, the extraction and processing of the metals often pose environmental effects and social challenges. Thus, this chapter on the REE business looks at how ESG rules affect things. It discusses their effect on protecting nature, dealing with people's rights, and organising systems in these industries.

16.1 The importance of ESG in the REE industry

REEs are the metallic elements from group 3 in the periodic tables including the lanthanides series. These include scandium and yttrium in the periodic table, as well as the 15 lanthanides elements namely lanthanum, cerium, praseodymium, neodymium, promethium, samarium, europium, gadolinium, terbium, dysprosium, holmium, erbium, thulium, ytterbium, and lutetium [1]. The market analysis reported by Mordor Intelligence TM stated that the global REE market size is forecasted to expand from 167.99 million tons to 206.25 million tons in the year 2023–28 due to continuous demand for REE in advanced technologies [2]. They have become pervasive in improving energy efficiency and in digital technologies such as hybrid and electric vehicles, smartphones as well as military equipment [3].

The REE industry's need for modern technology is changing significantly. This happens because essential ideas around the environment and social rules change how businesses run things. Environmental, social, and governance (ESG) has become a guide in helping companies act responsibly, sustainably, and ethically. It serves as one of the ways to make the companies more socially conscious and act responsibly when utilising the raw materials in their industries. Sustainability is a crucial matrix that enhances a company's image, ultimately

[1]Department of Chemistry, Faculty of Resource Science and Technology (FRST), University Malaysia Sarawak (UNIMAS), Malaysia
[2]Department of Psychology, Faculty of Social Science, Arts and Humanities, Lincoln University College (LUC), Malaysia

drawing more investors and gaining increased public attention. In short, a good ESG score will help a company's business growth.

However, with the growing popularity of environmental consciousness, people are shifting their interest to 'green technologies' from mining raw materials like petroleum. Nevertheless, green technologies usually utilise a lot of REE in their productions [4].

16.2   Environmental considerations

Thinking about the environment is crucial in making REE business go. ESG principles push changes towards being green and doing things rightly, which helps sustainability, too. The ideas in ESG rules change how the REEs industry gets resources, uses energy, and handles waste. Businesses in this area must use good ways for the environment, try new tech, and care about fixing nature damaged by their actions [5]. Adding ESG ideas into the environmental world helps with old nature problems about mining REEs and makes it a responsible and caring part of the worldwide market. By doing things that are good for the environment, companies protect nature. This helps their future and makes them strong against problems around the world, which change over time.

The success of the REE business depends on how far they consider environmental factors in the process. As global society slowly transitions away from fossil fuel-based energy, ESG plays a paramount role in ensuring sustainability. If ESG factors are not taken seriously in the REE business, mistakes from past operations of fuel-based industries are likely to be repeated with the new green technological advancements. Failing to learn from past mistakes with a lack of commitment to ESG rules will not only compromise ESG factors but also hinder the REE industry's ability to evolve in accordance with social demands and environmental expectations [6]. Thus, it is important for the REE business to abide by ESG rules to ensure the growth and the expansion of green technologies, that are aligned with global sustainability demands.

16.3   Sustainable mining practices

In the past, getting REEs always caused problems for nature. These problems involve damaging animal homes and making water dirty. Good practices for mining rare earths are encouraged by ESG rules, which make change happen. This change is about using things wisely and asking companies to lessen the damage they cause while getting resources. Using these rules well is essential for better technology [7]. This allows companies to reduce their effect on the environment a lot. A sustainable mining practice will not only reduce the negative impacts of mining on the environment, but it will also help in mitigating climate change-related risks [8]. Abiding by ESG rules will require REE businesses to act more responsibly in extracting REEs in a more environmentally conscious way.

Also, ESG rules make businesses use plans for cleaning up more than just getting resources. They try to restore environments and reduce the effects on creatures like animals or plants. Strategic planning in cleaning up after extracting resources will reduce the chance that endangered species of flora and fauna will be negatively impacted by the mining process. ESG rules will contribute to the introduction of environmentally friendly mining practices by REE businesses.

16.4   Energy transition

The energy-intensive nature of REE extraction and processing poses substantial environmental challenges. ESG principles guide the industry towards an essential energy transition, advocating for the adoption of renewable energy sources and energy-efficient technologies. This shift not only addresses the industry's carbon footprint but also aligns with broader global efforts to combat climate change. By reducing reliance on non-renewable energy sources, the REE sector can make significant strides in minimising its environmental impact and contributing to a more sustainable energy future. This paradigm shift addresses the need to embrace a more sustainable energy transition that will highly contribute to global efforts in combatting climate change. REE operations should take a more proactive approach by following ESG principles in reducing greenhouse gas emissions and reducing the dependency on non-renewable resources.

16.5   Waste management

ESG considerations extend comprehensively to waste management practices within the REE industry. This involves addressing the disposal and treatment of by-products and radioactive materials generated during REE extraction. ESG principles dictate that companies must implement robust waste management strategies that prioritise recycling and reusing materials wherever possible. By minimising the environmental impact of waste generated throughout the production process, the industry can mitigate its contribution to pollution and resource depletion.

Fuel-based industries are notorious for irresponsible waste management when extracting raw materials, but with ESG guidelines, the REE industry will need to prioritise the reuse and recycling of materials to facilitate the reduction of pollution. This not only will help environmental preservation, but also positively impact global sustainability.

16.6   Social responsibilities

Caring for our planet is now very important in the business of REEs. Working on environment, people, and leadership ways ensures good actions are done carefully for a better future. This article discusses the complicated aspects of social responsibility [7]. It examines how ESG rules shape community involvement, worker actions, and supply chain ethics in REE businesses. Highlighting the importance of every level of the supply chain in the REE industry adhering to ESG rules is crucial for moving towards a more sustainable and socially conscious direction. As social responsibility towards a greener Earth is not the responsibility of just one person, it must be embraced at every layer of the REE industry for the global society to witness a meaningful impact on the environment.

16.7   Community engagement

REEs business is connected to local people by nature, which causes concerns about helping society. ESG ideas are like a helper, saying that working with the people around is very important. Businesses are not just co-workers but critical parts of their work area [9]. The rules say that companies must include local people in decision-making. This means getting people in the community to talk freely, asking them for their thoughts, and using these ideas when making plans.

Furthermore, ESG rules require businesses to honour the cultural legacy. This means understanding the value of local customs and ways of doing things. The REE industry, which often works in areas full of different cultures, must be careful about its actions [10]. It must consider the long-term and cultural historical backgrounds of the communities involved. This way, businesses help protect cultural history. They also build a teamwork bond that goes past money deals. The ideas behind ESG need companies to help the general health of communities affected by mining actions. This is more than just creating jobs and helping the economy. Businesses are asked to put money into helping their neighbourhoods grow. This includes schooling, health care, and extensive building projects like roads or bridges. By following ESG rules, the REE industry changes its job from being just an economic thing to a good community member who helps make society better.

16.8   Governance structures

Leaders are essential in directing the REE business's path, especially when ESG or environment, social caring, and good leadership rules come into play. In this area, governance that considers ESG deals with being open and sharing information [5]. It also makes sure companies follow the rules set by lawmakers. This robust system guides businesses to act responsibly and sustainably in their practices through ethical decision-making processes.

Governmental bodies or any entities associated with REE business must strongly stress the magnitude of incorporating ESG rules in the REE industry. For instance, Agnese and colleagues [11] found that the origin of investors can predict the decisions of banks regarding activities related to ESG. This shows the significant impact that key stakeholders, such as major investors, can have on shaping the extent of ESG implementations and consequently contributing significantly to a country's economic development. Lawmakers in various countries must take the

initiative to create a business model that encourages both local and foreign investors to adhere to ESG rules. This helps to foster a sense of responsibility among
stakeholders to follow ethical conduct in conducting REE businesses.

16.9   Transparency and reporting

Good business practices in the REE sector stress clear and accurate reporting,
mainly focusing on transparency. Firms in this type of business should follow a
policy that openly shares detailed information about their effect on the environment, social projects, and how they are managed [12]. Transparent reporting is
essential for getting trust from people who matter, such as investors and watchdogs.
This includes local communities, too. In the REE field, transparent reporting means
sharing information about how we care for nature. It should explain actions to
protect the environment and offer details on existing social projects [7]. This
openness meets the increasing need for responsibility and helps shareholders to
make intelligent choices about their engagement with the business. People who put
money in the stock market are looking more and more for detailed reports on ESG.
These reports check how big companies treat the environment, social issues, and
business ethics. Companies that happily give this information edge out their competition by immediately showing they care about doing good practices [10]. Open
reporting goes beyond numbers. It also includes stories that make clear the journey
to sustainability in business, problems faced, and ways used for constant
improvement. Talking straight and openly with essential people can help the REE
industry build better connections. This leads to more involved decision-making that
everyone knows about, making it a friendlier place for all kinds of folks.

16.10   Innovation and technology

Innovation and technology are significant in the REE industry. This is especially
true as ESG ideas push them to have practices that are good for nature and do right
by people [10]. There are two essential parts in this area: new ways of getting
things out and technologies that can use them again. Both help lessen the harm it
does to nature or where living creatures live.

Reducing ecological footprints during REE extraction is extremely important
to contribute to the improvement of environmental health. Technological
advancements and innovations will help minimise the environmental harm that
REE extraction might pose. ESG rules will hold the REE industry accountable,
compelling them to follow more sustainable mining practices responsibly, thus
positively impacting the global environment.

16.11   Alternative extraction methods

Led by the idea of ESG, companies in rare earth metals are actively welcoming
creativity to look for new ways to extract different ores. For a long time, familiar

ways of getting things out have caused significant environmental problems. They can ruin places where animals live and make water dirty [13]. The new idea based on ESG wants to fix these problems by looking at ways of taking stuff out that are better for nature and last longer in the future. People are working hard to find new ways to reduce the impact of removing REEs from nature. These other ways try to cut down on breaking habitats, using too much water, and releasing bad things. By doing this, the industry follows what is expected from ESG and helps in a big world goal to lessen how bad resources hurt nature. Creating and using new ways to get resources shows how the industry always tries to improve [10]. The principles of ESG stress how important it is to keep up with the latest technology. This can help improve sustainability. By using new ways of getting REEs and being environmentally friendly, the REE industry shows it cares about taking steps to reduce harm to our planet.

16.12 Conclusion

In conclusion, including ESG rules in the REE business is very important to lead it towards a green and caring future. Caring about the environment leads people to use mining practices that are good for the planet, change energy sources, and responsibly manage waste. Community involvement, fair treatment at work, and good supply chain management are needed according to social duties. Governance systems are formed by openness, following rules, and making good moral choices. Also, new ideas and technology are supported to reduce harm to our planet. How well a company does in ESG can affect if they can join the market and how sure investors feel about it. As the REE business changes due to ESG goals, it works on environmental and social problems. It also shows itself as a good and fair part of the world market. The connection between ESG rules and the REE industry offers a forward-looking way for sustainability. This gives us a more robust future with more care about nature.

References

[1] Balaram, V. 'Rare earth elements deposits: Sources, and exploration strategies'. *Journal of the Geological Society of India*. 2023; 98: 1210–1216.

[2] Mordor Intelligence. *Rare earth metals market size and share analysis - growth trends and forecasts (2023–2028)*, 2 November 2023. https://www.mordorintelligence.com/industry-reports/rare-earth-elements-market.

[3] Costis, S., Mueller, K. K., Coudert, L., Neculita, C. M., Reynier, N., and Blais, J. F. 'Recovery potential of rare earth elements from mining and industrial residues: A review and cases studies'. *Journal of Geochemical Exploration*. 2021; 221: 1–55.

[4] Golroudbary, S. R., Makarava, I., Kraslawski, A., and Repo, E. 'Global environmental cost of using rare earth elements in green energy technologies'. *Science of The Total Environment*. 2022; 832: 1–12.

[5] Paat, A., Veetil, S. K. P., Karu, V., and Hitch, M. 'Evaluating the potential of Estonia as a European REE recycling capital *via* an environmental social governance risks assessment model'. *The Extractive Industries and Society*. 2021; 8(4): 1–8.

[6] Jowitt, S. M., Mudd, G. M., and Thompson, J. F. 'Future availability of non-renewable metal resources and the influence of environmental, social, and governance conflicts on metal production'. *Communications Earth & Environment*. 2020; 1(1): 1–8.

[7] Cherepovitsyn, A., Solovyova, V., and Dmitrieva, D. 'New challenges for the sustainable development of the rare-earth metals sector in Russia: Transforming industrial policies'. *Resources Policy*. 2023; 81: 103347–103355.

[8] Chen, H. M., Kuo, T. C., and Chen, J. L. 'Impacts on the ESG and financial performances of companies in the manufacturing industry based on the climate change related risks'. *Journal of Cleaner Production*. 2022; 380: 134951.

[9] Leruth, L., Mazarei, A., Regibeau, P., and Renneboog, L. 'Green energy depends on critical minerals. Who controls the supply chains?' *PIIE Working Paper*. 2022; 846: 1–35.

[10] Filho, W. L., Kotter, R., Özuyar, P. G., Abubakar, I. R., Eustachio, J. H. P. P., and Matandirotya, N. R. 'Understanding rare earth elements as critical raw materials'. *Sustainability*. 2023; 15(3): 1919.

[11] Agnese, P., Arduino, F. R., and Secondi, L. 'Does ownership structure affect environmental, social and governance activity? Evidence from the banking system of an emerging economy'. *Global Business Review*. 2022; 23(6); 1403–1423.

[12] Choudhury, B. K. 'Sustainability in metal industries', in S. K. Dutta, J., Saxena, and B. K. Choudhury (eds), *Energy Efficiency and Conservation in Metal Industries*, CRC Press, Boca Raton, FL, 2022, pp. 187–214.

[13] Lèbre, É., Stringer, M., Svobodova, K., *et al*. 'The social and environmental complexities of extracting energy transition metals'. *Nature Communications*. 2020; 11(1): 4823.

Chapter 17

ESG in innovative and experimental approach for small-scale renewable energy generations

Babu Sasi Kumar Subramaniam[1], Wai Yie Leong[2] and Muthu Manokar Athikesavan[3]

17.1 Introduction

The energy produced from fossil fuels, which emit significant amounts of carbon dioxide, is a serious environmental contaminant. Solar energy is expanding at the fastest rate because of the overwhelming need for renewable energy. The sun is the primary source of energy for Earth. All forms of energy, including biomass, water, and air, are derived from solar radiation. The total amount of solar energy produced is 3.8×1020 MW. The solar energy from the sun that is received in 30 minutes is equivalent to the world's annual energy requirements, according to Kreith and Kreider [1]. Unlike other energy sources, such as fossil fuels, which are utilized for a variety of purposes and release greenhouse gases into the atmosphere. Panwara *et al.* [2] stated that solar energy is a healthy, safe energy source that doesn't harm people or the environment. While high-temperature solar thermal applications are mostly utilized to provide energy, low-temperature solar thermal applications are used to heat houses and other buildings.

17.1.1 History of solar energy

Anderson [3] state that Archimedes and the Greek scientists repelled the Roman fleet 1800 years ago by using hundreds of polished shields that reflected and absorbed solar energy. As per Meinel and Meinel [4], Augustin Mouchot created a steam generator using minimal solar energy in 1878, but the French government disregarded it because of its higher cost. Krider and Kreith [5] stated that the solar collectors that Monsanto designed were made of reflective cone-shaped silver-metal plates with a diameter of 5.4 m, an area of 18.6 m^2, and a moving part weight of 1400 kg. And it was also stated that a large umbrella composed of 1788 glasses

[1]Department of Mechanical Engineering, Kings Engineering College, India
[2]Faculty of Engineering and Quantity Surveying, INTI International University, Malaysia
[3]Department of Mechanical Engineering, B.S. Abdur Rahman Crescent Institute of Science and Technology, India

reflected sunlight onto boiling water, producing steam. In 1913, Schumann created a 62-meters parabola that covered an area of 1200 m^2. It had a 37–45 kW steam capacity and ran continuously for 5 hours. Nevertheless, the First World War, the price of gasoline, and the fact that the receiver systems used two-axis mirror tracking temperatures between 100 and 1500 °C led to forced closures in 1915. Kalogirou [6] states that in the 1960s, a large number of Enterprises developed FPTs, which were used in water heating systems. These collectors had an absorption area of 4 m^2 and a capacity of 180 L/m^2. It was connected to a heater or thermal transmitter to continuously supply hot water. Malik *et al.* [7] report that the Monchot's complex supporting structure was installed and utilized a large aluminum surface coated in silver glass to concentrate solar energy needed to generate distillation water. Norton [8] state that solar drying is an additional solar energy application that is mostly utilized by the agriculture sector to reduce moisture content and lengthen storage periods. As per the findings of Lysen [9], in 2002, semiconductors other than silicon, cadmium compounds, and cupric sulfide were used to make PV cells. These technologies, which are connected to a photovoltaic grid, are utilized to store more than 15% of the local electricity and are used to store batteries independently of the power grid.

From the previously mentioned perspective, we often divide solar energy into concentrated and non-concentrated solar systems to construct a sustainable energy source for various energy applications. This chapter includes a basic introduction to thermal energy storage and a variety of solar technologies. Concentrated solar is the most important type of collector, which is briefly explained below.

(a) A solar thermoelectricity device effectively produces power by using thermo-electric materials to create a disparity in temperature between the two sides. A parabolic disc concentration is used to capture the light from the hot side. The formulas used to calculate the magnitude of the thermoelectric voltage caused by the temperature differential across the materials are

$$Z = \sigma S^2 / K$$

where σ is the electrical conductivity, S is the See beck coefficient, and K is the thermal conductivity (TC) of materials. According to Fan *et al.* [10], a single thermoelectric generator was tested and was able to generate 4.9 W of power at a temperature gradient of 110 K with a conversion efficacy of 2.9%. The STED model is too expensive at the market level. However, overall capability rose when combined with a hybrid system.

(b) A dye-sensitized solar cell (DSSC) is a type of photovoltaic (PV) cell that relies on a photoactive dye between the anode and electrolyte. In 1991, Michael Grätzel and Brian O'Regan developed the Grätzel cell in response to the need for a less expensive, more efficient alternative to traditional solar cells.

(c) A concentrated photovoltaic cell uses optics to concentrate the most solar energy onto the portion of the Photo voltaic matter that engenders electrons from photons. Movement current is produced by the hinge port, which accelerates electrons as they pass from the valence to the conduction bands. It is more suitable for solar farms than being utilized on roofs. To boost effective-ness, a concentrated solar panel was added to the hybrid system.

(d) Photovoltaic solar panels have photovoltaic elements that absorb solar energy and transform it into electrical energy using the photoelectric effect. The most popular photovoltaic materials are thin films consisting of semiconductors and mono crystalline silicon. The bulk of photovoltaic modules are constructed using wafer-based crystalline silicon, which converts light energy into elec-trical energy at a rate of up to 14%–20%. In 2025, a 23% increase in research development capability is predicted. Materials like copper-gallium-diesel and cadmium telluride were used to generate thin photovoltaic materials, and this manufacturing productiveness was quite high. The outside surfaces are sup-ported by glass, stainless steel, or plastic. The cost-effective solar energy usage increased despite the photovoltaic industry's accurate predictions for more productive nighttime and overcast conditions.

Concentrated collectors as compared to the other collectors were able to reach higher temperatures and better thermal efficiency and some of their noteworthy and distinctive varieties are detailed here.

(a) Parabolic trough collector coated by the reflective sheet of materials that absorbed the solar energy is reflected into the focus point of the receiver and penetrates the working fluid, where it is converted into thermal energy reaches a temperature between 50 and 400 °C. The receiver is completely covered in a double-ended glass tube to prevent heat loss, and a tracking system rotates the collector so that it is always at the focus point on the receiver and absorbs more solar radiation, as stated by Kalogirous [11].

(b) Linear Fresnel collector concentrates sunlight onto a common receiver using a linear tower glass. It is challenging to integrate a storage system into these elastic or flat-curved collectors, which can be seen as a parabolic through reflector.

(c) **Parabolic dish reflectors** are a point-focus collector; the parabolic dish reflectors concentrate solar light into a receiver. To reproduce the beam into the recipient, the dish is built with a glass reflector. a fluid that circulates through a receiver and absorbs solar radiation, turning it into thermal energy. The thermal energy can be used for household heating and cooling systems or transformed into electricity.

(d) **Solar tower** is a high radioactive energy input effort made possible by the mul-tiple mirrors reflecting sunlight into a receiver is referred to as a "heliostat field collector." The flowing working fluid in the receiver generates thermal energy.

Tracking system is required for the collectors to follow the sun and transfer solar light into the tiny receiver system, according to Siddiqui [12]. Single axis and double axis tracking are the two different varieties. Generally speaking, the elec-tronic system improved reliability and monitoring precision by employing sensors to trigger the motor control system automatically in response to observed solar radiation levels.

Dynamic response controls, which measure solar light and use that information to drive the tracking system, are likewise operated by computer-controlled machinery. Nikesh *et al.* [13] reported that tracking system increased energy storage by 20%–30% compared to non-tracking. Tiberiu and Liviu [14] state that the most efficient tracking system features a machine that is easy to operate and consumes less energy. The findings support the anticipated solution for large-scale photovoltaic structures.

Thermal storage systems are essential because collected energy is kept in storage for a range of heating applications. The following are some of the assumptions to consider before selecting a thermal energy storage: (i) The temperature must be kept at or above 100 °C; (ii) the specific heat capacity remains constant to reduce the heat losses.

Thermal energy storage systems are the most crucial components of an effective industrial heat and cooling recovery system and are categorized as follows:

- **Sensible heat storage**: the temperature will rise or fall during the heating or cooling process, but there are no phase changes. The stored energy depends on the characteristics of thermal conductivity and specific heat capacity.
- **Latent heat thermal storage**: the heat was either absorbed or released by the storage system, causing phase change materials, but no change in temperature. The SE is used mostly for heating appliances Panwara *et al.* [15]. A salt hydrate was employed by Chiu *et al.* [16] as a latent heat storage medium to store solar energy. It had a high thermal conductivity, a low cost, and higher storage. It also had strong storage abilities. The work is based on the building industry and includes phase change materials as a building component. High heat sensitivity and minimal heat and temperature changes are characteristics of the phase change materials construction element.
- **Thermo-chemical storage** is dependent on specific chemical and physical reactions, which result in the thermal energy. Throughout the process, the storage system performs an isothermal function. This system has more potential for future usage.

From the aforementioned vantage point, different solar technology types were explored. The lowest-cost method of generating additional energy is currently being used for small-scale heating applications, and the performance abstract of these types of collectors is described.

17.1.2   Flat plate collector

Flat plate collector is used in an appealing and economical way to dry agricultural food goods, which has the potential to replace fuel wood in major global developments. It is composed of a sheet of aluminum that is 5 mm thick, 2×1 m^2 in size, and painted black. On the outside of the enclosure, an insulated collector with a thickness of roughly 10 cm and glass wool with a thickness of 25 mm are utilized. These experiments were conducted to collect information on different atmospheric conditions and to evaluate the achievement of the collector under free and enforced convection approaches.

17.1.3   Evacuated tube collector

The evacuated tube collector experimented with the thermal efficiency at various flow rates with and without an absorber plate. The evacuated tube apparatus's 3.6 m^2 surface area and 30 different tube numbers. Due to a pressure difference, heat from the evacuated tube is dispersed into the manifold's channel. The evacuated tube collected heat, which increased solar energy absorption and decreased heat losses. The blower functions in a forced convection mode, bringing air from the atmosphere into the interior of the manifold. As an absorber plate, an aluminum block-covered sheet is positioned at the base of an evacuated tube. In the present study, the effectiveness of evacuated tubes with and without absorber plates was compared.

17.1.4   Phase change materials

The phase change materials are essential energy sources for effective use when there is a mismatch between energy supply and usage and are used to maintain a steady temperature in systems that store latent heat. The waste recovery, cooking, and steam supply industries were most frequently utilized. The intent of the current study is to use it as a stable flow with high energy for cooking applications. Two different phase-change materials, such as palmitic and stearic acids, were chosen for investigation in an attempt to improve the thermal conductivity properties of melting and freezing. By using differential scanning calorimetric analysis, latent heats were assessed to identify the variables that affect the melting and freezing properties as well as variations in temperature over time and during cycle activities.

17.1.5   Parabolic trough collector with thermal energy storage (phase change materials)

The phase change materials are an experimental work carried out by the parabolic trough with manually operated tracking system that collects solar energy and converts it into thermal energy using the working fluids of water .It is then stored into the phase change materials filled in the heat storage and the heat is then utilized for cooking purposes. The research aims to determine the thermal efficiency of the parabolic trough at 82° reached its peak thermal efficiency at 1 p.m. at a flow rate of 0.035 kg/s, 0.045 kg/s, and 0.065 kg/s were 67.8%, 62.4%, and 59.4%, respectively. At the same time, the storage energy was accomplished, and the utmost energy was 10,327 W, 9987 W, and 9370 W respectively. In the flow rate of water at 0.035 kg/s of cooking vessel at 6.00–8.00 p.m., energy is 10.9, 8.9, and 7.5 times greater than the typical dry food energy. The aforesaid overcome the flow rate of 0.035 kg/s was higher than other kg/s, as was the stearic acid used as a phase change material.

17.1.6   Parabolic trough Collector with thermal energy storage (phase change materials-nanofluids)

Based on a study of a cooking vessel with heat storage that employs waste engine oil and water as the working fluids at flow rates of 0.035, 0.045, and 0.065 kg/s, the current research was conducted. Heat was extracted from the parabolic trough and stored in a thermal storage device. The system was made up of rectangular boxes that held stearic acid phase change materials with 0.3 vol.% Al_2O_3 nanofluids. The waste engine oil medium's heat storage capacity was 0.33 times greater than that of the water at a flow rate of 0.035 kg/s, and it was also greater than the flow rates of 0.045 and 0.065 l/s. Heat was gathered in the thermal storage and then compelled into use and delivered to the cooking vessel via the collector, which reflected solar radiation to the receiver. The energy output of the cooking vessel waste engine oil and water, at a flow rate of 0.035 kg/s, was found to be approximately 12.4, 14, and 15.1, and 9.8, 10.5, and 11.5 times lower than that of the groundnut, ginger, and turmeric crops at different flow rates of 0.035 L/s, 0.045 L/s, and 0.065 L/s, as well as inferior to the storage tank. Last, this discovery has uses in food production, specifically in veggie drying.

17.2   Solar energy: the expanding frontier of clean, renewable power

Energy production from fossil fuels is a major environmental concern due to the significant amounts of carbon dioxide they emit. In response to the urgent need for renewable energy, solar energy is expanding rapidly. Solar energy, which originates from the sun, is fundamental to all forms of energy, including biomass, water, and wind. Remarkably, the solar energy received by Earth in just 30 minutes is equivalent to the world's annual energy requirements, as noted by Kreith *et al.* stated that unlike fossil fuels, which release greenhouse gases and contribute to environmental degradation, solar energy is a clean and sustainable source. It does not harm the environment or human health. Solar thermal applications are categorized into high-temperature systems, primarily used for electricity generation, and low-temperature systems, which are commonly used for heating homes and other buildings.

17.2.1   History of solar energy

Anderson and colleagues noted that Archimedes used polished shields to repel the Roman fleet with solar energy 1800 years ago. Meinel and team mentioned August Monset's steam generator powered by minimal solar energy in 1878, but it was rejected due to cost. Krider *et al.* described Monsanto's reflective cone-shaped solar collectors. Kreith *et al.* discussed Monochot's mirror-based helical reflections. Eneas *et al.* detailed an umbrella creating steam with sunlight. Schumann's 1913 parabolic system was impressive but closed in 1915 due to external factors. In the 1960s, flat plate collectors were developed for water heating with an absorption area of 4 m^2 and 180 L/m^2 capacity. Monchot's structure used silver-coated aluminum to generate distillation water. Solar drying is used in agriculture to reduce moisture. By 2002, PV cells were made from semiconductors other than silicon, cadmium compounds, and cupric sulfide. These technologies can store over 15% of local electricity and operate independently from the power grid.

Solar energy is divided into concentrated and non-concentrated solar systems to create sustainable energy sources for various energy needs, including Thermal Energy Storage.

From the aforementioned viewpoint, SE is often separated into concentrated and non-concentrated solar systems in order to build the sustainable energy source for diverse energy uses. This chapter includes a basic introduction of thermal energy storage and a variety of solar energy technologies. Concentrated solar is the most important type of the collector was briefly explained below.

(a) Solar thermal electricity uses thermo-electric materials to create a temperature difference, generating power. Light is captured with a parabolic disc concentrating sunlight. Formulas like $Z = \sigma S^2/K$ calculate thermoelectric voltage due to temperature difference. A study by Fan *et al.* [10] found a single generator produced 4.9 W at 110 K with 2.9% efficiency. Combining with a hybrid system increases overall effectiveness.

(b) Dye-sensitized solar cell harnesses PV-based photo generic sensitivity between anode and electrolyte, developed by Gratzel in 1991.

(c) Concentrated photovoltaic cells focus solar energy to generate electrons, producing current through a hinge port. It is ideal for solar farms and paired with a solar panel for enhanced efficiency.

(d) Photovoltaic solar panels use photovoltaic elements to absorb sunlight and convert it into electricity. Common materials include thin films of semiconductors and monocrystalline silicon. Most modules use wafer-based crystalline silicon, with an efficiency of 14%–20%. Research predicts a 23% increase in development capability by 2025. Materials such as copper–gallium–indium–selenide and cadmium telluride are used to create thin photovoltaic films. Despite accurate predictions for more efficient night and overcast energy production, the cost-effective use of solar energy has increased.

Concentrated collector reached higher temperatures, enhanced thermal efficiency, notable unique varieties are listed below.

a) **Parabolic trough collectors** use reflective materials to focus solar energy onto a receiver, converting it into thermal energy between 50 and 400 °C. The parabolic trough receivers are covered in glass tubes to prevent heat loss, with a tracking system to optimize solar radiation absorption.

b) **Linear Fresnel collectors** concentrate sunlight onto a common receiver using tower glass, but integrating storage systems can be challenging.

c) **Parabolic dish reflectors** concentrate sunlight onto a receiver using glass reflectors, converting solar energy into thermal energy for various applications.

d) **Solar towers** use multiple mirrors to focus sunlight onto a receiver, generating thermal energy through a working fluid.

A tracking system is essential for collectors to optimize solar light absorption. Electronic systems with sensors improve reliability and monitoring precision by automatically adjusting motor control in response to solar radiation levels. Dynamic response controls and computer-controlled machinery enhance efficiency, increasing energy storage by 20%–30%. It emphasizes the importance of an easy-to-operate machine that consumes less energy for efficient tracking systems in large-scale photovoltaic structures.

Thermal storage systems store collected energy for various heating applications. Important considerations for selecting a storage system include maintaining a temperature above 100 °C and constant specific heat capacity to minimize heat losses. Thermal energy systems are vital for industrial heat and cooling recovery, categorized based on their design.

- **Sensible heat storage** involves changes in temperature during heating or cooling without any phase changes, depending on thermal conductivity and specific heat capacity.
- **Latent heat thermal storage** uses phase change materials to absorb or release heat without temperature changes, commonly used in heating applications.
- **Thermochemical storage** relies on chemical reactions to store and release energy isothermally. Nanofluids improve heat transfer in these systems.

17.3   Literature review

Studying the following literature was highly beneficial for carrying out the work and for improving the performance of each experimental work.

17.3.1   Flat plate collector

The flat plate collector is one sort of collector converter that transforms solar energy into heat energy as needed. Numerous studies have looked at the effectiveness of this technique, which involves the absorber plate in a flat collector transferring heat to fluid flowing through pipes. One of the energy capturing devices is an flat collector, according to Rai [17]. While the solar energy is being transferred to the absorber plate inside the working fluids of the pipe or tube, the effect of effectiveness is being investigated. These collectors were subjected to experiments by Kostić and Pavlović [18], both with and without a reflector. The outcomes demonstrated that reflectors enhanced collector thermal energy when placed in an ideal position. Khatik *et al.* [19] conducted research on a flat plate collector with or without an absorber plate that absorbs solar energy, reflects it, stores it, and transmits it into the working fluids. It also minimizes radiation loss, which results in a rise in the capacity of the absorber plate, which is 5.12% greater than without the absorber plate. Rhushi *et al.* [20] completed the research employing a flat plate with a horizontal angle of 28°. The ability of a slope collector grew over the course of every hour of monitoring. Manjunath *et al.* [21] accomplished a study on the experimental and computational model of a flat plate collector. Conventional methods of working fluids were shown to possess strong agreement among experimentation and CFD modeling methods. Benli [22] evaluated the flat-plate collector's reverse, straight corrugated trapeze, and base type. When these investigations were carried out, the day and solar radiation were identical. The results showed that the flat plate collector had a higher heat transfer coefficient and an inferior stress drop. The effectiveness of a flat plate collector with inserted porous baffles at varying 50 cm^2 of surface area with thicknesses of 6 and 10 mm was examined by Bayraka and Oztopb [23]. Temperatures of the inside, exterior, absorber plate, and solar radiation were monitored at air flow rates of 0.016 kg/s and 0.025 kg/s. The experimental findings demonstrated that the thermal efficiency was better and the temperature difference was greater at an flow rates of 0.025 kg/s with 6 mm thickness compared to 0.016 kg/s. Sethi *et al.* [24] examined the effects of positioning an angled rough absorber plate in front of the flat plate collector to assess different operating parameters within the 3600–18000 Reynolds number range. An experiment by Yadav *et al.* [25] examined the erratic airflow across a rough and smooth rectangular duct surface. The front flow passage has an arc-shaped configuration that was intended to convey warm air on one side while providing proper insulation on the other three sides. The trials' findings showed that the maximal heat transfer coefficient and friction factor were 2.89 and 2.93 times bigger than those of the smooth surface, respectively. Chamolia *et al.* [26] studied the thermal efficiency of a flat plate collector using a double-pass heat exchanger. They also investigated several techniques, such as expanded surfaces, packed beds, corrugated absorbers, etc. To create a low-cost, highly effective solar collector, Chii-Dong *et al.* [27] tested a flat plate collector of the Thermal efficiency, angles, and resistance coefficient factors were examined. The greatest instantaneous thermal efficiency of 40% was attained with a low flow rate of 100 m^3/h and solar radiation on the collected area of 600 W/m^2 absorption patterns. The most effective solar collector is said to be an absorber tube solar collector. Mohammadia and Sabzpooshani [28] conducted research on a convectional-type flat plate collector with an inside-connected fin and baffles. According to the investigation, the upward single-pass type of flat plate collector had a higher output and greater effectiveness in contrast to the flat plate collector of the convectional type. Saxena *et al.* [29] conducted research on a simple solar air warmer. The granular carbon covering used in the model serves as the absorption medium. Four different configurations were considered in relation to the model's effectiveness. Bahrehmand and Ameri [30] experimented on an analytical model of the single- and double-glass flat plate collector. A triangular fin on a two-glass collector was shown to be more efficient than a rectangular fin on a single-glass collector. Additionally, it evaluated the heat balancing equations for various collector parts, including the rectangular and triangular fins with different widths, lengths, and shapes, as well as the tin metal sheet in the absorber plate. Al-Kayiem and Tadahmun [31] conducted an experiment on a rectangular duct with three different absorber plate angles: 30°, 50°, and 70°. The analysis revealed that the inclination angle of 50° had the highest heat transfer and Nusselt number, and likewise, in order to lower heat losses and optimize double-pass puissance. Ravi and Saini [32] explored the implementation of several techniques, such as fins and corrugated or grooved surfaces. Sarath *et al.* [33] carried out studies using convectional flat-type and parabolic concentrators. Results showed that the concentrator collector performed 15.3% better than the conventional flat plate. Reviewing the literature has a noteworthy favorable effect on the experiment analysis.

17.3.2   Evacuated tube collector

A flat plate collector is employed only in commercial heating systems because of its higher heat losses, and 80 °C is the maximum temperature limit. Evacuated tube collectors are among the best ways to use solar energy. The review of the literature is highly helpful for designing and improving performance. Subramanian *et al.* [34] created the evacuated tube collector to examine the drying ginger under meteorological circumstances. It showed up to this conclusion when compared to natural sun drying, which was reduced from 13 to 6 hours. Lamnatou *et al.* [35] improved the evacuated collector's functioning with a heat pipe for solar drying. The warm air from the collector's output reaches the temperatures needed to dry vegetables by passing through the heat pipe. It also demonstrated the effectiveness of the collector for drying large amounts of substance without using a preheater. Wang *et al.* [36] conducted an experiment using ten evacuated collector fins connected to separate circular parabolic concentrators. An evacuated tube collector heat exchanger coupled with copper tubes. The surrounding air is warm, with the assistance of a concentrator following a slow exchange. Used the model analysis to determine that the heat exchanger exit air temperatures were greater than 200 °C. Kazeminejad [37] derived an equation for the heat transfer coefficient of the series. and parallel types of ETCs. In its study, a number of variables were examined, and it was determined that the parallel is the best option for serious types. Morrison *et al.* [38] examined water as a cheap working fluid with a higher heat transfer coefficient. Shah and Furbo [39] performed tests on the vertical tube evacuated collector. The absorber is a tubular structure that fully absorbs solar radiation in every direction. The flat plate collector was compared to these thermal energy values and found to be higher. Kim and Seo [40] experimented with the numerical analysis of the various four types of evacuated tube absorption patterns. The most effective solar collector is said to be an absorber tube solar collector. Shah and Furbo [41] examined the CFD model of the evacuated tube collector under various operating conditions with connections made either horizontally or vertically in the manifold channel. The results demonstrated that only minor modifications were permitted throughout the process. Budihardjo and Morrison [42] studied the collector of a flat plate and an evacuated tube. The results indicated that the 30-number evacuated was less efficient than a flat plate of a two-parallel type. Ma *et al.* [43] looked at the heat transfer coefficient and efficiency factor of the parabolic trough collector, copper, or fin-type absorber tube. The findings revealed the absorber's nonlinear heat loss coefficient. Yadav and Bajpai [44] studied the presentation of an evacuated tube collector in the presence of parallel and counter-flowing airflow. When compared to the parallel flow rate, the counter flow rate had a higher flow rate and was more effective Hayek *et al.* [45] examined the parabolic trough collector's overall efficiency when using both the heat pipe and the glass tube. Compared to glass tubes, heated pipes have an effectiveness that was 15%–20% increased [46]. When the setup was completed and the temperature reached its peak level of 55.7 °C above ambient, it saved 50% more energy than an open solar dryer. Bala *et al.* [47] used the evacuated tube collector for the drying of the mushrooms. The dryers were constructed from flat plate collectors, and each drying tunnel unit included three fans and a 40-W solar module. Water served as the operational fluid when Zhijian Liu *et al.* [48] tested the heat transfer coefficient 915 times on the ETC. The results were recorded and analyzed using a digital logic computer. Saravanakumar and Mayilsamy [49] investigated several sensible heat storage materials using a solar collector at weather-related testing in both convection modes. Carried out evacuated tube collector tests on dried Muscat grapes. At ambient temperatures between 29.5 °C and 33.2 °C, the evacuated tube collector chamber generated a maximum temperature and collector output of 87 °C and 74 °C, respectively. The highest thermal efficiency of the dryer was found to be 29.92%. Rajagopal *et al.* [50] utilized natural flat collector techniques to dry copra, extracting additional moisture from the flat collector. Gumus and Ketebe [51] experimented with drying ogbono and corn at various 110–130 °C temperatures. Under all favorable conditions, it found that the most effective drying temperature for producing soft, evenly dried maize and ogbono was 110 °C. Loha *et al.* [52] were successful in absorbing the characteristics of ginger slices using a flat plate collector. The experimentation was carried out at 45 °C, 50 °C, 55 °C, and 60 °C drying temperatures at air velocity (1.3 m/s), and its effects on ginger's thermal conductivity and its moisture properties were studied.

17.3.3 Phase change materials

Chaudhari *et al.* [53] stated that solar energy is one of the main sources of heat energy and can be stored as sensible and latent heat storage. In practice, latent heat outperforms sensible heat. Phase-change materials are substances that store latent energy and absorb and release heat when melting and freezing. Phase change materials are available in a range of thermal properties and were employed for a diversity of energy storage relevance, including solar cooking, industrial steam, and residual thermal energy systems. Fewer phase change materials can store a large thermal efficiency, which provides steady heat for a long time. Although low thermal conductivity is a phase-change flaw, numerous experimental techniques are suggested to raise material thermal conductivity. The latent heat storage was calculated using a DSC, which is based on the thermal parameters of the freezing and melting of phase change storage. The studied literature was helpful in guiding the investigation for the following study. Harikrishnan and Kalaiselvam [54] investigated the TiO_2 nanofluid-palmitic acid phase change materials. Thermal energy storage is unaffected, despite these minor changes having a considerable effect on differential scanning calorimetry results. According to the test results, nanofluid PCMs for solar water heating systems were more dependable, stable, and capable of transferring heat more effectively. Harikrishnan and Kalaiselvam [55] experimented using CuO nanofluids and oleic acid PCMs to store thermal energy for cold devices, which contain CuO nanoparticles in oleic acid that range in size from 1 to 80 nm. The study showed that CuO nanofluids had a greater thermal conductivity than olive oil. So CuO-oleic acid nanofluids were suggested for cool energy-saving applications. Velraj *et al.* [56] suggested that phase-change materials might be employed for high thermal energy storage during the freezing and melting processes. Because phase-change materials with high heat storage were dependent on thermal conductivity and heat transfer rate, Fang *et al.* [57] looked at using a DSC test to analyze the phase change materials temperature and latent heat storage for air conditioning systems. Tyagi *et al.* [58] examined the temperature of the melting, boiling, and latent heat for inorganic salt phase change materials using the DSC test. The analysis concluded that these phase change materials are most appropriate for built applications and the most suitable energy storage for low-temperature applications. Rahimi *et al.* [59] explored the phase change behavior through testing in heat exchangers with finned tubes or bare tubes. In the bare tube exchanger, the flow rate during the solidification process was effectively raised while the melting time was decreased. Xia and Zhang [60] investigated the behavior characteristics of phase change materials acting as acetamide for solar applications, which are mostly utilized for storing low thermal energy. In comparison to graphite composites, the thermal properties were 10% better at acetamide. Nallusamy *et al.* [61] used three distinct strategies to optimize heat transmission for latent heat storage and maximize the use of heat. Rashid *et al.* [62] described the spherical capsule holding the phase change materials placed into the bed of thermal storage, which analyzed the features for the freezing and melting processes. Putra *et al.* [63] conducted research on the flow rate of beeswax/CuO nanoparticles used in phase change for low-energy storage applications. The results showed an increase in thermal conductivity and a decline in latent heat. Jesumathy *et al.* [64] found paraffin wax to be the best phase change material for latent heat storage. The heat transfer of paraffin wax ranged from 40 to 70 °C, and it was raised to 58–60 °C by the influence of a heat exchanger, reflecting that the phase change material and the latent heat storage were 210 kJ/kg. Thirugnanam and Marimuthu [65] conducted experiments on the phase change material (paraffin wax) used in the double pipe heat exchanger, carried out by the two different flow rates. Finally, it was determined that phase change material is an appropriate technology for storing energy recovery systems. The work focused on the freezing, melting, and latent heat storage of palmitic and steric acid as phase-change materials for use in culinary applications. According to the existing literature, only a few studies on the freezing and melting characteristics of this palmitic and steric acid phase change material have been carried out. The comparative experimental study was fully investigated in terms of time and cycle operations.

17.3.4   *Parabolic trough collector*

Senthil Manikandan *et al.* [66] experimented with various variables, including the collector's thermal energy, which increases with the efficiency of the tracking mechanism for the continuous reflecting sheet at the solar radiation focus point. Similarly, Helwa *et al.* [67] studied the tracking mechanism for the continuous reflecting sheet at the solar radiation focal point to enhance the output of the parabolic collector receiver. Basil Okafor [68] contrasted different kinds of collectors, and the parabolic trough collector attained the highest temperature and thermal energy of 86.5 °C, or 31.53%. Additionally, it has been stated that up to 74% of this energy is used to meet economic needs. Hadi Ali [69] experimented with a parabolic trough collector and gained a temperature of 90 °C at a 30 °C environmental temperature with a receiver focal length of 47.02 cm. Balakrishnan *et al.* [70] evaluated a parabolic trough collector receiver setup with a metal tray covered in black paint and no glass cover. The system's highest temperature reached was 104 °C. Ronge *et al.* [71] examined an experiment conducted on a stationary parabolic trough collector with and without a tracking system, which resulted in an ability of 27.6% and an output of 41.2% in about 11–12 hours. On the other hand, it performed at 43.6% overall and 53.1% in 11–12 hours in the Thermal storage.

17.3.5   Parabolic trough collector with solar cooker

The amount of energy required to cook is growing every day. In India, fuel sources, including firewood, LPG, and other power sources, are frequently utilized. The fuel that is now on the market is considered to be scarce and pricey. The best way to cook food is to use solar power because it poses no environmental risks. Noman *et al.* [72] created a polished parabolic trough collector with a focus ratio of 9.867, indicating that, under stationary settings for solar radiation absorption, it produced a thermal energy range of 50%–30%. Asmelash *et al.* [73] scrutinized the parabolic trough collector for drying soybeans with a solar dryer. A 30 mm copper tube was used to transfer the heat from the parabolic trough collector to the solar cooker. The system reached its maximum efficiency of more than 6% while the temp temperatures were discovered to be 191 and 119 °C for its absorber tube, respectively. The effect of increasing solar radiation on a parabolic trough collector with higher thermal energy and energy production at the intake and exit was examined by Shukla and Khandal [74]. Chaudhary *et al.* [75] compared the feats of a regular cooker and a solar cooker. A solar cooker with an exterior glazing surface painted black retained 32.3% more heat than a regular cooker, and phase change materials retained 28% more heat. A solar cooker constructed by Wollele and Hassen [76] required 45 minutes to cook 1 kg of rice at 355 K and 421 W of solar radiation. When a solar cooker was placed on an insulated storage tank, it was attained in 40 minutes. Kumar De *et al.* [77] detailed the methodology for energy-efficient cooking and food preservation techniques that preserve nutritious energy while protecting the environment by lowering CO_2 and harmful emissions through the use of a stovetop experiment conducted without a pressure cooker. The outcome demonstrated that heating 1 liter of dry rice used 1.5 MJ of energy and 2640 s, while maintaining the food's nutritional value and protecting the environment by lowering CO_2. Yahuza *et al.* [78] conducted research on the Box type solar cooker using 1.5, 1, and 0.5 liters of water. The highest cooking temperatures, which varied according to the increase in solar radiation, were 81.6 °C and 81.7 °C.

17.3.6   Thermal energy storage

Panchal *et al.* [79] examined the use of solar energy in sensible and latent heat storage materials for heat storage during the day and night. An experiment by Saini *et al.* [80] included parabolic trough-aided phase change materials for heat storage with solar cooking. When the thermal oil was compared to water, its temperature rose by 10–24 °C and its heat production increased by 19.45% to 30.38%. Mawire *et al.* [81] investigated the use of indirect cooking in an oil/pebble-bed parabolic trough energy storage simulation model. The findings showed that a variable flow rate's rapid rate of power extraction and temperature decline make it suitable for cooking. A temperature higher than the constant flow rate was consistently maintained for the variable flow rate.

17.3.7   Nanofluids in thermal energy storage

As per Farhana *et al.* [82], nanofluids have the capability to augment the solar collectors' energy storage. This study examined the thermal energy of various nanoparticles in the base fluids used in the investigation, including TiO_2, CuO, ZnO, and Al_2O_3. Saxena and Gaur [83] looked into the effects of nanofluids in different kinds of collectors. Solar energy was significantly increased with the inclusion of nanofluids. Hussein *et al.* [84] investigated ZnO/Ethylene Glycol-Pure Water at 1.0–4.0% volume concentration in a parabolic trough collector thermal conductivity. The maximum thermal energy in an experimental scenario was found to be 62.87% with a flow rate of 0.045 kg/s. Norouzi *et al.* [85] found that the parabolic trough receiver and the flow rate of the working fluids were diminished by an increased parabolic trough temperature of more than 60% of the fluid temperature, and the thermal energy was 17%. Rizal *et al.* [86] showed a solar cooker can produce 16 liters of hot water at 40–60 °C for 4 hours and 60 liters at the same temperature for 5 hours. Norouzi *et al.* [87] tested the ability of a parabolic trough collector to absorb more solar radiation by using a tracking system for heat-carrying Al_2O_3 nanofluids. Investigations were conducted on the impact of a number of parameters, including heat output, thermal energy with corresponding flow rate, and nanoparticle concentration. Elarem *et al.* [88] used an evacuated tube equipped with a thick fin arrangement for an experiment for rapid heat transfer of the paraffin wax phase change materials, CuO nanoparticles. An analysis was done on the modification to the system-act metrics. Singh and Gaur [89] examined the evacuated tube collector discharge with and without a dryer. In both scenarios, bottle gourd, ginger, and tomato were the average convective heat transfer coefficients and were 305.4%, 3.8%, and 153%, respectively, and thermal energy was 14.22%–27.99% higher than the dryer without an evacuated tube, respectively. Experiments utilizing parabolic trough energy storage to increase solar cooker efficiency have been published in the literature. By adopting phase change materials-Al_2O_3 nanofluids as the storage medium and water and waste engine oil as the working medium, the experiment performed on parabolic trough energy storage was improved.

17.4   The experiment setup

17.4.1   Flat plate collector

An air flow rate channel with a single path for a flat plate collector connection as shown in the figure 17.1.The experiment's configuration of absorber plates, blowers, glass covers, drying chambers, temperature sensors, chimneys, and glass wool insulation. The collection measured (2 × 1 × 0.10) meters. The collector's edges were covered by a 10-cm-thick aluminum plate. The entire system was insulated with glass wool materials that were 25 mm thick to avoid heat losses. The absorber plate is made of 5 mm-thick metal that has been coated in black to improve heat absorption and reduce heat losses. Through the 50-mm-thick holes in front of the flat collector, air can gradually exit through the upper side of the absorber plate and reach the surface. The overall design is to keep the maximum temperature constant. The testing took place in February 2017 between 8 a.m. and 6 p.m., and the corresponding temperature measurements for the forced and natural convection methods were taken to achieve the collector efficiency. The experiment was run under average atmospheric air conditions.

17.4.2   Evacuated tube collector

The setup comprises 30 vacuum tubes, each measuring 0.06 m, 0.03 mm, and 1.5 m in length, respectively. The evacuated tube collector has a 3.6 m^2 surface area, and one of its ends is in length. The blower was connected to the front of the manifold and rotated continuously, sucking air from the atmosphere. The long, double-walled evacuated tube collector has an external surface that is translucent to collect more solar radiation and an interior with a selectively absorbed coating that prevents heat loss over a longer period of time. Figure 17.2 shows the schematic view of the evacuated tube collector. The ambient air that enters and exits the inner and outer circular manifolds becomes heated as a result of absorbing more solar radiation.

Figure 17.1   Experimental setup for flat plate collector with pre-heating chamber

Figure 17.2   ETC arrangements

17.4.3   Phase change materials

It is necessary to prepare the phase-change materials for encapsulation before the experiment begins. One beaker containing 150 g of powdered palmitic acid was heated to 60 °C using a magnetic stirrer. The polyethylene components have a diameter of 75 mm. Water is added to the liquid palladium acid to heat it to 80 °C. The change in temperature was recorded during the charge and discharge processes and in the region beneath the DSC to assess the latent heat for the selected phase change materials. Water is used as a working fluid, and its temperature ranges from 20 °C to 90 °C. Electrical stirrers were utilized to keep the temperature within the tank consistent. Maintaining a steady temperature for the working fluids is essential before testing the phase-change materials. During the freezing and melting processes, four K-type thermocouples were inserted within the phase change material encapsulation to monitor temperature from the inner surface to the center every 5 seconds. In the current study, to identify the latent heat storage characteristics under various thermal load circumstances, all sensors and transducers were connected to an 80-channel data recorder (Agilent 34972A, USA).

17.4.4   Parabolic trough-assisted phase change heat storage for solar cooker

Figure 17.3 depicts the experimental setup for phase change material assisted by a parabolic trough collector. It consists of an 80-L storage tank composed of mild steel of thickness 2.5 mm, and its outside surface was completely insulated using thick glass wool materials (0.05 m), a parabolic trough receiver, and manual tracking. The storage tank is built on its inner side which houses a rectangular box (40 cm × 40 cm × 4 mm thick) parallel to the left and right sides of the storage tank. It was filled with phase change materials—0.3% volume concentration of Al_2O_3 nanofluids—to enhance the temperature of the working fluids at the exit of the tank. The system consisted of two tracks: the working fluids in the storage tank are pumped into the solar cooker through a check valve; in the other route, the transfer liquid is first made in the thermal storage at room temperature and sent by a check valve into the parabolic trough collector receiver, where it absorbs heat and stores it in the storage tank. The experiment was conducted continuously. The entire system is sealed with insulating materials, including glass wool. Throughout the working hours, flow rates of 0.035 kg/s, 0.045 kg/s, and 0.065 kg/s were used to measure the temperatures of the storage tank and parabolic collector, with their specifications indicated in Table 17.1. An 80° 14′ 15.42″ East, Latitude 13° 4′ 2.78″ North, was the location of the planned and built small parabolic trough enhanced thermal energy storage in Chennai, India.

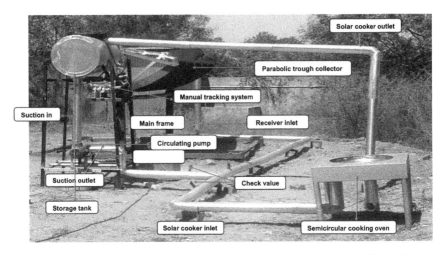

Figure 17.3   Schematic arrangement of parabolic trough-assisted phase change materials heat storage for cooking vessel

Table 17.1   Specification of PTC, reflector sheet, and receiver tube

S. No	Properties	Dimensions	S. No	Properties	Dimensions
	Parabolic Trough Collector			**Receiver tube**	
1	Collector aperture area	4.5 m²	1	ETC diameter	0.051 m
2	Aperture width	1.5 m	2	Receiver tube OD	0.047 m
3	Length-to-Aperture ratio	0.642	3	Receiver tube ID	0.043 m
4	Rim angle	67.8°	4	Thickness of the tube	0.04 m
5	Coating absorptance	0.944	5	Length (cover tube)	3 m
6	Coating emittance	0.9	6	Materials	AISI Type 304 SS
7	Mirrors reflectivity	0.91	7	Specific heat capacity	0.5 J/g-°C
8	Concentration ratio	13.1	8	Thermal conductivity	16.2 W/m-K
9	Slope error	rad ± 12	9	Melting point	1400–1455 °C
10	Specularity	rad ±13	10	Solids, liquids	1400 °C
	Reflector Sheet				
1	Specular reflectance	94%	4	Water vapour transmission	Negligible
2	Hemispherical reflectance	94%	5	Operating temperature	−40 to +90 °C
3	Nominal thickness	0.1 mm	6	Temperature difference	6 to 8 °C.

17.5   Preparation of solutions

17.5.1   Phase change materials with nanofluids

For drying, the cooking food needs a temperature between 50 °C and 90 °C. Here is the stearic acid, which has a melting and boiling point of 69.3 °C, 361 °C is used as a phase change material, and the addition of Al_2O_3 nanoparticles (sizes of 10 and 40 nm) revealed that a higher concentration of metal oxides significantly increases the working fluid's temperature, which increases thermal energy. Enhancing nanofluid stability in terms of viscosity, thermal conductivity, and heat transfer characteristics has been the focus of recent studies. Because they have a greater potential to improve heat transfer, nanofluids are more appropriate for application in real heat transfer processes.

Figure 17.4 (a) illustrates the solid phase change materials that have been encased with a 0.3% stearic acid seal and Al_2O_3 Nanofluids. Of these, 75% of the containers contain Al_2O_3 nanofluids in stearic acid phase change materials, while the remaining 25% expands during the heat transfer operations of charging and discharging. We bought homogeneous aluminum oxide. nanoparticles and stearic acid phase change materials from SWASCO Laboratories in Mumbai. The nano-powder was combined with phase-change materials at a determined ratio of 0.3%. The substance was heated and then poured into rectangle boxes before coating and sealing. The digital representation of phase change materials containing 0.3 vol.% Al_2O_3 nanofluids are displayed in Figure 17.5. Longer dispersion stability and homogeneous dispersion were seen with nanoparticles at all concentrations. The produced nanofluids can therefore aid in boosting thermal conductivity. The produced Al_2O_3 nanoparticles used in phase change materials are displayed in Tables 17.2 and 17.3.

Figure 17.4   (a) Encapsulation (b) digital image of the phase change materials – 0.3% volume Al_2O_3 nanoparticles

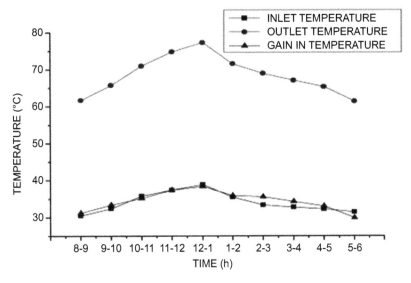

Figure 17.5   Temperature of the collector during natural convection mode

Table 17.2   Thermophysical properties of the stearic acid (phase change materials)

S. No.	Properties	Value/type
1	Melting point °C	69.6
2	Density (g/cm^3)	0.847
3	Boiling point	376.1
4	Unsalable	WA, Ether
5	Points of freezing	55–7 °C
6	Chemicals structure	$CH_3(CH_2)_{16}COOH$

Table 17.3   Thermophysical properties of the Al_2O_3 nanoparticles

S. No.	Properties	0.3vol.%Al_2O_3	Uncertainty
1	TC (W/m/K)	0.669	0.001
2	latent heat (kJ/kg)	1916	0.12
3	SPC (J/kgK)	4.026	0.01
4	Viscosity (mm^2/S)	4.026	0.002
5	Density (kg/m^3)	1.034	0.001
6	Proportion of energy (kJ/kg)	0.175	0.01

Table17.4   Thermodynamics properties of the waste engine oil, water

S. No.	Properties	Waste engine oil	water
1	Boiling temperature	400 °C	99.974 °C
2	Critical temperature	235 °C	373.946 °C
3	Critical density	940 kg/m^3	0.322 g/cm^3
4	Melting heat latent	389	334 kJ/kg
5	Latent evaporating heat (at100 °C)	145.6 btu/lb	40.657 kJ/mol
6	Temperature of melt(at101.325kPa)	10–40 °C	0 °C = 32 °F
7	Specific heat capacity	2483 J/kgK	4184 Jkg^{-1}K^{-1}
8	Kinematic viscosity	5 × 10^6 kg/ms	1 × 10^6 kg/ms
9	Thermal conductivity	0.1314 W/mK	0.6

17.5.2   Working medium

George *et al.* [90] examined the cooling and lubricating characteristics of many multi-grade SAE 20 W-50 engine oils using five samples. The upshot demonstrated the greater specific heat capacity of oil with lower internal energy and less viscosity had superior heat transfer properties. The thermodynamic properties of water and waste engine oil as working fluids are displayed in Table 17.4.

17.6   Assessment of uncertainty

Sabatelli *et al.* [91] stated that the parameter is connected to a measurement's result and represents the distribution of values that can be appropriate to the measurement. Uncertainties originate from a measurement that considers the sensor, data logger, and measuring device uncertainty, according to Michaelides *et al.* [92]. Following testing, a transient mathematical model of the collector is compared to the data gathered under a variety of operating situations. Every 5–10 minutes, test results are measured using the transient energy equation. The following equation is used to find the collector concert. For quasi-dynamic model as

For quasi-dynamic model as

$$\eta = C_1 P_1 + C_2 P_2 + \cdots + C_m P_m \tag{17.1}$$

Where P_1, $P_2 \cdots P_m$ are quantities, while $C_1, C_2 \cdots C_m$ are characteristic constants and m for number of parameters. For dynamic system test method, according to Morrison *et al.* [38],

$$Q_u = \eta_0 [K_0 \cdot B \cdot GB + K_0 \cdot D \cdot GB] - a_0 (T - T_a) - a_1 (T - T_a)^2 - C \left(\frac{dT}{dt} \right) \tag{17.2}$$

where η_0, a_0, a_1, C the coefficients are K_0, B, GB are determined by the test measured data's correlation.

For mathematical model, $(T_m - T_a)/Gt$ is the ratio of temperature differential to solar radiation with regard to time, which is used to determine the collector efficiency. The formula to find the mean temperature, T_m, is $T_m = (T_1 + T_2)/2$. Thus, the collector efficiency is expressed as

$$\eta = F_m(\tau\alpha) - \frac{F_m U_L(T_m - T_a)}{Gt} \tag{17.3}$$

where F_m is the mean collector efficiency factor and its volatility uncertainty was 0.12.

17.7 Model calculations

17.7.1 The flat plate collector and evacuated tube collector efficiency

Hematian and Bakhtiari [93] stated that the thermal efficiency of the collector is the proportion of incident solar absorption on the absorber plate to the beneficial thermal gain.

$$\eta_{coll} = \frac{P_{useful}}{P} \times 100 \tag{17.4}$$

$$P_{useful} = mc_p(T_2 - T_1)\text{Watts} \tag{17.5}$$

$$P = I_\beta \times A_C \tag{17.6}$$

$$m = \rho s V \text{ in Kg/sec} \tag{17.7}$$

where P_{useful} is the practical heat gain by the collector in Watts, P is the solar absorbed by absorber plate W/m^2 and m is the flow rate in Kg/sec, C_p is at specific heat capacity of pressure in J/Kg °K. T_1, T_2 are inlet and outlet temperature of collector tray. I_β is the precise measurement of SR made by a solar meter with accuracy ± 1 W/m^2. A_c is the region of the flat plate collector and V is the outlet velocity of the air, measured by a speed meter with a precision of 0.1 m/s. ρ is the density of the air in Kg/m^3 and S is the drying chamber's area in m^2.

17.7.2 Heat output and potency of parabolic trough collector

Kalogirou [11] argued that in steady-state conditions, the energy absorbed by the working fluids is minus the heat loss from the environment, is the rate of useable heat of the solar collector. The rate of useful vigor from the parabolic trough collector of area A_C can be obtained

$$Q_u = A_C[G_i(\tau\alpha) - U_L(T_p - T_a)] = mC_P(T_0 - T_i) \tag{17.8}$$

$G_i(\tau\alpha)$ = the solar radiation absorbed by receiver (W/m^2), $U_L(T_p - T_a)$ = the product of the overall heat lo coefficient represents T_p, T_a are receipt, ambient temperature in °C.
The thermal efficiency of storage tank is attained by

$$\eta_{th} = \frac{mC_p(T_0 - T_i)}{A_C[G_i(\tau\alpha)]} \tag{17.9}$$

17.7.3   *Energy losses for parabolic trough collector*

According to Kalogirou *et al.* (2004), heat losses to the environment from all thermal energy storage various heat transfer modes

$$Q_t = \frac{(T_p - T_a)}{R_L} = ULA_C(T_p - T_a)$$

$$R_L = \frac{1}{ULA_C}$$

(17.10)

UL = Overall heat coefficient based on collector area (W/m^{2}k).

17.7.4   *Energy losses for parabolic trough collector receiver*

The heat lost from the parabolic trough collector receiver to glass

$$Q_{loss} = A_C h_{cpg}(T_p - T_a) + \frac{A_c \sigma \left(T_p^4 - T_a^4\right)}{\left(\frac{1}{\varepsilon_p}\right) + \left(\frac{1}{\varepsilon_g}\right) - 1}$$

(17.11)

where h_{cpg} = heat transfer coefficient between the receivers to glass covers (W/m^2 k). ε_p, ε_g are infrared emissivity's of absorber plate, glass cover.

17.7.5   *Heat storage for thermal energy storage*

Malik *et al.* [7] the thermal energy of fluids (Q_s) at steady state is depended upon the temperature differential (ΔT_s), specific heat capacity and mass of quantity as illustrated in the below equations.

$$Q_s = mC_p \Delta T_s$$

(17.12)

An energy equilibrium of the storage tank gives

$$mC_p \frac{dT_s}{dt} = Q_s + Q_r + Q_t$$

(17.13)

where Q_s, Q_r = rate of engrossed, mislaid solar energy in the storage tank.

17.7.6   *Utilized energy output for cooking vessel*

Kalogirou [11], the fluid temperature

$$\text{Film Temperature } T_f = \frac{(T_p + T_a)}{2}$$

(17.14)

$$\text{Reynolds Number } R_e = \frac{md}{A\mu} = \frac{\rho UD}{\mu m}$$

(17.15)

$$\text{where } m = \sqrt{\frac{UL}{K\delta}}$$

(17.16)

Nusselt's Number $N_u = 0.0158(R_e)^{0.8}$ (17.17)

$$N_U = \frac{h_C D}{K}$$ (17.18)

$h_C = Actual\ heat\ transfer\ coefficient\ in\ W/m^2 K$

$Useful\ heat\ output(Q_u) = h_c A_c (T_p - T_a)$ (17.19)

17.8 Measuring device

17.8.1 Flat plate and evacuated tube and parabolic trough collector

K-type thermocouple sensors, which had an accuracy and resolution of 0.1–450 °C, were used to take temperature readings at numerous locations. The A Fluke 59 Max+ Infrared Pyrometer is used to accurately measure the flow rate because it is made of precision laser technology. Use the AcuRite 613 Indoor Humidity Monitor to measure the air's humidity. A vane anemometer of the Testo 410-1 type is used to compute the air velocity of the experiment at speeds between 0.4 and 20 m/s with an accuracy of 0.1 m/s and a working temperature range of 10–50 °C. The working fluid flow rate is measured using the SI-3601 Oval Gear Flow Meter, and the value is confirmed using the check valve. A MAX6675 data collection device is attached to monitor the complete setup.

17.9 Result and discussion

17.9.1 Flat plate collector

The natural and forced convection modes of the solar collector's strength were tested over a number of days, with the average intake and outlet temperatures being recorded every hour. The ambient air drawn in by the air blower passes through the absorber plate of the collector and exits through the insulated collecting tray.

17.9.1.1 Natural convection

These analysis methods evaluate the collector's efficiency without the aid of external equipment. Traditionally, air is introduced into the collector through the absorber plate and exhausted through an insulated collection tray, the temperature of which is monitored every hour. The temperature of the absorber plate during the day at the gain in, input, and outflow is explained in Figure 17.5. Between 12 and 1 p.m., there was the greatest temperature difference, with temperatures of 78 °C and 48 °C; between 2 and 3 p.m., the temperature naturally decreased as a result of fair passing through the flat plate collector absorbing the solar energy. This shows that if the solar radiation is lowered, the temperature at the collector output will also reduce. The effectiveness of the flat plate collector at diverse flow rates throughout the period is depicted in Figure 17.6. Compared to low flow rates,

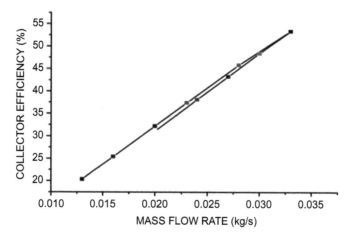

Figure 17.6   Collector Efficiency during various mass flow rate

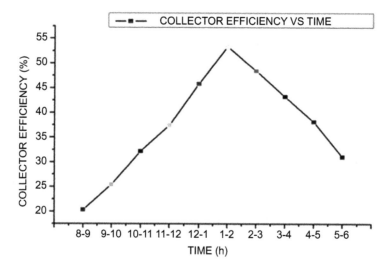

Figure 17.7   Collector efficiency during natural mode

which do not absorb as much solar radiation and result in greater heat losses, the flow rate at 0.035 kg/s does absorb more solar radiation while still achieving better collector efficiencies. Because of the great temperature difference between the collector's input and output at a flow rate of 0.03 during the middle of the day, between 1 and 2 p.m., the thermal efficiency was higher in natural convection.

The collector efficiency during the daytime in natural convection mode at a flow rate of 0.035 kg/s is shown in Figure 17.7. Due to the fact that less solar energy was being absorbed and the incoming air temperature was lower during the early phase, the effectiveness was lower. As the air temperature rose over time as a result of the flat plate collector's enhanced solar radiation absorption the air's moisture content decreased, and the variation in temperatures between the input and exit increased. The solar radiation was acquired during midday, which increased the collector efficiency to its highest level and the change in collector efficiency was observed every hour.

17.9.1.2 Forced convection

The same process used to evaluate the collector efficiency for natural convection should be used for forced convection. The air is moving over the absorber plate at a 25% quicker rate with forced convection, which causes less solar radiation absorption and greater heat dissipation; hence, the efficiency of forced convection will be reduced.

In front of the chamber is installed a blower (15 V, 1 amp, and 8 W): The flat plate collector performed better during the middle of the day (Figure 17.8) due to the greatest differential between the intake and outlet of the collector, as well as the flow rate and more solar radiation that were present at the time. The collector's maximum outlet temperature was measured to be close to 70 °C, and it peaked up to midnight with a starting temperature of 48 °C. Following that, its temperature gradually dropped and was lower than that of natural convection. In the figure above, the flat plate collector's inflow, outflow, and associated rise in temperature are all readily visible. Only 48% of the maximum thermal efficiency, as shown in Figure 17.9, was attained at midday because it was dependent on the heat-absorbed flow rate of air and solar radiation. Because heat is dispersed over the collector due to heat absorption and the air's maximum speed, forced convection collector efficiency is lower than natural convection. The collector efficiency spontaneously decreased after mid-day.

At a flow rate of 0.035 kg/s, Figure 17.10 shows that natural convection outperformed forced convection due to the larger temperature difference between the

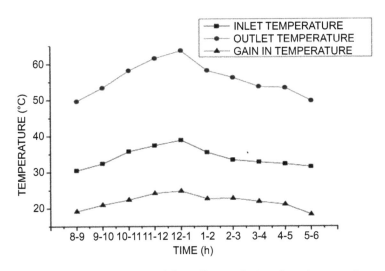

Figure 17.8   The temperature of the collector during forced convection mode

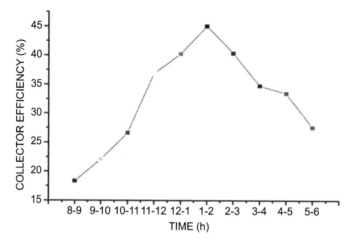

Figure 17.9   Compare the collector efficiency during the period of time

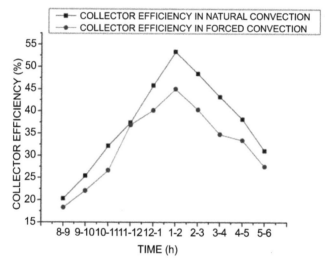

Figure 17.10   Compare the collector efficiency during the natural and forced convection

interior and exterior of the collector during the day. Natural convection had an18.6% higher collector efficiency than forced convection. A flat plate collector's natural mode was developed and tested, and it performed better than the forced convection mode.

17.9.2   Evacuated tube collector

Evacuated tubes with and without absorber plates were used for the experiment's setup. The experiment was done in March 2017 between the hours of 8:00 a.m. and

6:00 p.m., and the ambient temperature was from 30.2 °C to 39.9 °C. The evacuated tube collector with and without an absorber plate served as the foundation for the study. Figure 17.11 depicts the temperature during the time periods with and without the absorber plate. The gain outlet, and inlet temperatures were 67, 108, and 39 °C for the absorber plate during the period of 12–1 p.m. on the experiment day and 58, 98, and 36 °C for without the absorber plate, respectively, and in between 2:00 and 3:30 p.m., the temperature naturally dropped because of reduced sun exposure, cool air, and increased heat losses. The heat of the collector was impacted by the reduction in solar energy at the end of the day. As a result of the absorber plate's ability to retain solar energy for a prolonged duration, there was less heat loss when using one compared to without an absorber plate.

Figure 17.12 displays the change in the air's relative humidity during the duration of the experiment's day. Initial air humidity levels ranged from 52% to 64%, but when the sun's rays intensify in the afternoon, the humidity drops to about 39% for areas with absorber plates. Figure 17.13 exemplifies the discrepancy of solar radiation with regard to time. The evacuated tube originally attained solar radiation of 675 W/m^2, which then gradually increased to 880 W/m^2 at midnight, when it reached its maximum level, and then gradually fell to finally reach 730 W/m^2.

Figure 17.14 shows the evacuated tube collector with absorber plate, which reached its maximum temperature at midday and was 13.6% higher than it would have been without the absorber plate This is because the absorber plate initially absorbs heat, which is stored and endures for a long time before being distributed. As a result, the output temperature of the absorber plate was higher.

Figure 17.15 explains the different flow rates of air with respect to the collector efficiency. Low flow rates that absorbed more solar radiation also caused more heat losses, whereas high flow rates that absorbed more solar radiation but couldn't last for a long time caused more heat losses to occur. When compared to the flow rate of 0.032 kg/s, they received more solar radiation and had lower heat losses. So, it

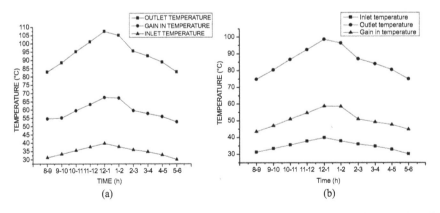

Figure 17.11   Temperature during the period of evacuated tube collector (a) with and (b) without absorber plate

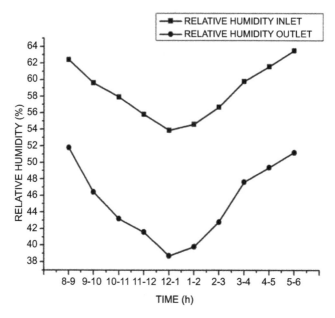

Figure 17.12   Relative humidity of air during the period of evacuated tube collector

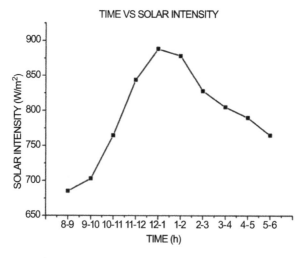

Figure 17.13   Solar radiation during the period of evacuated tube collector

was able to collect with greater capacity. Natural convection was more effective in the middle of the day, between 1 and 2 p.m., when there was a maximum air temperature differential between the collector's inlet and output at a flow rate of 0.032 kg/s. The efficiency was reduced due to the early period's slow solar energy and cool incoming

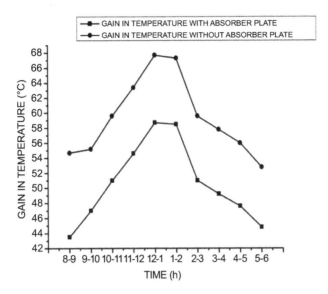

Figure 17.14   Gain in temperature with and non-absorber plate evacuated tube collector

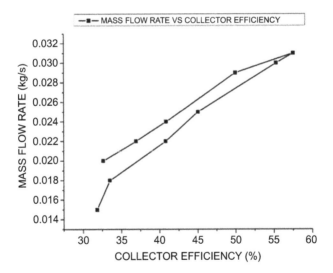

Figure 17.15   Collector efficiency at variant flow rates of the evacuated tube collector

air. As the air temperature grew as a result of the flat plate collector's enhanced solar radiation absorption, the difference between the input and output temperatures widened with time. The collector efficiency reached its peak during the midday hours, as shown in the figure, when the most solar radiation was captured.

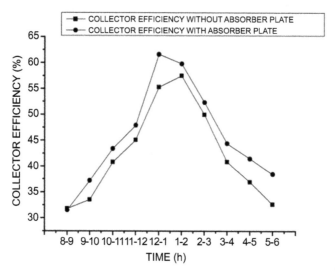

Figure 17.16   Collector efficiency with and without absorber plate of the evacuated tube

At the same time, the collector efficiency was fluctuating hourly, as indicated in Figure 17.16 displays the collector efficiency for an absorber plate with or without an evacuated tube. The results indicate that an evacuated tube collector with an absorber plate efficiently stores more energy than a collector without an absorber plate. In cold or cloudy weather, it was also employed to raise the temperature of the collector's input and output. Evacuated tubes, both with and without absorber plates, were very efficient in the beginning. However, after gradually increasing up until 12–1 p.m., gradually increasing efficiency at 1–2 p.m., and then gradually declining due to reduced solar radiation and increased energy losses.

17.9.3   Phase change materials

The studies explain the properties of palmitic and stearic acid in the phase-change materials during freezing and melting periods during working hours. The phase-change material during melting and freezing procedures was influenced by the temperature at which it was heated or cooled over time. The section explains the time-being change and cycle of operation of the freezing and melting processes of stearic and palmitic acids.

17.9.3.1   Melting and freezing characteristics of the phase change materials with respect to time

Figure 17.17 shows the time-melting or freezing effects of palmitic and stearic acid due to their phase change materials. The temperature increments from 60.28 °C to 62.5 °C and 52.8 °C to 54.5 °C for the time period of 9–24 minutes for both palmitic and stearic acid were shown to be because of increased thermal conductivity, which allows faster heat transfer and reduces the amount of time needed for the phase change materials to complete the charging process. Figure 17.17(b) illustrates that increased thermal conductivity allows for a faster rate of heat transfer in phase-change materials, shortening the discharge process. The temperature decreases at a steady rate from 60.25 °C to 60.1 °C, 54.9 °C to 54.79 °C for a duration of 15 min for both palmitic and stearic acid.

17.9.3.2   Effect of cycling operation for melting and freezing uniqueness of the phase change materials

Figure 17.18 depicts the behavior of the phase-changing stearic and palmitic acids over the cycle period in regard to melting and freezing temperatures. Figure 17.18 depicts stearic acid phase change materials at steady-state temperatures at

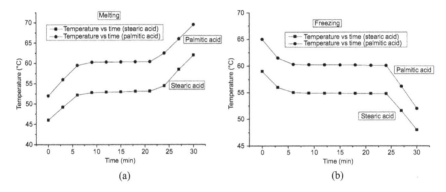

Figure 17.17   Effect of palmitic and stearic acid phase change materials for (a) melting and (b) freezing

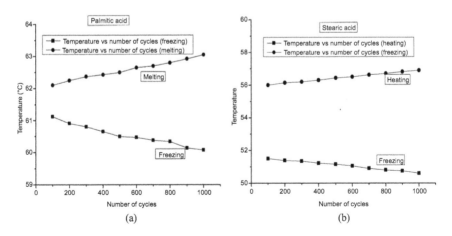

Figure 17.18   During the effect of the cycling operation (a) palmitic (b) stearic phase change materials

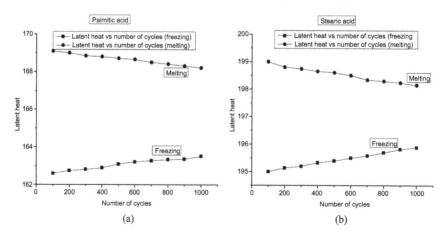

Figure 17.19   Effect of latent heat for the palmitic and stearic acid phase change materials (a) melting and (b) freezing

increment stages from 62.37 °C to 62.8 °C and decrement stages from 60.8 °C to 60.34 °C. During the stages, the cycle operation takes place between 300 °C and 500 °C for melting and freezing processes, and in palmitic acid, the temperature growths and decrements are in a stable range from 56.2 °C to 56.7 °C and 51.34 °C to 50.8 °C, respectively. The phase change materials carry out an entire charge or discharge process, but it will take more time since they will store more energy throughout the charging and discharging time, causing the heat load to remain constant for a long period of time.

17.9.3.3   Latent heat for melting and freezing characteristics of the phase change materials

Similarly, the consequence of the cycling process of palmitic and stearic acid on the latent heat for melting and freezing is presented in Figure 17.19. For the melting and freezing of palmitic acid under steady-state conditions, the cycling operation from 300 to 800 exhibited a drop in lateness at 168.85–168.42 °C and an increase at 162.82–163.34 °C, and in the stearic acid, a decrease in latent heat at 198.74–198.3 and a rise at 195.2–195.7 °C, respectively. Small deviation decreases and increases occur during the melting and freezing processes as a result of the heat that phase-change materials have absorbed and retained.

17.9.4   *Parabolic trough collector with Thermal energy storage*

In February and March of 2018, the experimentation began. The presentation of the parabolic trough collector receiver and thermal storage is evaluated and analyzed using the working fluids of water at flow rates of 0.035 kg/s, 0.045 kg/s, and 0.065 kg/s.

17.9.4.1 The achievement of parabolic trough collector receiver at flow rate of 0.035 kg/s

When the experiment began at 8:00 p.m., there was no sunlight, and moisture was present. As a result, both the environmental and parabolic trough collector temperatures were the same. The parabolic trough collector receiver temperature will be affected by solar radiation as well as the surrounding environment. At 9.00 p.m., the ambient and the parabolic trough collector-outlet temperatures were only marginally different. Over time, the difference widened due to increased solar radiation and decreased moisture and heat losses. At the test time, the parabolic trough collector receiver reached a temperature of 104.9 °C with a corresponding heat gain and efficiency of 1802 W,67.2% at a flow rate of 0.035 kg/s for the working fluid of water. The solar radiation was observed to be 590 W/m^2 at the ecological temperature of 37.6 °C. The accomplishment of the parabolic trough with temperature, heat gain, and ability at the flow rate of 0.035 kg/s is indicated during the day of 8:00 a.m.–5:00 p.m. The thermal efficiency of the parabolic trough collector at a flow rate of 0.035 kg/s is illustrated in Figure 17.20. After 2:00 p.m., there is a decline in solar radiation and an adjustment to energy losses as a result of the values. From 2 o'clock to 4 o'clock, the heat gain and thermal efficiency rapidly declined.

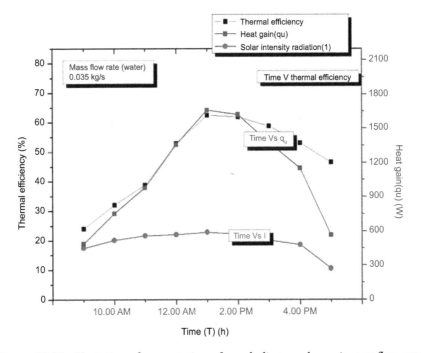

Figure 17.20 Variation of presentation of parabolic trough receiver at flow rates of 0.035 kg/s

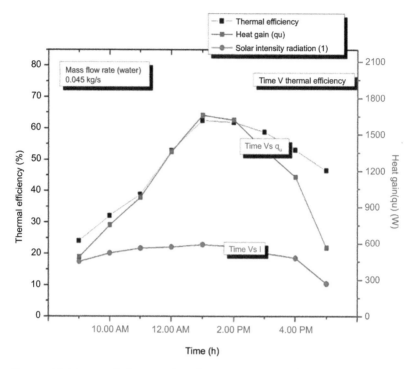

Figure 17.21   parabolic trough collector receiver at flow rates 0.045 kg/s

17.9.4.2   The implementation of parabolic trough receiver at flow rates of 0.045 kg/s

The procedure will proceed exactly as before, except the flow rate will now be 0.045 kg/s as shown in figure 17.21. The ambient energy levels are similar during the process's period but in the beginning, there is no change in the receiver outlet. Instead of 10:00 a.m.–1:00 p.m., that increases solar radiation, reduces heat losses, and increases heat gain and thermal efficiency.

17.9.4.3   The presentation of the parabolic trough collector at flow rate of 0.065 kg/s

Figure 17.22 depicts the parabolic trough receiver's energy output and thermal efficiency as working fluids in the water at a flow rate of 0.065 kg/s. Energy gain and thermal efficiency were lower at flows of 0.065 kg/s compared with 0.035 kg/s and 0.045 kg/s due to the working fluid's high flow rate, which resulted in lower radiation absorption and significant energy losses.

17.9.4.4   The storage tank at flow rate of 0.035 kg/s

The parabolic trough collector receiver collects the output energy and deposits it into a storage tank. 12 rectangular boxes are in the storage tank to hold stearic acid for accumulated energy.

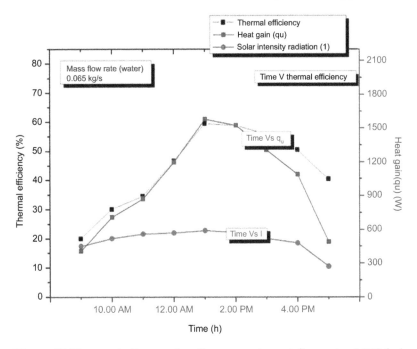

Figure 17.22   parabolic trough collector receiver at flow rates 0.065 kg/s

The storage phase change materials system's key advantage is that it delivers consistent energy. At flow rates of 0.035 kg/s, 0.045 kg/s, and 0.065 kg/s, the efficiency of a storage tank and a parabolic trough collector was compared. Even though the working fluid and storage tank temperatures were almost equal to the parabolic trough collector's primary input, it was discovered that convection and an increase in solar radiation that was stored inside phase change materials caused those temperatures to slightly rise over a period of minutes. The SE increased until 1 p.m. and declined until 5 p.m. After 6 o'clock, when there is neither atmosphere nor strong radiation, it rapidly and significantly decreases. Figure 17.23 shows that the storage tank attained its highest inlet temperature, sun intensity, and heat gain, respectively, at 1 o'clock in the afternoon at a flow rate of 0.035 kg/s of water. The parabolic trough collector receiver performs 9.5 times lower than storage tank heat gain up to 5 p.m.; storage tank output gradually declined and in each hour was 0.9 more decreased.

Figure 17.24 shows the discrepancy in the solar radiation, heat gain, and outlet temperature of the storage tank at the flow rate of 0.045 kg/s. Moreover, 0.035 kg/s was similar to 0.045 kg/s but with slightly less energy. The storage tank outlet temperature of 90.7 °C, 590 W/m^2 of solar intensity, and 9987 W of heat gain were all obtained at 1 o'clock in the afternoon. The parabolic trough receiver's energy was six times lower than the storage tank. The storage tank production gradually declines after 5 p.m. as a result of the lack of solar radiation and the steady energy loss of 0.93 times that takes place per hour.

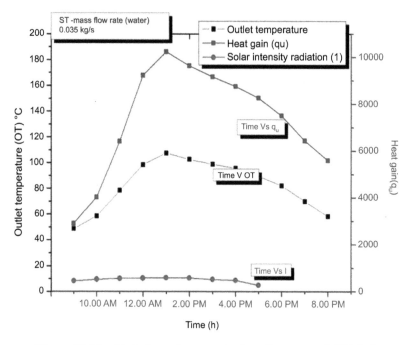

Figure 17.23   Variation of storage tank at flow rates .0.035 kg/s

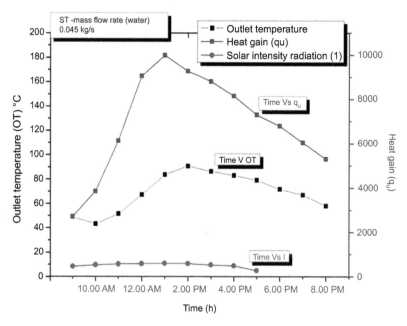

Figure 17.24   Variation of storage tank at flow rates 0.045 kg/s

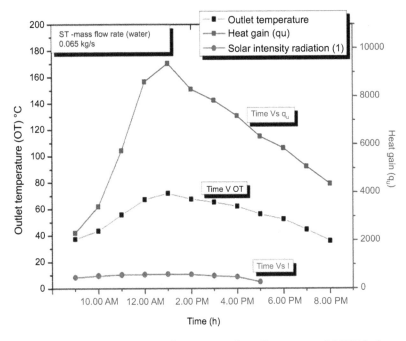

Figure 17.25   Variation of storage tank at flow rates of 0.065 kg/s

Figure 17.25 illustrates the storage tank's thermal efficiency at a flow rate of 0.065 kg/s. The image showed that the SR was 590 W/m^2 at 1 p.m., which corresponds to the storage tank reaching its maximum outlet temperature and a heat gain of 72 °C and 9370 W, respectively. The heat output of the storage tank was five times more effective than that of the parabolic trough receiver. Due to the lower sun radiation and hourly heat losses of 1.08 times, the storage tank outputs gradually decreased until 5 p.m.

17.9.5   The parabolic trough with solar dryer integrated with energy storage using PCMs-Al$_2$O$_3$ nanofluids

The goal of the device was to use phase change materials (Al$_2$O$_3$ nanofluids) for energy storage to measure the mean heat output and capability of the parameters in a solar dryer at flow rates of 0.035 kg/s, 0.045 kg/s, and 0.065 kg/s for groundnut, ginger, and turmeric crops by using working fluids of water and waste engine oil.

17.9.5.1   Comparing heat output for crops (0.035 L/s)

At different hours, the crops' moisture content decreased due to the recommended dryer's increased thermal efficiency. When compared to alternative drying techniques, a faster drying rate helped retain heat because of its solar dryer.

The heat output, heat removal rate, and humidity loss of solar-dried crops using water and waste engine oil, with a flow rate of 0.035 L/s, are shown in Figure 17.26. Groundnuts absorbed 6 times less agricultural energy in waste engine

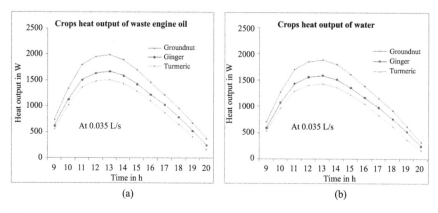

Figure 17.26   (a) and (b) Heat output of crops with respect to time (0.035 L/s)

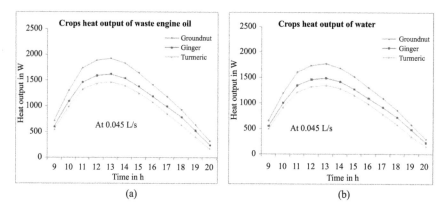

Figure 17.27   (a) and (b) Heat output of crops with respect to time (0.045 L/s)

oil and 4.6 times less than in water. This is due to heat losses being transferred into the surrounding air, resulting in less energy being stored in the tank.

17.9.5.2   Comparing heat output for crops (0.045L/s)

In Figure 17.27, the heat output of crops is compared to the working fluids of water and waste engine oil at 0.045 L/s. The solar dryers at 1 p.m. at a flow rate of 0.045 L/s for waste oil and water were 11008 W and 7977 W, respectively. The engrossed vigor of the solar dryer was found to be around 6.4.2 times lower than that of the storage tank when compared to groundnut, ginger, and turmeric. This was attributed to greater losses of radiation and modes of heat transfer in the surrounding environment.

17.9.5.3   Comparing the heat output of crops (0.065 L/s)

The heat output of crops grown on waste engine oil and water mediums at a rate of 0.065 L/s is shown in Figure 17.28. The results of the study indicate that the absorbed heat for the drying crops were lower than the flow rates of 0.035 L/s and 0.045 L/s due to the higher volume flow rate occurring at a high heat loss. For ground nut, ginger, and turmeric, heat output was found to be 5.2, 6.2, and 8.2 times lower than waste engine oil and 5.3, 5.9, and 5.6 times slower than water, respectively.

17.9.5.4   Contrasted with the crop efficacy (0.035 L/s)

An examination of crop yields at a rate of 0.035 L/s in waste engine oil and water media is shown in Figure 17.29. These values decreased after 1 p.m. as an end result of localized heat loss and decreased solar exposure. After output at 7.00 and 8.p.m., crops at a rate of 0.035 L/s at 6.00 p.m. were 54%, 45%, and 41% for waste engine oil, 51%, 43%, and 39% for water medium, respectively. For all crops, the yield of waste engine oil was about 4% greater than that of the water medium.

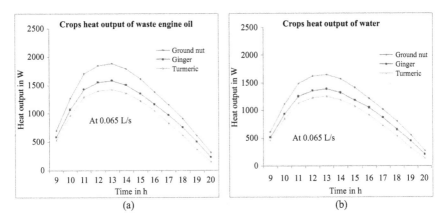

Figure 17.28   (a) and (b) Heat output of crops with respect to time (0.065 L/s)

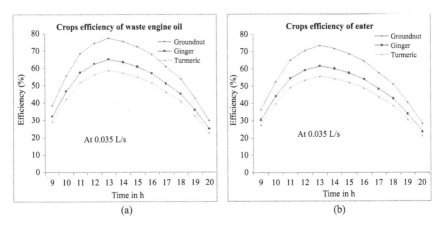

Figure 17.29   (a) and (b) Effectiveness of the crops with respect to time (0.035 L/s)

17.9.5.5   Comparing the potency of the crops (0.045 L/s)

The effectiveness of the harvests in the water and waste engine oil at a flow rate of 0.045 L/s is shown in Figure 17.30. The ratings decreased after 1 p.m. due to a decrease in solar radiation and environmental reheat losses. Following an 8% reduction at 7:00 and 8:00 p.m., crops' respective efficiency at 0.035 L/s flow rate at 6:00 p.m. was 49%, 41%, and 37% for waste engine oil, 48%, 40%, and 36% for water. All crops together produced only 3% more waste oil overall than in the water medium.

17.9.5.6   Comparing the capability of the crops (0.065L/s)

Figure 17.31 illustrates the crop efficiency at a flow rate of 0.065 L/s. The groundnut, ginger, and turmeric crops showed 35%, 30%, and 27% efficiency in the waste engine oil at 9:00 a.m., and 32%, 27%, and 24% efficiency in the water. The crops 'respective percentages at 1 p.m. were 64%, 54%, and 49% for water and 71%, 59%, and 54% for waste engine oil. After 1 p.m., it lost value due to a decrease in solar

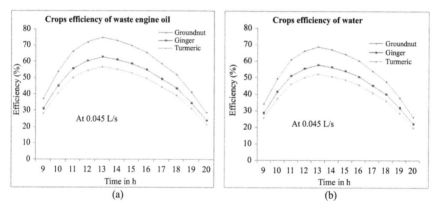

Figure 17.30   (a) and (b) Effectiveness of the crops with respect to time (0.045 L/s)

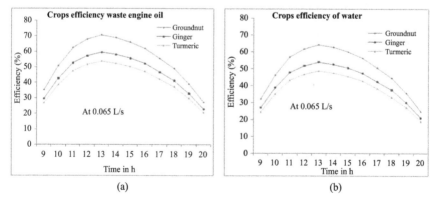

Figure 17.31   (a) and (b) Effectiveness of the crops with respect to time (0.065 L/s)

energy and ambient temperature loss. Upon collecting the crops at 5:00 p.m., the water showed 51%, 42%, and 38% efficiency, whilst the waste engine oil demonstrated 56%, 47%, and 42% efficiency.

17.10 Conclusion

17.10.1 Flat plate collector

The flat plate collector converted solar energy from light into thermal efficiency by moving air over the gap between the absorber plate and the collector. In the renewable energy heating technology, the highest and lowest temperatures of the air in the flat plate collector for forced convection mode were 98.6 °C and 39.9 °C, and the natural convection was 107.6 °C and 67.7 °C, respectively, at the flow rate of 0.035 kg/s in the period of 1–2 p.m. because more solar radiation was received and heat losses were reduced.

17.10.2 Evacuated tube collector

The objective of the research work is to examine the thermal efficiency of the evacuated tube with and without an absorber plate with respect to time at various air flow rates. At the flow rate of 0.032 kg/s attained, the maximum outlet and inlet temperatures of the evacuated tube collector without the absorber plate were 98.6 °C and 39.9 °C, respectively, and those with the absorber plate were 107.6 °C and 67.7 °C. Similarly, the gain temperature of the evacuated tube with the absorber plate is 13.6% higher than that of the tube without the absorber plate.

17.10.3 Phase change materials

The heat transfer for solar energy storage was tested in an experimental investigation of phase-change materials. The study of the repercussions showed that the deviations were minimal in the phase change materials. The study of the consequences showed that the deviations were minimal in the phase change materials as compared to standard latent heat storage and phase change material characteristics during the melting and freezing processes. The temperature for melting and freezing for both palmitic and stearic acid rose due to the thermal conductivity of phase change materials, which allowed for faster heat transfer, a heated surface area, and a reduced distance for absorbing or releasing energy during the melting and freezing processes.

17.10.4 Parabolic trough collector with thermal energy storage

The experiment's findings to determine the storage tank and parabolic trough collector receiver are the outlet temperature, heat gain, and thermal efficiency of a working fluid of water at flow rates of 0.035, 0.045, and 0.065 kg/s, respectively. The conclusion demonstrated that the parabolic trough collector achieved maximum outlet temperature, heat gain, and thermal efficiency of 104.9°, 1802 W, and

67.2% at flow rates of 0.035 kg/s, 102.5°, 1658 W, and 62.4% at flow rates of 0.045 kg/s, and 97.4°, 1578 W, and 59.4% at 0.065 kg/s, respectively, which corresponded to the atmospheric and solar radiation of 37.6 °C and 590 W/m^2, respectively. The parabolic trough receiver used to gather energy formed the basis for the storage tank's capacity. The experiment showed that the storage tank's maximum output temperature and heat gain were 107.5 °C and 10237 W, respectively. These temperature increases are almost 5% and 10% for 0.045 kg/s and 0.065 kg/s, respectively. In the aforesaid findings, flow rates of 0.035 kg/s were superior to 0.045 kg/s and 0.065 kg/s.

17.10.5 The parabolic trough with solar dryer integrated with energy storage using Phase change materials-Al_2O_3 nanofluids

At flow rates of 0.035, 0.045, and 0.065 L/s, the effectiveness of the solar dryer for drying groundnut, ginger, and turmeric crops for water and waste engine oil was successfully established. At flow rates of 0.035 L/s, the waste engine oil produced 0.33 times more energy than the water medium. By using phase change materials of 0.3 vol.% Al_2O_3 nanofluids as thermal energy storage, the process increased charging and discharging by 25–35% with usable heat and efficiency.

References

[1]   Kreith, F., and Kreider, J. F. (1978). *Principles of Solar Emerging*. New York: McGraw-Hill.

[2]   Pawar, S. R., Mohite, V. R., and Vivekananda, N. (2021). Review of thermal energy storage systems and their applications. *International Journal of Innovations in Engineering Research and Technology*, 2(6), 1–10.

[3]   Anderson, B. (1977). *Solar energy: Fundamentals in building design*. New York: McGraw-Hill.

[4]   Meinel, A. B., and Meinel, M. P. (1976). *Applied Solar Energy: An Introduction*. Reading, MA: Addition-Wesley.

[5]   Krider, J. F., and Kreith, F. (1977). *Solar Heating and Cooling*. New York: McGraw-Hill.

[6]   Kalogirou, S. (1997). Solar water heating in Cyprus, current status of technology and problems. *Renewable Energy*, 10, 107–1012.

[7]   Malik, M. A. S., Tiwari, G. N., Kumar, A., and Sodha, M. S. (1985). *Solar Distillation*. New York: Pergamon Press.

[8]   Norton, B. (1992). *Solar Energy Thermal Technology*. London: Springer-Verlag.

[9]   Lysen, E. (2003). Photovoltaics: An outlook for the 21st century. *Renewable Energy World*, 6(1), 43–53.

[10] Fan, H., Singh, R., and Akbarzadeh, A. (2011). Electric power generation from thermoelectric cells using a solar dish concentrator. *Journal of Electronic Materials*, 40, 1311–1320.

[11] Kalogirou, S. A. (2004). Solar thermal collectors and applications. *Progress in Energy Combustion Science*, 30(3), 231–295.

[12] Sorensen, B., Doble, M., Broussely, M., *et al.* (2008). *Renewable Energy Focus e-Mega Handbook*. San Diego, CA: Academic Press.

[13] Nikesh, D., Watane, R., and Dafde, A. (2013). Automatic solar tracking system. *International Journal of Scientific and Engineering Research*, 4(6), 93.

[14] Tudorache,T., and Kreindler, L. (2010). Design of a solar tracker system for PV power plants. *Acta Polytechnica Hungarica*, 7(1), 23–39.

[15] Panwara, N. L., Kaushikb, S. C., and Kotharia, S. (2011). Role of renewable energy sources in environmental protection: A review. *Renewable and Sustainable Energy Reviews*, 15, 1513–1524.

[16] Chiu, J. N., Martin, V., and Setterwall, F. (2009). A review of thermal energy storage systems with salt hydrate phase change materials for comfort cooling. 11th International Conference on Thermal Energy Storage, Stockholm, Sweden.

[17] Rai, G. D. (2013). *Non-Conventional Sources of Energy*. New Delhi: Khanna Publishers.

[18] Kostić, L. T., and Pavlović, Z. T. (2012). Optimal position of flat plate reflectors of solar thermal collector. *Energy and Buildings*, 45, 161–168.

[19] Khatik, J. S., Yeole, S. V., and Juned, A. R. A. (2014). Comparative experimental study of absorber plate with absorber plate having concavities. *International Journal of Research in Engineering and Technology*, 3, 13–15.

[20] Rhushi, P. P., Byregowda, H. V., and Gangavati, P. B. (2010). Experiment analysis of flat plate collector and comparison of performance with tracking collector. *European Journal of Scientific Research*, 40, 144–155.

[21] Manjunath, M. S., Karanth, K. V., and Sharma, N. Y. (2012). A comparative CFD study on solar dimple plate collector with flat plate collector to augment the thermal performance. *International Journal of Mechanical and Mechatronics Engineering*, 6(10), 2246–2252.

[22] Benli, H. (2013). Experimentally derived efficiency and exergy analysis of a new solar air hater having different surface shapes. *Renewable Energy*, 53, 58–67.

[23] Bayraka, F., and Oztopb, H. F. (2012). *Arib Hepbastic was investigated the performance of the initial energy and exergy of solar air hater inserted porous baffles*.

[24] Sethi, M., Varun, G., and Thakur, N. S. (2012). Correlations for solar air heater duct with dimpled shape roughness elements on absorber plate. *Solar Energy*, 86, 2852–2861.

[25] Yadav, S., Kaushal, M., Varun, G., and Siddhartha, S. (2013). Nusselt number and friction factor correlations for solar air heater duct having protrusions as roughness elements on absorber plate. *Experimental Thermal and Fluid Science*, 44, 34–41.

[26] Chamolia, S., Chauhana, R., Thakura, N. S., and Saini, J. S. (2012). A review of the performance of double pass solar air heater. *Renewable and Sustainable Energy*, 16, 481–492.

[27] Chii-Dong, H., Ho-Ming, Y., and Tsung-Cing. (2011). Collector efficiency of upward-type double pass solar air heaters with fins attached. *International Communications in Heat and Mass Transfer*, 38, 49–56.

[28] Mohammadia, K. K., and Sabzpooshani, M. (2013). Comprehensive performance evaluation and parametric studies of single pass solar air heater with fins and baffles attached over the absorber plate. *Energy*, 57, 741–750.

[29] Saxena, A., Agarwal, N., and Ghansyham , S. (2013). Design and performance of a solar air heater with long term heat storage. *International Journal of Heat and Mass transfer*, 60, 8–16.

[30] Bahrehmand, D., and Ameri, M. (2015). Energy and exergy analysis of different solar air collector systems with natural convection. *Renewable Energy*, 74, 357–368.

[31] Al-Kayiem, H. H., and Tadahmun, Y. A. (2015). On the natural convection heat transfer in a rectangular passage solar air heater. *Solar Energy*, 112, 310–318.

[32] Ravi, K. R., and Saini, R. P. (2016). A review on different techniques used for performance enhancement of double pass solar air heaters. *Renewable and Sustainable Energy*, 56, 941–952.

[33] Sarath, K. K., Siva, K. R., and Govindraj, M. (2012). Design advanced solar water heater. *Indian Journal of Mechanical Engineering and Research*, 2, 31–41.

[34] Subramanian, C. V., Neelamegam, P., and Umayal Sundari, A. R. (2014). Drying kinetic of Muscat grapes in a solar drier with evacuated tube collector. *IJE Transactions B: Applications*, 27(5), 811–818.

[35] Lamnatou, C., Papanicolaou, E., Blessiotis, V., and Kyriakis, N. (2012). Experimental investigation and thermodynamic performance analysis of a solar dryer using an evacuated –tube air collector. *Applied Energy*, 94, 232–243.

[36] Wang, P. Y., Guan, H. Y., Liu, Z. H., Wang, G. S., Zhao, F., and Xiao, H. S. (2014). High temperature collecting performance of a new all-glass evacuated tubular solar air heater with U-shaped tube heat exchanger. *Energy Conversion and Management*, 77, 315–323.

[37] Kazeminejad, H. (2002). Numerical analysis of two-dimensional parallel flow flat-plate solar collectors. *Renewable Energy*, 26(2), 309–323.

[38] Morrison, G. L., Budihardjo, I., and Behnia, M. (2004). Water-in-glass evacuated tube solar water heaters. *Solar Energy*, 76(1–3), 135–140.

[39] Shah, L. J., and Furbo, S. (2004). Vertical evacuated tubular-collectors utilizing solar radiation from all directions. *Applied Energy*, 78(4), 371–395.

[40] Kim, L. Y., and Seo, T. (2007). Thermal performances comparisons of the glass evacuated tube solar collectors with shapes of absorber tube. *Renewable Energy*, 32(5), 772–795.

[41] Shah, J., and Furbo, S. (2007). Theoretical flow investigations of an all glass evacuated tubular collector. *Solar Energy*, 81(6), 822–828.

[42] Budihardjo, I., and Morrison, G. L. (2009). Performance of water-in glass evacuated tube solar water heaters. *Solar Energy*, 83(1), 49–56.

[43]   Ma, L., Lu, Z., Zhang, J., and Liang, R. (2010). Thermal performance analysis of the glass evacuated tube solar collector with U-tube. *Building and Environment*, 45(9), 1959–1967.

[44]   Yadav, A., and Bajpai, V. K. (2011). Thermal performance of one-ended evacuated tube solar air collector at different air flow rates: experimental investigation. *International Journal of Ambient Energy*, 33, 35–50.

[45]   Hayek, M., Assaf, J., and Lteif, W. (2011). Experimental investigation of the performance of evacuated tube solar collectors under eastern Mediterranean climatic conditions. *Energy Procedia*, 6, 618–626.

[46]   Medugu, D. (2011). Performance study of two designs of solar dryers. *Archives of Applied Sciences Research*, 2, 136–148.

[47]   Bala, B. K., Morshed, M. A., and Rahmen, M. F. (2009). Solar drying of mushroom using solar tunnel dryer. *Proceedings of the International Solar Food Processing Conference*.

[48]   Liu, Z., Liu, K., Li, H., Zhang, X., Jin, G., and Cheng, K. (2015). Artificial neural networks-based software for measuring heat collection rate and heat loss coefficient of water-in-glass evacuated tube solar water heaters. *PLoS ONE*, 10(12), e0143624.

[49]   Saravanakumar, K., and Mayilsamy, P. T. (2010). Forced convection flat plate solar air heaters with and without thermal storage. *Journal of Scientific & Industrial Research*, 69, 966–968.

[50]   Rajagopal, T., Sivakumar, S., and Manivel, R. (2014). Development of solar dryer incorporated with evacuated tube collector. *International Journal of Innovative Research in Science, Engineering and Technology*, 3(3), 2655–2658.

[51]   Gumus, R. H., and Ketebe, E. (2013). The effect of temperature on drying rate of agro food: Corn (Maize) and Ogbono (Irivingia Gabonnensis). *IOSR Journal of Engineering (IOSRJEN)*, 3(3), 36–42.

[52]   Loha, C., Reeta, D., Choudhury, B., and Chatterjee, P. K. (2012). Evaluation of air drying characteristics of sliced ginger (Zingier officinale) in a forced convective cabinet dryer and thermal conductivity measurement. *Journal of Food Processing Technology*, 3(6), 1000160.

[53]   Chaudhari, V. N., Rathod, M. K., and Chaudhari, K. A. (2013). Stearic Acid as Phase Change Material: Thermal Reliability Test and Compatibility with some Construction Materials. *seeds*, 4(5).

[54]   Harikrishnan, S., and Kalaiselvam, S. (2013). Experimental investigation of solidification and melting characteristics of nanofluid as PCM for solar water heating systems. *International Journal of Emerging Technology and Advanced Engineering*, 3, 628–635.

[55]   Harikrishnan, S., and Kalaiselvam, S. (2012). Preparation and thermal characteristics of CuO–oleic acid nanofluids as a phase change material. *Thermochimica Acta*, 533, 46–55.

[56]   Velraj, R. V. S. R., Seeniraj, R. V., Hafner, B., Faber, C., and Schwarzer, K. (1999). Heat transfer enhancement in a latent heat storage system. *Solar Energy*, 65(3), 171–180.

[57]   Fang, G. Y., Li, H., and Xu, X. B. (2003). Study on thermal properties of a new phase change cool storage material. *International Journal on Architectural Science*, 4(4), 147–149.

[58]   Tyagi, V. V., Panwar, N. L., Rahim, N. A., and Kothari, R. (2012). Review on solar air heating system with and without thermal energy storage system. *Renewable and Sustainable Energy Reviews*, 16(4), 2289–2303.

[59]   Rahimi, M., Ranjbar, A. A., Ganji, D. D., Sedighi, K., and Hosseini, M. J. (2014). Experimental Investigation of Phase Change inside a Finned-Tube Heat Exchanger. *Journal of Engineering*, 2014(1), 641954.

[60]   Xia, L., and Zhang, P. (2011). Thermal property measurement and heat transfer analysis of acetamide and acetamide/expanded graphite composite phase change material for solar heat storage. *Solar energy materials and solar cells*, 95(8), 2246–2254.

[61]   Nallusamy, N., Sampath, S., and Velraj, R. J. R. E. (2007). Experimental investigation on a combined sensible and latent heat storage system integrated with constant/varying (solar) heat sources. *Renewable energy*, 32(7), 1206–1227.

[62]   Rashid, F. L., Al-Obaidi, M. A., Dulaimi, A., Mahmood, D. M., and Sopian, K. (2023). A review of recent improvements, developments, and effects of using phase-change materials in buildings to store thermal energy. *Designs*, 7(4), 90.

[63]   Putra, N., Prawiro, E., and Amin, M. (2016). Thermal properties of beeswax/CuO nano phase-change material used for thermal energy storage. *International Journal of Technology*, 7(2), 244–253.

[64]   Jesumathy, S. P., Udayakumar, M., and Suresh, S. (2012). Heat transfer characteristics in latent heat storage system using paraffin wax. *Journal of Mechanical Science and Technology*, 26, 959–965.

[65]   Thirugnanam, C., and Marimuthu, P. (2013). Experimental analysis of latent heat thermal energy storage using paraffin wax as phase change material. *International Journal of Engineering and Innovative Technology (IJEIT)*, 3(2), 372–376.

[66]   Senthil Manikandan, K., Kumarasan, G., Velraj, R., and Iniyan, S. (2012). Parametric study of solar trough collector system. *Asian Journal of Applied Sciences*, 5(6), 384–393.

[67]   Helwa, N., Bahgat, A. B. G., El Shafee, A. M. R., and El Shenawy, E. T. (2000). Maximum collectable solar energy by different solar tracking systems. *Energy Resources*, 22(1), 23–24.

[68]   Basil Okafor, E. (2013). Performance evaluation of a parabolic solar cooker. *International Journal of Engineering and Technology*, 3(10), 923–927.

[69]   Hadi Ali, M. (2015). Studying and evaluating the performance of solar box cookers (Untracked). *International Journal of Computational Engineering Research*, 5, 2250–3005.

[70]   Balakrishnan, M., Claude, A., and Arun Kumar, D. R. (2012). Engineering, design and fabrication of a solar cooker with parabolic concentrator for heating, drying and cooking purposes. *Archives of Applied Science Research*, 4(4), 1636–1649.

[71] Ronge, H., Niture, V., and Ghodake, M. D. (2016). A review paper on utilization of solar energy for cooking. *Imperial International Journal of Eco-Friendly Technologies (IIJET)*, 1, 121–124.

[72] Noman, M., Wasim, A., Ali, M., *et al.* (2019). An investigation of a solar cooker with parabolic trough concentrator. *Case Studies in Thermal Engineering*, 14, 100436.

[73] Asmelash, H., Bayray, M., Kimambo, C. Z. M., Gebray, P., and Sebbit, A. M. (2014). Performance test of parabolic trough solar cooker for indoor cooking. *Momona Ethiopian Journal of Science*, 6(2), 39–54.

[74] Shukla, S. K., and Khandal, R. K. (2016). Design investigations on solar cooking devices for rural India. *Distributed Generation and Alternative Energy Journal*, 31(1), 29–65.

[75] Chaudhary, A., Kumar, A., and Yadav, A. (2013). Experimental investigation of a solar cooker based on parabolic dish collector with phase change thermal storage unit in Indian climatic conditions. *Journal of Renewable and Sustainable Energy*, 5(2), 023107.

[76] Wollele, M. B., and Hassen, A. A. (2019). Design and experimental investigation of solar cooker with thermal energy storage. *AIMS Energy*, 7(6), 957–970.

[77] Kumar De, D., Nathaniel, M., Nath De, N., and Ajaeroh Ikechukwu, M. (2014). Cooking rice with minimum energy. *Journal of Renewable and Sustainable Energy*, 6(1), 013138.

[78] Yahuza, I., Rufai, Y., and Tanimu, L. (2016). Design, construction and testing of parabolic solar oven. *Journal of Applied Mechanical Engineering*, 5(4), 1000212.

[79] Panchal, H., Patel, J., and Chaudhary, S. (2019). A comprehensive review of solar cooker with sensible and latent heat storage materials. *International Journal of Ambient Energy*, 40(3), 329–334.

[80] Saini, G., Singh, H., Saini, K., and Yadav, A. (2016). Experimental investigation of the solar cooker during sunshine and off-sunshine hours using the thermal energy storage unit based on a parabolic trough collector. *International Journal of Ambient Energy*, 37(6), 597–608.

[81] Mawire, A., McPherson, M., and Van den Heetkamp, R. R. J. (2008). Simulated energy and exergy analyses of the charging of an oil–pebble bed thermal energy storage system for a solar cooker. *Solar Energy Materials and Solar Cells*, 92(12), 1668–1676.

[82] Farhana, K., Kadirgama, K., Rahman, M. M., *et al.* (2019). Improvement in the performance of solar collectors with nanofluids—A state-of-the-art review. *Nano-Structures & Nano-Objects*, 18, 100276.

[83] Saxena, G., and Gaur, M. K. (2018). Exergy analysis of evacuated tube solar collectors: a review. *International Journal of Exergy*, 25(1), 54–74.

[84] Hussein, H. M. S., El-Ghetany, H. H., and Nada, S. A. (2008). Experimental investigation of novel indirect solar cooker within door PCM thermal storage and cooking unit. *Energy Conversion and Management*, 49(8), 2237–2246.

[85]   Norouzi, A. M., Siavashi, M., Ahmadi, R., and Tahmasbi, M. (2021). Experimental study of a parabolic trough solar collector with rotating absorber tube. *Renewable Energy*, 168, 734–749.

[86]   Rizal, T. A., Amin, M., Widodo, S. B., *et al.* (2022). Integration of phase change material in the design of solar concentrator-based water heating system. *Entropy*, 24(1), 57.

[87]   Norouzi, A. M., Siavashi, M., and Oskouei, M. K. (2020). Efficiency enhancement of the parabolic trough solar collector using the rotating absorber tube and nanoparticles. *Renewable Energy*, 145, 569–584.

[88]   Elarem, R., Alqahtani, T., Mellouli, S., *et al.* (2021). Numerical study of an evacuated tube solar collector incorporating a nano-PCM as a latent heat storage system. *Case Studies in Thermal Engineering*, 24, 100859.

[89]   Singh, P., and Gaur, M. K. (2021). Heat transfer analysis of hybrid active greenhouse solar dryer attached with evacuated tube solar collector. *Solar Energy*, 224, 1178–1192.

[90]   George, N. J., Obianwu, V. I., Akpan, A. E., and Obot, I. B. (2010). Lubricating and cooling capacities of different SAE 20W–50 engine oil samples using specific heat capacity and cooling rate. *Archives of Physics Research*, 1(2), 103–111.

[91]   Sabatelli, V., Marano, D., Braccio, G., and Sharma, V. K. (2002). Efficiency test of solar collectors: uncertainty in the estimation of regression parameters and sensitivity analysis. *Energy Conversion and Management*, 43(17), 2287–2295.

[92]   Michaelides, I. M., Kalogirou, S. A., Chrysis, I., *et al.* (1999). Comparison of the performance and cost effectiveness of solar water heaters at different collector tracking modes, in Cyprus and Greece. *Energy Conversion and Management*, 40(12), 1287–1303.

[93]   Hematian, A., and Bakhtiari, A. A. (2015). Efficiency analysis of an air solar flat plate collector in different convection modes. *International Journal of Green Energy*, 12(9), 881–887.

Chapter 18

ESG and SDG in healthcare

Palaniswamy Mohankumar[1]

The spread and development of infectious illnesses are frequently linked to envir-onmental degradation caused by humans, which modifies biodiversity and, in turn, host–pathogen dynamics. Utilizing natural resources wisely, disposing of trash properly, and safeguarding the environment are crucial steps toward promoting the health and welfare of people, pets, plants, and the ecosystem. These actions can also stop the spread of pathogens. A recent example is the COVID-19. Global supply chain interruptions brought on by the COVID-19 epidemic, the Russo-Ukrainian war, and the Israel–Palestine conflict have created major economic dis-turbances in recent times, making stock markets all across the world very sensitive and volatile. Apart from financial success, investors and fund companies are increasingly taking environmental, social, and governance (ESG) performance into account when developing a sustainable finance strategy. This indicates that stake-holders expect the businesses they put their money into to be sustainable, socially conscious, and successful. A Royal Bank of Canada market survey indicates that a growing number of investors think that capitalizing on businesses that perform well in terms of ESG can lower investment risks and boost the return on investment. The unlawful sewage release by one of Xiaomi's suppliers in 2018 violated regulations pertaining to environmental protection, which had an impact on Xiaomi's intention to list on the Hong Kong stock exchange [1]. This shows how a company's pros-perity does not ensure its long-term viability as a corporation, as environmental and social problems can have an impact. Consequently, more and more corporate choices are taking ESG factors into account. Nowadays, listed firms are keen to enhance their own ESG capacities by adhering to established ESG assessment frameworks to consistently cultivate a sustainable business image within the industry.

Concerns about how corporate governance and social responsibility affect nonfinancial aspects of the business, like employee perks and board operations, are growing among investors. The world's economic operations and quality of life have already been severely impacted by global climate change, which is made worse by the rapid industrialization that has increased energy demand. Global efforts are

[1]School of Health Sciences, Hindustan Institute of Technology and Science, India

underway to find new technologies and strategies to either reverse or modify the course of global warming. Previous research has demonstrated that the growth of industry and the number of homes and cars have all contributed to an alarming rise in GHGs and global warming. For this reason, countries all over the world promote ESG, with one of the main goals being the reduction of carbon emissions [2].

According to a Bank of America analysis, US and Western European companies adhering to sustainable development principles had a 20% improvement in their capitalization-to-earnings ratio between January 2007 and August 2019 alone. Simultaneously, conventionally employed elements—like the presence of physical assets, financial performance, and market share—are becoming less and less significant in determining a company's worth. Even still, intangible assets—like intellectual property and brand value—are becoming more and more significant; their proportion in determining the worth of S&P 500 index businesses rose from 30% in 1998 to 68% in 2018 [3].

18.1   Environmental social governance

The Paris Agreement of 2015, the Sustainable Development Goals set forth in the UN 2030 agenda, and the European Green Deal of 2019 have all brought these issues greater prominence and attention. Any financial activity that takes into account the affects or elements of the environmental (E), social (S), and governance (G) domains is considered sustainable finance. The "E" in this paradigm stands for all actions pertaining to environmental problems, including managing waste and pollution, conserving biodiversity, using water, and producing greenhouse gas (GHG) emissions. "S" pertains to working conditions both inside and outside the organization, employee rights, and workplace treatment, whereas "G" covers management-related matters such as executive choices and board composition [4]. Promoting investment choices and business assessments that are in line with the objectives of a low-carbon, environmentally friendly, and sustainable economy is, in essence, the aim of sustainable finance.

Research has indicated that the incorporation of ESG factors into a company's valuation model enhances its nonfinancial metrics, including stakeholder values, market acceptance, customer happiness, and debt costs. Numerous studies assert that a firm's equity premium and value significantly increase upon the integration of ESG elements into valuation and investing decisions. The unanswered question is whether attempts to increase ESG disclosure would boost the potential of the company to create wealth for its shareholders and turn a profit, or if ESG disclosure is only meant to enhance its reputation [5].

18.1.1   ESG in the past

The acronym ESG refers to the combined effects of a business's operations on the environment and society. To include the discovery, appraisal, and management of sustainability-related threats and possibilities with regard to all stakeholders and the environment, strong and transparent governance is required. ESG is not a brand-new idea; it first appeared in the 1960s [6]. The foundation of its structure is the idea of corporate social responsibility (CSR), which is an optional business endeavor that aims to manage the company in a morally and environmentally responsible way.

The United Nations launched the idea of ESG in 2004 with the goal of progressively modernizing the corporate environment with improved sustainability. In addition, ESG development is supported by investments that are ethical, sustainable, and environmentally friendly. Consequently, investors use ESG as a benchmark to evaluate the conduct of companies and their potential for future financial success. As demonstrated by Volkswagen's 2010 emissions crisis, which affected the company's stock price and resulted in billions of dollars in capital losses, investors may steer clear of businesses whose activities may be a sign of risk factors by adhering to ESG norms [1]. To accomplish the sustainability aims and objectives in the industries, a customized ESG framework is crucial in addition to generalized ESG frameworks. Ecologically speaking, supply chain enterprises should stop utilizing, abusing, and destroying natural assets.

With the introduction of the first CSR Framework by Malaysian corporations in 2006, the country has become a significant sample for ESG research. In Malaysia, the first report on sustainable development was released in 1987 [5]. Teoh and Thong discover that Malaysian companies do not seem to be very involved in CSR, and they seem to be more concerned with their workers and the financial success of their products than with how these things affect society and the environment. The CSR reporting framework, which became necessary for all Malaysian companies in 2006, and the Sustainability Framework, which demanded ESG disclosures from the companies in 2015, have further reinforced this endeavor. The government's endeavor to include CSR in its Tenth Malaysia Plan serves as evidence of the government's commitment to guaranteeing the execution and prosperity of CSR. Legally speaking, the Companies Act of 2016 also made CSR disclosures mandatory, which helped Malaysian companies get exceptionally high CSR scores. In addition, there appears to have been a rise in ESG disclosures since the FTSE4Good Bursa Malaysia Index was introduced in 2014 and the SDGs were adopted and implemented in January 2016. This highlights the goals of these initiatives, which include lowering information asymmetry, enhancing transparency, and offering nonfinancial voluntary disclosures that aid in investor decision-making. With nearly 97% of the top 100 Malaysian companies disclosing their corporate sustainability performance, compared to the global average of 72%, the country has recently been named as the leading pioneer in CSR reporting worldwide. Malaysia is among the few countries of its kind globally that implemented the Malaysian Code for Institutional Investors in 2014. This code establishes and unifies regulations to include sustainability concerns in investment analysis portfolios. The present ESG framework is a wide-ranging, dynamic concept that incorporates actions linked to corporate governance and sustainability with CSR concepts. The policies that make up current ESG practices, also referred to as ESG 1.0, were put into place without first assessing their effects using quantitative techniques.

18.1.2   Role of SME in ESG

Listed firms must establish a contractual agreement with logistics service providers (LSPs) to facilitate transportation management, warehousing, and freight operations for general company operations. Still, a number of the LSPs are small and medium-sized businesses (SMEs) that lack the resources—both financial and human—to completely adhere to the most recent ESG measuring standards. For example, the majority of SMEs do not have formal data organization for carbon and greenhouse gas emissions. The Trade and Industry Department of Hong Kong reports that more than 340,000 SMEs are involved in a variety of business sectors, making up 98% of all company establishments and employing 45% of all private sector workers. Three-quarters of Hong Kong's commercial entities are SMEs in the shipping and transportation industries [1].

18.1.3   Role of healthcare in ESG

Promoting universal health for illnesses that are primarily preventable and/or curable in their early stages is the main objective of health professionals. Whether it takes the form of long-term maintenance, therapeutic treatments, or preventive care, providing general health care results in pollution and a large carbon footprint. It is our ethical and moral duty as healthcare professionals to control how our actions affect the environment and to make sure that we do it in a sustainable way. Three main factors contribute to CO_2 emissions in the healthcare industry: (i) patient and healthcare worker travel to and from care facilities; (ii) the production, distribution, and purchase of supplies and ancillary goods along the supply chain; and (iii) waste generation and management, including the handling of single-use plastics (SUPs), which pose a serious environmental risk and need immediate attention [7]. Given the massive amounts of SUP personal protective equipment (PPE) used during the ongoing COVID-19 epidemic, the SUP load is more relevant than ever. The present surge in the usage of SUPs emphasizes how challenging it is to put sustainable health care practices into reality since patient safety and receiving the best care frequently take precedence over environmental effects.

To achieve ESG in the long run, four specialized high-quality care shall be provided: preventive care, optimized care, integrated care, and ownership of care. Its measures and outcomes are mentioned in Table 18.1. The four categories of high-quality patient-centered care will unavoidably lead to two major accomplishments in environmental sustainability: (i) decreased demand for expert interventions, decreased patient travel between consultations, and longer-lasting restorations all contribute to a decrease in CO_2 emissions overall. (ii) Lesser the requirement for procurement, lowers waste generation overall. For disorders that may be prevented, sustainable healthcare is easily attained with a strong emphasis on high-quality, patient-centered treatment. The whole healthcare team, care managers, commissioners, regulatory agencies, and patients as co-managers and co-creators of their own care work together to provide high-quality healthcare.

Table 18.1 Healthcare in ESG [7]

Type of care	Health		Environmental outcomes
	Measures	Outcomes	
Preventive care—Patient-specific assessment and management	Health education—Promotion of health care, consumption of food, drink, and tobacco Hygiene measures—Proper toothbrushing and usage of fluoride toothpaste, handwash technique, basic hygiene Healthy diet—Reduced intake of sugar, acidic drinks, alcohol, and tobacco	Healthy body, health mind No unnecessary treatment Reduced cost Professional satisfaction Improved patient quality of life	Reduced CO_2 emission—Less journey to clinic or hospital Reduced waste—Less clinical waste
Optimized care—Combination of knowledge, skills, and experience	Best practice—Evidence-based treatment High-quality treatment—Usage of quality equipment and materials	No unnecessary treatment Reduced cost Refined clinical outcomes Better care	Reduced CO_2 emission—Less journey to clinic or hospital Reduced waste—Less packing and clinical waste
Integrated care—Structured treatment, integrated services, and patient participation	Structured treatment plan—Active therapist and patient participation, joint decision-making in treatment Treatment appointments—Shared family appointments	Healthy body, health mind No unnecessary treatment Reduced cost Professional satisfaction Improved patient quality of life	Reduced CO_2 emission—Less journey to clinic or hospital
Ownership of care—Individual and team ownership in offering care, professional development, and satisfaction achievement	Provide best practice—Practice and maintain a focused professional development Be an example—Make difference and inspire others Clinical governance— Improve the quality of service and conduct regular audits Professional involvement—Be part of a professional society, international research participation	Professional satisfaction Improved patient quality of life	Reduced CO_2 emission—Less journey to clinic or hospital Reduced waste—Less packing and clinical waste

18.1.4   Role of climate change in ESG

Human health is being adversely impacted by climate change in a number of ways, both directly and indirectly. Empirical research is necessary to address how climate change affects public health. Current tactics by government, NGOs, industry, and academics sometimes rely on insufficient information or advice developed in other circumstances. Research funding in the social sciences must be increased, and multidisciplinary cooperation is crucial. Indirect health effects such as malnutrition due to an increase in food costs are a consequence of climate change. Direct health effects of climate change include an increase in the occurrence of heat strokes due to heat waves. Both are moderated by the environmental and social determinants of health (SEDH).

In light of the fact that the world's urban population is anticipated to experience a record-breaking rise from [8] to 6.3 billion in the year 2050. It is predicted that the health risks linked to climate change will be catastrophic for the urban poor. It is anticipated that Asia and Sub-Saharan Africa will contribute up to 70% of the total number of urban people globally in the year 2050. In underdeveloped nations, internal migration is frequently seen as a reaction to climate change. Due to the ongoing deterioration of soil and significant shifts in rainfall and temperature trends, agricultural operations in rural regions are consistently at risk. As a result, there is increasing pressure to explore alternate economic opportunities, leading to rural–urban migration.

One of the most significant and clear consequences of climate change is the increase in average surface temperatures. The Paris Agreement aimed to decrease global warming to below 2 °C, preferably below 1.5 °C. However, with the present greenhouse gas emission data, current estimates suggest that the average temperature worldwide could increase by 3–5 °C by the end of the century. Cities are especially susceptible to the consequences of excessive heat during summertime due to the urban heat island (UHI) phenomenon. According to popular belief, the UHI effect incidences are caused by a greater number of concrete buildings, restricted vegetation and greenery, and higher population counts. In cities with temperate and tropical climates, where there is little vegetation or greenery, a large portion of the daytime heat is caught by concrete buildings. The UHI effect is the explanation for why temperatures tend to be higher inside cities than they are in the outskirts [8].

In summer 2022, Europe saw a heatwave never seen before, reaching as far north as Scotland and reaching a record-breaking maximum temperature of 35 °C. The average annual temperature of Scotland from 1961 to 1990 was 7 °C, indicating a cold environment. A substantial body of research indicates that elevated temperatures are linked to detrimental health outcomes, such as elevated rates of hospitalization and mortality. For instance, Wales and England had almost 3000 additional fatalities during the UK's record-breaking warm summer of 2022 [9]. Global warming, population aging, and urbanization together may significantly raise the likelihood of harmful health impacts from heat if appropriate action and adaptation are not taken.

Guidelines on Heat-Health Action Plans (HHAPs) were released by the WHO Regional Office for Europe with the intention of assisting nations and areas in developing, enhancing, and putting into practice heat-health measures to avoid negative heat-related effects. By 2019, there were about 20 European nations that had put in place local, regional, or national heat-health warning systems and/or HHAPs. Studies have revealed a rise in death rates in cold regions including Scotland, Sweden, Estonia, Finland, and Russia during periods of high temperature.

18.1.5   Monitoring ESG

The increasing depletion of environmental resources has been damaging to society throughout time. The annual amount of material extracted worldwide has expanded substantially, rising from 30 billion tons in 1970 to 70 billion tons in 2010. Despite a weakening global market and declining population as of 2000, material extraction has grown. Industry estimates indicate that in 2030 and 2050, respectively, 125 billion and 180 billion tons of resources will be required to power the global market. This is because the middle class is growing, especially in emerging countries, and the global population is growing. Overall, 300–335 billion tons are predicted to be recovered as raw materials by 2030 [10]. In recent times, professionals and academics have come to view governance as essential to successful natural resource planning. Indeed, in the last two decades, a great number of management entities and administrations have adopted more participatory, collaborative, and multifaceted governance frameworks in reaction to the obvious shortcomings of the conventional top-down, bureaucratic, and usually government-led administration approaches.

The stability of biological sources must be preserved by increasing the functioning of governance institutions in countries where environmental degradation occurs despite enormous foreign investment in inventive approaches to attain ecological goals and ground-level activities. A governance system's capacity to accomplish its aims may be limited by a few, but in most cases, resolving several interconnected governance issues is a prerequisite for the system's success. Conflicts among interested parties, a lack of funding to carry out particular strategies, the total lack of government support for particular green projects, or a failure to incorporate or utilize indigenous knowledge to gain a deeper understanding of ecological systems are all examples of governance issues. Degradation of the environment has continued globally, despite significant efforts in the previous few years to identify and address governance issues.

"Storages of elements that exist in the natural environment that are rare and commercially useable in production or consumption, either in their raw forms or after minimum processing," is how the World Trade Organization defines natural resources in its annual report [10]. Natural resources may be broadly classified into two groups: renewable resources and nonrenewable resources. A few examples of renewable resources include land, trees, and water; nonrenewable resources include minerals, fossil fuels, and diamonds. Researchers have devised many classifications, including fuel and nonfuel, stealable and nonstealable, and point and diffuse, to measure the effect of nonrenewable resources on violence.

18.1.6   ESG assessment

The idea of an ESG rating has arisen within the context of sustainable finance and its goals to categorize and evaluate businesses' positions and attitudes with respect to sustainability objectives. Investors, the firms themselves, and the general public all value these objectives. A multitude of rating agencies have arisen in response to the growing demand for identifying and assessing the sustainability of firms, particularly their ESG performance. This demand stems from investment opportunities as well as the need to comply with European Commission (EC) laws.

The unifying strategy employed by these organizations requires evaluating the three main pillars: E (Environmental), S (Social), and G (Governance). Subsequently, they contain significant good or negative actions or efforts conducted by corporations inside each of these pillars. However, as Lopez *et al.* noted, there is no one definition of sustainability and no widely accepted technique for determining it [4]. The new Corporate Sustainability Reporting Directive (CSRD) is one of the initiatives to encourage finance for sustainable projects that have been implemented recently. This directive seeks to extend sustainability reporting requirements to a wider variety of businesses, irrespective of their size, and expands on the principles previously outlined by the Nonfinancial Reporting Directive (NFRD). To guarantee that businesses give accurate, transparent, and comparable information, efforts are being made to create a new set of regulations and requirements with the new CSRD. Additionally, the CSRD covers all major corporations, listed or not, regardless of employee count, whereas the NFRD's purview was restricted to big listed companies with more than 500 workers. Due to this extension, the number of enterprises covered by sustainability reporting and the sharing of sustainability data has increased from 11,000 under the previous mandate to approximately 50,000.

There is still a lack of standardization and statutory sustainability reporting requirements, particularly for unlisted small and medium-sized organizations and microenterprises, even if the new CSRD broadens the scope of covered companies. It is noteworthy that SMEs, the bulk of which are unlisted, are responsible for 64% of the adverse environmental effects, including energy usage, greenhouse gas emissions, and waste disposal. For this reason, even with the new CSRD, the exclusion of unlisted SMEs and microenterprises is still an issue. SME participation in required sustainability reporting and the ESG rating process makes sense given that SMEs make up 99% of all firms in the European Union (EU) and microenterprises account for 92.7% of all businesses. SMEs are also major contributors to the European economy. For SMEs to stay up to date with advancements in sustainable finance, obtain funding assistance for the shift to a sustainable economy, and adjust their business plans appropriately, their involvement is imperative.

When making investments in or collaborating with SMEs, banks and bigger corporations need access to pertinent sustainability data from small businesses to comprehend the risks and possibilities involved. This emphasizes the necessity of making ESG rating for SMEs obligatory as opposed to optional. SMEs are also forced to commit time and resources to data collection, analysis, and reporting due to the requirement of ESG performance disclosure. As a result of their data availability, transparency policies, and implementation of sustainable management frameworks in their operations, larger organizations typically obtain more favorable ESG ratings than smaller ones. This phenomenon is known as the "company size bias," which may be lessened or eliminated [4]. The list of Level 1, Level 2, and Level 3 ESG assessment prioritization are provided in Table 18.2.

Table 18.2   Suggested Level 1, 2, and 3 ESG assessment prioritization in logistics [1]

Level 1	Level 2	Level 3
E	Carbon and greenhouse gas emissions	Emission control, nonrecycled products, awareness workshop
	Climate change	Environment protection facilities, climate change education and related goals
	Packaging	Green materials, reduce the use of materials, pallet and waste management
	Product quality	Cargo disposal procedure, green packaging, product handling instructions, quality control, storage conditions
	Renewable energy	Paperless operation, recycling waste water, renewable electricity
	Reverse logistics	Recycling packaging material, waste handling, return and recall policy
	Sustainable sourcing	Efficient sourcing, CSR monitoring, green thinking, accreditation
S	Diversity, equity, and inclusion	Diversity of employees, equal employment, antiharassment and antidiscrimination policy
	Fair labor practices	Job specifications, structured career path, work insurance, labor law awareness training
	Human rights in supply chains	Complaint mechanism, code of conduct for workers, human rights awareness training and auditing
	Supply chain visibility	Traceability system, inventory audit, transport and warehouse management, information sharing
	Workforce health and safety	Standard operating procedures, escape routes, employee safety, ergonomic workplace, work from home
	Work-life balance	Recreational activities, flexible working hours, allowances, psychological counseling services
G	Anticorruption	Fair recruitment, antibribery, background check, regular audit
	Business ethics and integrity	Law compliance, ethical guidelines, privacy policy
	Data and transparency	Standard data management, use of genuine antivirus, data confidentiality, and recovery
	ESG metrics, analytics, and compensation goals	Sustainability vision, development strategy, and related certifications, Disclosure of ESG metrics
	Risk management	Risk monitoring and remedies, information leakage prevention, crisis response plan
	Supplier and customer relationships	Supplier risk management and satisfaction, regular visits to customers, customer relationship management

To determine which firms are leading and trailing their respective industries in terms of ESG, two significant ESG measuring databases—KLD and ASSET4—are examined [1]. On one side, KLD offers comprehensive CSR evaluations that cover seven important stakeholder characteristics: community, employee relations, diversity, environment, human rights, product sustainability, and corporate governance. However, ASSET4, a different popular database, has continuously offered thorough CSR data for Russell 1000 Index companies. Analysts gather the yearly ESG data and then combine the qualitative and quantitative data into a single study. Measurement-wise, research on ESG often selects an annual score for the environment, a score for society, and a score for governance, and then builds a CSR index by giving each of the three categories equal weight. The two approaches to measuring ESG that were previously discussed assess CSR to produce ESG indexes. However, there are some minor differences between the basis and purpose of CSR and ESG. Compared to CSR, the basis of ESG is more focused, with an emphasis on the collaborative assessment of the three aspects of governance, society, and environment to create the ESG index.

Amid the uncertainty surrounding the concept of sustainability, the International Organization of Securities Commissions (IOSCO) has disclosed that several indicators, frameworks, and techniques are used by rating agencies and data suppliers to analyze and assess the ESG performances of corporations. After that, they give stakeholders ESG data that are appropriate for the individual firms. According to Billio *et al.*'s research, this variation in rating criteria causes various rating agencies to hold divergent opinions on the ESG performance of the same firms [4].

For example, corporations are rated by Morgan Stanley Capital International (MSCI) according to how they handle and are exposed to ESG opportunities and risks. Refinitiv, on the other hand, scores firms based on many factors, including their ESG performance and amount of disclosure, as well as their contentious business practices. The Financial Times Stock Exchange (FTSE) Russell, on the other hand, takes into account variables such as the degree to which a company is exposed to ESG risks, the significance of those risks to its operations, and the caliber of its risk management capabilities. Conversely, Sustainalytics considers the exposure level to ESG risks in relation to the industry mean for similar risks. It also makes a distinction when allocating weights between acceptable risks and peculiar risks that are unanticipated or unimportant in the context of the company's industry. In conclusion, because different rating agencies employ different calculation techniques, the categories created under each ESG pillar, the techniques used to determine the ratings for each individual pillar, and the total ESG scores differ.

The majority of the data sources used by rating agencies and the firms themselves are not publicly available, as underscored by IOSCO. Rating agencies also utilize distinct data products. One example of this is that according to research, just 49% of private corporations submitted their Scope 1 and 2 emissions in the 2021 Carbon Disclosure Project (CDP) Questionnaire [4]. Additionally, the information utilized to compile the ratings comes from a variety of sources, such as publications from nongovernmental groups, yearly reports, and corporate surveys. Therefore, to obtain the pertinent ESG data, rating agencies mostly rely on disclosures provided by the corporations themselves.

Apart from the difficulties posed by varying methodology and information sources, there exist concerns pertaining to the transparency of sustainability statistics. Even though transparency is a critical component of ESG ratings, a study found that many businesses raised their ratings by boosting their disclosure without significantly raising their ESG performance. Moreover, US operating corporations have significantly lower disclosure rates for ESG performance and data. Only 57% of the 1000 biggest firms in the Russell 1000 Index revealed their Scope 1 and 2 emissions, according to a recent 2022 study [4]. On the other hand, all of the enterprises that were subject to the European Union's more stringent guidelines and regulations disclosed their output of greenhouse gases.

18.2 Copernicus satellite observations

An innovative strategy is presented in recent research, with an emphasis on the environmental (E) pillar, to overcome two significant shortcomings of the existing ESG grading system. First, a fresh approach to directly gathering environmental data from Copernicus satellite observations is put out to evaluate the E pillar of major corporations' ESG ratings. The new data-collecting technique only uses raw, observable data from one source—a satellite database. Refer to Figure 18.1. Utilizing satellite imaging is recommended by researchers Yang and Broby for waste management, natural resource management, water pollution, and air

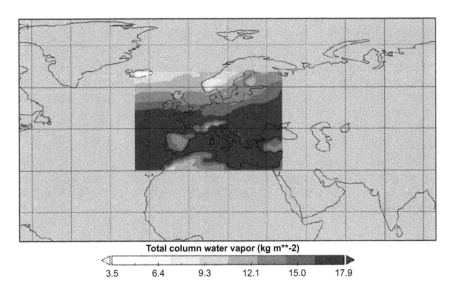

Total column water vapor (kg m**-2)

3.5   6.4   9.3   12.1   15.0   17.9

Figure 18.1   Satellite image of the water vapor in Europe [4]

Table 18.3   Variables measured from the Copernicus satellite

Measured variables	Services
Carbon dioxide (CO_2)	Copernicus Atmosphere Monitoring
Carbon monoxide (CO)	Service (CAMS)
Formaldehyde (CH_2O)	
Leaf area index (LAI)	
Methane (CH_4)	
Nitrogen dioxide (NO_2)	
Nitrous oxide (N_2O)	
Sulfur dioxide (SO_2)	
Volatile organic compounds (VOC)	
Water vapor	
Ozone (O_3)	Copernicus Climate Change Service (C3S)
Volumetric soil water	

pollutant emissions tracking. Second, the neural network (NN) is designed to determine the correlation between the environmental data and the final ESG ratings by feeding it with both the firms' declared ESG rating data and environmental data from satellite imaging. This methodology facilitates the assessment and quantification of SMEs' environmental performance. The European Commission (EC) oversees the Copernicus program, which aims to provide end users with assessments, calculations, and projections from satellites and on-location for a variety of uses, such as traffic control, farming preparation, environmental sustainability, and urban infrastructure administration. Copernicus consists of six services namely Atmosphere Monitoring Service, Climate Change service, Land Monitoring Service, Marine Environment Monitoring Service, Emergency Management Services, and Security Service, each providing databases for various subjects and goals [4]. The names of measured variables under the Copernicus satellite are mentioned in Table 18.3.

Utilizing a neural network (NN), the suggested technique presents a novel solution. This neural network is calibrated to compute E ratings for branches of major corporations, taking into account their ecological factors. Copernicus satellite observations are used to obtain the required input data for the neural network. These observations cover 14 distinct categories, such as air pollution, ground water use, devegetation, and greenhouse gas emissions. The findings indicate that the neural network model successfully corresponds to the fluctuations in the sustainability data of SMEs. It reliably captures the annual changes in emissions, pollutants, and other sustainability measures, and assigns E scores correspondingly. One interesting finding is that in contrast to the consistent ESG scores that rating agencies usually award, the NN model gives different branches of the same organization varied E values. This strategy of assigning diverse ESG scores to branches is considered fairer and more instructive. It enables the examination of a company's operation at multiple branches and locales in terms of sustainability.

18.2.1 Barriers to ESG

ESG advocates encourage businesses and investors to support environmental conservation by implementing strategies like energy conservation, waste minimization, and recycling. Businesses must fulfill their social obligations, which include defending human rights and labor rights and enhancing workplace safety. Furthermore, companies should include excellent corporate governance principles in their operations and actions, such as the board of directors' autonomy in functioning, safeguarding of investors' rights and concerns, and executive remuneration. A greater number of businesses have released reports on sustainable development and actively shared pertinent details regarding their efforts to promote ESG in recent years, making it possible for investors to assess and track the success of the enterprise and its earnings.

There are initiatives in motion to harmonize and synchronize the comprehension of sustainability. The European Union (EU), for example, is making an effort to classify and identify economic operations that fit with its goals to meet the targets mentioned in the European Green Deal and the 2030 agenda. The EU created the EU Taxonomy, a categorization system that defines and enumerates activities that may be considered sustainable economic activity, to accomplish this. This makes it possible for interested parties to comprehend the requirements that undertakings need to fulfill to comply with the taxonomy. Additionally, stakeholders have the ability to evaluate the ways in which their actions support the six environmental objectives of the EU, the most important of which being adaptation and mitigation of climate change. Finally, stakeholders have the ability to assess if the actions they do follow the "do no harm" concept, which states that even if they support certain goals, they should not interfere with other goals at the same time [4].

In addition to the problems with the present rating procedure outlined above, the IOSCO suggests that the way input data is currently obtained might create conflicts of interest and result in nonobjective ESG rating conclusions. Furthermore, it is frequently essential to wait for sustainability or yearly reports to be released before gathering the required ESG data from businesses. Investors' requirement for more urgent analysis may not coincide with this waiting and data-collecting period. These differences in taxonomy, data sources, and techniques lead to a loss of precision in ESG ratings and difficulty comparing the ESG performances of various organizations. Consequently, this reduces the validity and precision of the rating outcomes.

An array of interdependent governance challenges often affects a natural resource management governance system's capacity to deliver social, environmental, and other desired outcomes rather than a single challenge. Studies suggest that the main obstacle preventing organizations from achieving the desired results in natural resource governance is a lack of human, financial, technological, and knowledge-based resources. Stakeholders' lack of participation in the decision-making process is one of the issues with governance. According to case studies conducted in Canada, Australia, Madagascar, and Chile, a lack of connection among stakeholders significantly impeded the effectiveness and success of planning and implementation activities [10]. For example, Oakley found that the lack of incentive programs to encourage regulatory parties to cooperate on sustainability concerns and the overall inadequacy of collaborative partnerships limited the overall capability to accomplish this.

Insufficient leadership on major resource challenges, a lack of financial and human resources to support implementation activities, a lack of mechanisms for resolving conflicts, and a lack of decision-making authority are among the characteristics of governing structures reported in developing countries. Nevertheless, studies on the governance frameworks for natural resource management in developed nations have frequently revealed issues with policy coherence and stakeholder institutions' alignment of objectives. The list of potential ESG barriers mentioned in the literature is presented in Table 18.4.

Table 18.4   Possible barriers to ESG adoption in enterprises [2]

S. No.	Barriers
1	Business model not fully aligned with sustainable development
2	Close supplier relationships
3	Deficiency in sustainable resource
4	Difficulty in converting awareness into action
5	ESG assessment barriers
6	Fear of failure
7	Financial constraints
8	High-cost barriers
9	High-customers' requirements
10	Improper carbon offsetting
11	Inadequate academic research support
12	Insufficient transparency
13	Lack of common definitions
14	Lack of efficient IT management system
15	Lack of government support
16	Lack of long-term strategies and planning
17	Lack of organizational integration
18	Lack of professionals and expertise
19	Lack of standards for measuring ESG performance
20	Lack of top management support
21	Lack of zero emission-related laws and certifications
22	Limited ESG data and lack of long-term comparability
23	Low-carbon technological barriers
24	Low return rate on investments
25	Market competition pressure
26	Social concerns
27	Stakeholder pressure
28	Strategic information sharing and communication
29	Unorganized waste management
30	Unwillingness to implement green innovation technology

18.2.2   Factors to consider

When making decisions, decision-makers may undervalue or overlook certain risks and dangers associated with an enterprise's sustainable development [11]. Here are a few possible risks and dangers:

Financial risks: Investing a substantial amount of money may be necessary for an organization to operate sustainably, particularly in the beginning. This may put more strain on the company's financial resources and raise the possibility of declaring bankruptcy.

Knowledge transfer risks: The foundation of sustainable growth is typically the transfer of expertise and wisdom that cannot be purchased on the open market. However, this knowledge could diminish and have a big impact on the development's sustainability and success if important experts depart the organization.

Reputational risks: Sustainability is linked to ethical business practices, social policies, and environmental stewardship. Violations of these guidelines, however, could harm the company's reputation and make it more difficult to sign new business deals and draw in external funding.

Legislation-related risks: Enforcing laws and regulations is essential to stability. However, unnecessarily strict regulations or changes in the law could result in higher costs and decreased profitability.

Risks of losing a competitive edge: Compared to conventional development, sustainable development may be priced higher. This may result in higher costs for goods and services, which would lower the company's ability to compete and boost sales.

Dangers associated with external environment interaction: Sustainable development necessitates engagement with suppliers, customers, governments, and the general public. Sales may decline and confidence may be lost if any of these groups' expectations are not met.

18.2.3   Implementing ESG

Researcher Martin claims that three distinct types of violence—civil war, ethnic struggle, and insurgency—can be caused by a lack of resources. In reference to the first, the current discourse indicates that renewable resources hardly never lead to international conflict. According to researchers Llamosas and Sovacool, the only renewable resource that has been thoroughly examined in relation to international conflict is water. To develop and run military sites as well as increase agricultural output, water is essential. According to these scholars, the natural competition between large countries makes resource-rich areas more vulnerable to conflict. Bayramov thinks that competition for limited resources, such as oil, has intensified with the advent of growing nations such as China, India, and Russia. Bond and Basu describe a new threat to prospective global safety in the international community as the rapidly depleting natural reserves, particularly oil, and the unequal distribution of these resources between the global north and global south.

The rising demand for organizations with strong ESG scores has led to considerable developments in the financial industry. Investing in socially conscious companies has grown significantly over the last few decades, both as a trend in the mutual fund industry and as a central theme in global financial studies. SRI, or socially responsible investing, is generally understood to be an investment method that identifies and rewards firms with high levels of CSR, with this indication being evaluated using ESG indicators.

Empirical evidence indicates that investors incentivize ESG-compliant companies, whereas inadequately disclosed ESG factors may signal unique risks. Inadequate ESG disclosure from businesses may lead to poorly chosen investments in high-risk industries that have the potential to harm the ecosystem or mistreat the workers. When a company incorporates ESG considerations into its investment decisions, investors will have more information to base their choices on, than just financial performance. Businesses that comply with ESG standards are deemed to have more sustainable development and environmental concerns, have stronger governance, and experience less volatility in their revenues. By 2030, the UN advises all businesses to publish their ESG practices. For businesses to actively participate in ESG disclosures that benefit both their shareholders and their business value chain, governments must provide a variety of tax incentives to encourage the implementation of ESG. Directors should thoroughly disclose the company's policies and ESG implementation in its annual report, according to the Malaysian Code of Corporate Governance, which is part of the program [5].

A recent study suggests that encouraging responsible investing leads to increased participation in ESG initiatives and enhanced business success. The government should make serious attempts to increase investor confidence in companies that participate in ESG initiatives by implementing the framework and legislative changes. Regulators must encourage businesses with less of a competitive advantage by providing them with financial support, tax breaks, or training.

18.3   Sustainable Development Goals

The United Nations (UN) endorsed "Transforming our World: The 2030 Agenda for Sustainable Development" at a special conference on September 25, 2015. This agenda aims to finish the global developmental goals that were started by the Millennium Development Goals (MDGs). The Agenda for Sustainable Development, including 17 Sustainable Development Goals (SDGs), 169 associated goals, and 230 indicators, aims to build upon the progress initiated by the MDGs while placing a significantly stronger emphasis on environmental, social, and economic sustainability. While the MDGs had three distinct health-related objectives, SDG3 is the sole direct health-related aim. Nevertheless, the other SDGs also have an indirect influence on health [12]. The list of SDG goals, indicators, and their corresponding summary are provided in Tables 18.5 and 18.6.

18.3.1   Role of Academia in SDG

Recently an analysis was done by mapping scholarly articles from 2015 to 2021 [15]. The study aimed to achieve three objectives: (1) to evaluate the amount and

Table 18.5 Sustainable Developmental Goals (SDG) and its summary

	Goals	Summary
SDG 1:	No poverty	End poverty in all its forms everywhere
SDG 2:	Zero hunger	End hunger, achieve food security and improved nutrition, and promote sustainable agriculture
SDG 3:	Good health and well-being	Ensure healthy lives and promote well-being for all at all ages
SDG 4:	Quality education	Ensure inclusive and equitable quality education and promote lifelong learning opportunities for all
SDG 5:	Gender equality	Achieve gender equality and empower all women and girls
SDG 6:	Clean water and sanitation	Ensure availability and sustainable management of water and sanitation for all
SDG 7:	Affordable and clean energy	Ensure access to affordable, reliable, sustainable, and modern energy for all
SDG 8:	Decent work and economic growth	Promote sustained, inclusive, and sustainable economic growth, full and productive employment, and decent work for all
SDG 9:	Industry, innovation, infrastructure	Build resilient infrastructure, promote inclusive and sustainable industrialization, and foster innovation
SDG 10:	Reduced inequalities	Reduce inequality within and among countries
SDG 11:	Sustainable cities and communities	Make cities and human settlements inclusive, safe, resilient, and sustainable
SDG 12:	Responsible consumption, production	Ensure sustainable consumption and production patterns
SDG 13:	Climate action	Take urgent action to combat climate change and its impacts
SDG 14:	Life below water	Conserve and sustainably use the oceans, seas, and marine resources for sustainable development
SDG 15:	Life on land	Protect, restore, and promote sustainable use of terrestrial ecosystems, sustainably manage forests, combat desertification, halt and reverse land degradation, and halt biodiversity loss
SDG 16:	Peace, justice, and strong institutions	Promote peaceful and inclusive societies for sustainable development, provide access to justice for all, and build effective, accountable, and inclusive institutions at all levels
SDG 17:	Partnerships for the goals	Strengthen the means of implementation and revitalize the global partnership for sustainable development

caliber of research produced during this time frame; (2) to examine the development of popular subjects by analyzing keywords; and (3) to evaluate the geographical distribution of the top worldwide research organizations and countries regarding the third Sustainable Development Goal (SDG). To accomplish these goals, open-source bibliometric databases, and mapping and visualization programs were employed. The bibliographic data was obtained from "The Lens" database, an open-source archive of scholarly literature including several research disciplines. The database collaborates with PubMed, CrossRef, Microsoft Academic Graph,

Table 18.6   Health-related goals and its Indicators of SDG-3

SDG—health-related goals [13]	Indicators [14]
3.1 Maternal mortality	3.1.1 Maternal mortality ratio
	3.1.2 Skilled birth attendance
3.2 Neonatal and child mortality	3.2.1 Under-5 mortality
	3.2.2 Neonatal mortality
3.3 Communicable diseases	3.3.1 HIV incidence
	3.3.2 Tuberculosis incidence
	3.3.3 Malaria incidence
	3.3.4 Hepatitis B incidence
	3.3.5 Neglected tropical diseases prevalence
3.4 NCDs and mental health/well-being	3.4.1 Noncommunicable disease mortality
	3.4.2 Suicide mortality
3.5 Substance abuse	3.5.1 Substance abuse coverage
	3.5.2 Alcohol use
3.6 Road traffic accidents	3.6.1 Road injury mortality
3.7 Reproductive health	3.7.1 Family planning need met and modern contraceptive methods
	3.7.2 Adolescent birth rate
3.8 Universal health coverage and access to medicines	3.8.1 Universal health coverage service index
	3.8.2 Financial risk protection
3.9 Pollution and Contamination	3.9.1 Air pollution mortality
	3.9.2 WaSH mortality
	3.9.3 Poisoning mortality
3.a Tobacco control	3.a.1 Smoking prevalence
3.b Research and development of vaccines and medicines	3.b.1 Vaccine coverage
	3.b.2 Developmental assistance for research and health
	3.b.3 Essential medicines
3.c Health financing and Health workforce	3.c.1 Health worker density
3.d Management of national and global health risks	3.d.1 International health regulation capacity

DOAJ, ORCID, and other organizations. Compared to subscription-based options that do not promote open science, its academic repository networks offer more thorough access. The study found that the United States, Switzerland, and the United Kingdom were the most productive nations, and the top two institutions were the World Health Organization and the University of London (refer to Tables 18.7 and 18.8).

18.3.2   Role of COVID-19 in SDG

The aims of SDG-3, which had seen notable worldwide advancements before the pandemic, have been significantly impacted by COVID-19. For example, before COVID-19, the rate of death for mothers and children decreased, the incidence of HIV among adults aged 15–49 worldwide decreased, and the rate of childhood immunization increased. The mortality rate for children under five also decreased,

Table 18.7 Top five articles with the most citations in SDG-3 [15]

Title	Year	Journal title	First author's country	Impact factor	Citations
The Effect of Multiple Adverse Childhood Experiences on Health: A Systematic Review and Meta-Analysis	2017	The Lancet. Public health	England	18.953	1809
High-Quality Health Systems in the Sustainable Development Goals Era: Time for A Revolution	2018	The Lancet. Global health	Netherlands	9.684	1088
The Lancet Commission on Global Mental Health and Sustainable Development	2018	Lancet	United Kingdom	22.226	1028
The State of US Health, 1990–2016: Burden of Diseases, Injuries, and Risk Factors Among US States	2018	JAMA	United States	14.891	794
National, Regional, and Worldwide Estimates of Stillbirth Rates in 2015, with Trends from 2000: a Systematic Analysis	2016	The Lancet. Global health	Netherlands	9.684	637

Table 18.8 Top five most productive institutions [15]

Institution name	Country	No. of documents	Citations
University of London	United Kingdom	98	3412
World Health Organization	Switzerland	37	1475
Liverpool School of Tropical Medicine	United Kingdom	30	1100
University College London	United Kingdom	22	1067
Harvard University	United States	21	1553

going from 76 deaths per 1000 live births in 2000 to 38 deaths per 1000 live births in 2019. The rate of neonatal mortality, or deaths in the first 28 days of life, also roughly halved, going from 30 deaths per 1000 live births in 2000 to 17 deaths per 1000 live births in 2019.

Prenatal care services, accessibility to reproductive health services, and family services were all interrupted by COVID-19 and the containment measures it imposed in the majority of nations. Furthermore, maternity staff were transferred to

COVID-19-affected and treated wards due to a rapid spike in the volume of COVID-19 patients and a scarcity of healthcare personnel in most nations. This naturally resulted in a reduction of attention to the needs of mothers and children. Unfortunately, only 228,000 more infant fatalities and 11,000 more maternal deaths were recorded in South Asia in 2020. This was partly because the majority of healthcare resources were directed toward containing the COVID-19 pandemic's destabilizing effects. During the crisis, individuals suffering from HIV/AIDS also saw significant delays in getting preventative and treatment services, including referrals and diagnostic testing. Comparably, the COVID-19 epidemic has hampered the prompt identification and treatment of newly diagnosed cases of TB as well as their follow-up. During the COVID-19 pandemic, the situations of substance users and addicts worsened relative to others, making them one of the most susceptible groups. Similarly, plenty of research shows that restriction methods, such as confinement and lockdown, have a variety of negative effects on mental health and psychological harm, including emotional discomfort, despair, stress, mood swings, irritability, sleeplessness, post-traumatic stress disorder, and anxiety. According to a recent WHO survey, there may have been a significant rise in the global suicide death rate in 2019 as a result of the unprecedented rise in mental illness and its devastating effects. As a result, several nations chose to incorporate mental health services into their COVID-19 plans for 2021. The majority of healthcare services, including promotion, prevention, treatment, and rehabilitation, have been disrupted by a lack of resilience and a lack of readiness in many health systems across the world, especially for the most vulnerable populations, including women, children, people with noncommunicable diseases, addicts, and those suffering from HIV/AIDS and tuberculosis. All governmental sectors, development partners, and civil societies must demonstrate their solidarity and cooperation in fighting the COVID-19 pandemic, current crises like noncommunicable diseases and climate change, and any potential future crises using the two approaches of the whole of government and the whole of society to maintain sustainable social, economic, health, and well-being relations.

18.3.3   Role of medical waste in SDG

The appropriate disposal of waste produced by healthcare institutions can have a direct influence on both human health and the environment, making medical waste management an essential practice in healthcare administration. Pharmaceuticals, research institutes, medical educational institutions, and other healthcare facilities that perform medical operations all produce waste materials that are classified as medical waste. Neglected or incorrectly handled medical waste has detrimental effects on the environment and public health. It is more expensive to remove the effects of this waste management than it is to create and put into place straightforward waste management techniques that do not need a big financial commitment. A considerable amount of medical waste is thrown in waste dumps and unauthorized dumpsites due to the carelessness of healthcare personnel at hospital facilities, and the inadequate, not accessible waste treatment and removal facilities [16].

It should be indicated that not all healthcare waste is possibly dangerous. Moreover, 75%–80% of the waste created by healthcare facilities that do not come

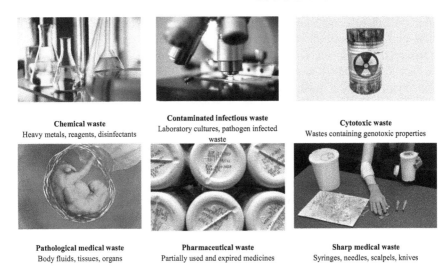

Figure 18.2   Common hazardous medical waste

into contact with patients' bodily fluids or infected patients is comparable in content to domestic waste. A fraction of healthcare waste, 10%–25% is designated to be harmful and can be a danger element for both the environment and the well-being of people. Figure 18.2 displays the collection of hazardous healthcare-related waste. There is a shortage of contemporary and appropriately equipped medical waste processing facilities that prevents efficient waste management. Additionally, it is possible that a large number of waste management and healthcare professionals were not properly trained in the treatment and disposal of medical waste, which results in the loss of important possibilities to address the issues.

18.3.4   Role of recycling in SDG

Significant worries have been raised about elevated arsenic (As) concentrations in groundwater in various regions of the world, particularly in low- and middle-income nations including Bangladesh, China, India, Nepal, and Vietnam. The world's population—more than 85% of it—gets their drinking water from groundwater. One example of a geogenic contaminant is arsenic, which comes from the minerals in rocks. As pollution in groundwater affects more than 230 million people. Of them, developing nations in South Asia report 78% of this exposure. According to recent research, As has been found in the groundwater of Brazil, Argentina, Bolivia, Chile, Bangladesh, China, Cambodia, Nepal, Vietnam, Mexico, and Bolivia. Arsenic exposure and pollution in the human community is one of the main obstacles to achieving the "Good health and well-being" goals. Humans can become infected by drinking water that has been tainted with arsenic or by eating food crops that have bioaccumulated arsenic. Numerous issues with the human body, including those related to the immune system, cardiovascular, neurological, hepatic, reproductive, and renal systems, are caused by arsenic.

Parkinson's disease, melanosis, leucomelanosis, keratosis, gangrene, cutaneous irritation, and cancer can all result from long-term As exposure. Some of the latest techniques that are available for the removal of As from the As-contaminated water are chemical oxidation and precipitation, coagulation and flocculation, electrocoagulation, biological treatment, adsorption, membrane filtration, nanoparticles, and ion exchange. Implementing such techniques shall improve the SDG-3 ratings [17].

Apart from recycling the As infected water, even waste water recycling contributes more to the ESG. Some traditional processes of recycling include the adsorption process, ion exchange, coagulation or flocculation, membrane technologies, and biological process. With the development in technologies, the latest procedures like advanced oxidation process, membrane bioreactor, ultrasound technique, hydrodynamic cavitation technology, and advanced green technologies are used. Each process has its own way of impact and contribution to the environment, finance, and society [18].

18.3.5   Role of AI in SDG

At the center of development efforts, cities today generate 80% of the world's gross domestic product (GDP) and release 75% of greenhouse gases linked to energy usage. Additionally, 70% of city dwellers lack access to at least one essential utility, such as power, water, or housing. According to projections, 70% of people on Earth are expected to reside in urban settings by 2050. By that time, it is also predicted that 90% of population growth, 80% of wealth gains, and roughly 60% of energy consumption will occur in urban areas worldwide, leaving developing nations and emerging economies with challenges in the areas of sustainability, environmental protection, and natural resource management. Resource conservation, sustainable consumption and production, conservation of energy, ecosystem monitoring, preservation of the environment, and natural resource management are just a few of the urgent urban environmental issues that artificial intelligence (AI) and sustainable Big Data innovation can help with. Additionally, cities can benefit from digital advancements through AI and Big Data solutions by increasing the productivity of their natural resources and human capital, building climate resilience, and creating decent jobs for the youth while also developing newer, cutting-edge technologies [19].

The digital age was brought about by the third industrial revolution (IR 3.0), which is most known for the widespread use of technological advances to automate industry, and the expansion of the internet. The world is currently on the verge of a fourth momentum known as the Industrial Revolution (IR 4.0). This orderly transformation, which can be classified by scientific and technological advancements that are upending industries and erasing national boundaries, is distinct in its velocity, possibilities, and impact on systems. It is now commonly acknowledged that AI, or software engines fueled by the Big Data revolution, is the primary driver behind the IR 4.0 Revolution [20]. The different initiatives or projects implementing AI and Big Data, along with its developer or publisher, and its expected outcomes are provided in Table 18.9.

Apart from AI and Big Data, the digital health concept also enhances the existing healthcare approach toward SDG. Digital health includes telemedicine,

Table 18.9 AI and Big Data for monitoring and implementing SDG [19]

Initiative	Developer	Outcome
All of Us	National Institutes of Health	Assess a patient's medical history as soon as possible, then treat them appropriately in the least amount of time
Aqueduct	World Resources Institute	Map and assess the water hazards in each region, both now and in the future
Azure FarmBeats	Microsoft Corporation	Utilize a vegetation and water index based on satellite images and evaluate the health of a farm
Big Data for Social Good	GSM Association	Use the large data capabilities of cellular providers to handle global emergencies, such as diseases and natural disasters
Central Africa Regional Program for the Environment	USAID	Forecast for the Democratic Republic of the Congo's potential deforestation
Digitization of Healthcare	Dash et al.	New biomarkers and clever therapeutic intervention techniques
Improved early warning systems for natural hazards monitoring	Ministry of Environment and Natural Resources of El Salvador	Provide El Salvador with warning and management mechanisms from natural hazards
IoT in Healthcare	Shameer et al.	Forecast the status of the patient's health and the rate at which the pathological state will develop
Managing Information for Natural Disasters (MIND)	Pulse Lab Jakarta	Encourage efficient information management and logistical planning in the wake of natural catastrophes
Mapping urban surfaces and cooling our cities	Global Cool Cities Alliance & World Resources Institute	Recognize changes in the urban surface to reduce the effects of urban heat through heat mitigation techniques
National Oceanic and Atmospheric Administration (NOAA)—Environmental Big Data	NOAA	Urban environmental protection, ecosystem monitoring, and restoration
National Oceanic and Atmospheric Administration (NOAA)—Global Hydro Estimator	Microsoft Corporation & NOAA	Provide global rainfall estimates in 15-min intervals at ~4 km resolution
Predictive Road Crash	Waze	Drivers are forewarned to use particular caution in the designated dangerous locations
Pulse Lab Jakarta	UN Global Pulse & United Nations	Finding Potential Positive Deviants (PDs) in Different Rice-Producing Areas of Indonesia
PulseSatellite	UN Global Pulse	Find and count the buildings in refugee communities, then estimate the size and distance between roofs to create a slum mapping dataset
Visualization and Analytics of Distribution Systems with Deep Penetration of Distributed Energy Resources (VADER)	Office of Energy Efficiency & Renewable Energy, US	Predicts the behavior of distributed energy resources and the use of power

Table 18.10 Benefits and potential drawbacks of implementing SDG [21]

SDG	Direct benefits	Indirect benefits	Potential drawbacks
No poverty	Provide tools to help better identify and comprehend the requirements of poor people	Economic growth	Segregation may lead to inequality
Zero hunger	Provide solutions to those who, in times of need, are unable to discover sources of food	Better food production and access Reduced wastage of food Strong supply chain	Low-income countries may left behind
Good health and well-being	Enhanced observation and management of health	Improved air quality Minimal traffic Online healthcare	Privacy breach while sharing patient data
Quality education	Increased academic accessibility in times of disaster like pandemics and syndemics	Enhanced educational programs	More time spent in using smart devices
Gender equality	Strengthened economy and a community that is safer and healthier	Women empowerment Participation of women in societal activities	Difficulty in accepting the change in stereotype
Clean water and sanitation	Tracking and controlling water and wastewater resources throughout real time	Better water distribution and management Safe drinking water Water recycling	Competition between corporations may lead to decreased service provision to urban poor
Affordable and clean energy	Encouragement of decentralized energy systems that incorporate renewable energy sources and electric automobiles	Energy saving Demand optimization Promotion of green energy	Access to smart technologies may widen the gap in community
Decent work and economic growth	Streamlined and secure banking and financing for small and medium-sized businesses	Better working environment Increased productivity	In absence of proper management, inequality may rise

Industry, innovation, infrastructure	Boost the pace of creative thinking	Increased productivity / Promotion of green energy / Energy saving	Integration of technology may become costly
Reduced inequalities	Decreased transaction costs that people of developing cities may receive remittances more easily	Reduced urban–rural disparity / Service accessibility	Unregulated technological advancement may result in discrimination
Sustainable cities and communities	Enhancing urban governance	Promotion of green energy / Energy saving	Resources from one end may be depleted for the sake of other
Responsible consumption, production	Help to improve governance of waste and wastewater	Better natural resource management	Smart technology may consume more energy
Climate action	Enhanced simulation of the effects of climate change	Integration of renewable energy	Green energy may not be sufficient to power hungry urban
Life below water	Tracking the contamination of water bodies and oceans	Reduced water pollution	Not identified
Life on land	Healthy life for all	Protection of natural resources	Light emitted from cities may affect animal, human, and astronomical society
Peace, justice, and strong institutions	More chances for public involvement	Transparency in governance / Reduced crime	Authoritative regimes could manipulate
Partnerships for the goals	Improved channels for promoting stakeholder interaction and cooperation	Trust among the stakeholders	Not identified

electronic medical records, Cloud-based systems, mobile health popularly known as mHealth, wireless health devices, virtual reality, and open-source software [13].

18.3.6   Recommendations

The direct and indirect benefits of SDG implementation and its potential drawbacks are mentioned in Table 18.10. As per the 2021 study [22], on a scale of 5, SDG-1, 3, 5, 7, 10, and 15 have progressed to 3 out of 5. SDG-8, 9, 12, and 14 have progressed beyond three. While SDG-2, 4, 6, 11, 13, 16, and 17 have progressed only up to 2 and a few below 3. Hence, this study put forward some recommendations for the successful achievement of SDGs.

- Reliable data infrastructure must be built at the national level because AI uses both structured and unstructured data to carry out data analysis and advanced decision-making without the need for human oversight.
- Governments must encourage research and innovation to support the growth of startups, Big Data solutions, AI technologies, and so on.
- The commercial sector collects a significant portion of the public sector's Big Data. For this reason, strong public–private collaborations are essential to maximizing the benefits of Big Data and AI.
- The SMEs must be given incentives to invest in Big Data, AI and related facilities, Big Data platform services, data protection and online safety, and research activities.
- Governments must take action to entice foreign investment in platform-based services, IT and data facilities, Big Data, and AI-related services.
- E-commerce and Big Data-based digital services may grow and improve their local exports of goods, job opportunities, and efficiency in addition to city economic growth, by eliminating regulatory and logistical obstacles as well as nontariff trade barriers.
- For the purpose of Big Data and AI-based regional and national growth and development, major intergovernmental bodies such as ASEAN, BIMSTEC, and SAARC must be enhanced and properly accessed.

18.4   One Health

A strategy that emphasizes the interdependencies across the fields of public health, veterinary medicine, and environmental sciences is known as the One Health approach. This concept is ingrained in all three of these fields together. To achieve optimum health outcomes for humans, animals, and ecosystems, it proposes a coordinated and collaborative approach to establishing programs, community-based activities, and policies. The purpose of this strategy is to achieve optimal health outcomes for all three groups. The strategy may be adopted on local, national, and worldwide levels and highlights the necessity of coordination, collaboration, as well as communication across sectors. The One Health concept has been regarded as a tool to grasp the complexity underlying health problems including zoonotic illnesses, which are diseases that may be transmitted from animals to human beings, and antimicrobial resistance, which arises when microorganisms develop resistance to antimicrobial medicines. The COVID-19 pandemic has also strengthened this concept, considering its likely animal origin and capacity to transfer between human beings and animals. There have been some problems with adopting the One Health model in practice. Challenges such as the issue of disciplinary silos, the lack of participation of the environment sector, the lack of finance, and the lack of understanding and commitment of policymakers. There have been some problems with adopting the One Health model in practice. Challenges such as the issue of disciplinary silos, the lack of participation of the environment sector, the lack of finance, and the lack of understanding and commitment of policymakers.

The One Health method takes into consideration the three different aspects of health: human health, animal health, and environmental health. There has been discussion in the academic literature around One Health implementation. According to the findings of the research, One Health concepts have been somewhat integrated across sectors. Concerns with the separation of data collection and processing, communication, and the absence of a unified concept of One Health are reported in the majority of research. According to the findings of the research, there are deficiencies in terms of financing, coordination, and administrative issues on both the regional and national levels. Additionally, there are obstacles in terms of policymakers' understanding and readiness to execute One Health initiatives.

Threats to human health include food contamination, antibiotic overuse, environmental degradation, and newly developing infectious illnesses. These trends are a result of globalization and fast urbanization. As a result, there are now more complications and inconsistencies pertaining to food safety, animal health, public health, and environmental health. Rapid urbanization and economic expansion are possible in a developing country with a huge population. These changes frequently result in squalid living conditions, elevated stress levels, infectious disease transmission, and congested cities [23]. Factors such as increases in energy use, transportation, by-product production, and intensification of economic activity led to increased air pollution. These variables endanger the health of animals in addition to having a negative effect on human health by increasing the risk of respiratory and cardiovascular disorders. For this reason, it is essential to implement methods that reduce these dangers to people, pets, and the environment.

The One Health strategy is centered on creating and executing policies, laws, strategies, and research including cross-sectoral cooperation to enhance public health outcomes. Its accomplishments are contingent upon solid institutional frameworks, collaborations, governance, and a One Health-oriented culture. Important elements include fostering mutual trust and understanding, combining all parties' resources, and promoting in-depth investigation and collaboration. One Health is an integrated, multisectoral, transdisciplinary strategy that strengthens our ability to address urgent public health issues that pose a threat to civilization. The underlying idea of the concept is the connection between health and the biosphere's interactions with humans, animals, plants, and environments. The Food and Agriculture Organization, the World Organization for Animal Health, and the United Nations Environment Program are quadripartites that have recently widely embraced One Health, which has been acknowledged by the World Health Organization (WHO) as the preferred approach to addressing health issues requiring complex solutions [6].

18.4.1   Healthcare in the past

The unintentional discovery of penicillin's antibacterial properties by Alexander Fleming in 1928 and the detection of a sulfonamide dye's comparable properties by Gerhard Domagk in 1932 led to the development of novel medicinal treatments. Although a variety of medications have been developed via further research, viral infections are still hard to treat. However, vaccine-based preventative methods really go back more than a century before bacterial treatment, as noted by Edward Jenner, who established the value of immunization against smallpox toward the end of the 18th century.

Concerns about the wellness of the public, community, and the ecosystem have been raised by the recent development of three corona virus illnesses. It was either bats or civet cats that first developed this severe acute respiratory illness in late 2002. Because of quarantine and isolation, the pandemic, which had spread to 26 countries and resulted in 8096 cases and 774 fatalities, was contained by 2003. Despite being shown to spread from camels to people, the Middle East respiratory illness is still present today. In contrast, SARS-CoV-2 spread quickly throughout the world's population and turned into a pandemic, albeit its exact origin is still unknown.

Safe and very powerful vaccinations against a broad range of viral illnesses have been created during the first half of the 20th century. The development of vaccinations against the 2019 corona virus sickness, which is brought on by the corona virus that causes severe acute respiratory syndrome, is examples of ongoing work in progress. Its DNA sequencing was completed in the final days of 2019—just days after the illness was initially reported. Thousands of lives were saved because of the rapid genetic mapping and the first-ever application of the effective messenger RNA technique in vaccine manufacture, which reduced the time needed for research and development from a decade or more to less than a year.

The severe harm that COVID-19 caused to every aspect of society highlights how urgently plans for strengthening a One Health strategy against the possibility of subsequent pandemics need to be made. All four major organizations—the Food and Agriculture Organization, the World Health Organization, the World Organization for Animal Health, and the United Nations Environment Programme—agree that further research is necessary to fully comprehend how the most recent coronaviruses crossed the animal-human divide and quickly enabled human-to-human transmission [24]. The main concern is how to stop future viral pandemics from spreading among wildlife and how to respond to them.

18.4.2   One Health in China

The information that follows offers a thorough examination of China's current health-related policies, emphasizing their applications and functions in furthering One Health administration. Four main systems are included in China's One Health

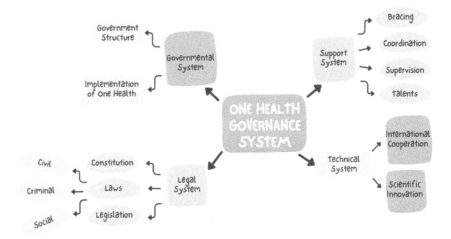

Figure 18.3   China's One Health governance framework

framework: the legal, governmental, technological, and logistic support systems [23]. A simple diagrammatic representation is presented in Figure 18.3.

Legal system—Regarding the legislative components of the One Health concept, there are three policies that function as fundamental tools for social system governance. In outlining the fundamentals of government, the Chinese Constitution places a strong emphasis on the welfare of the people as a whole, encompassing both individual and communal well-being. In addition to the constitution, national legislation and administrative rules are examples of legislative artifacts. The right to health is prioritized by national legislation in fields including criminal, civil, and social law. The "Criminal Law" is periodically amended to address different public health issues, especially those pertaining to criminal activity that jeopardizes public health and safety. The "Civil Code," which went into effect on January 1, 2021, has laws pertaining to health, particularly as they relate to individual rights and well-being. Social policies that tackle environmental issues—which are strongly tied to the One Health idea—such as the amended "Environmental Protection Law" of 2015, seek to advance social welfare. Adding to the above mentioned, the State Council of China also enacted laws such as "Basic Healthcare and Health Promotion Law of the Peoples Republic of China," "Law on the Prevention and Control of Infectious Diseases," "Food Safety Law," and "Drug Administration Law," which provides additional support to the One Health concept.

Governmental system—With this system, social governance is established in a hierarchical structure that distinctly delineates the legal responsibilities and jurisdictional boundaries of all governmental departments. It is beneficial to define roles at every level of government to avoid problems like local departments taking on duties that are outside the scope of their original role.

Technical system—The cornerstone of this governance paradigm for One Health is multidisciplinary technology innovation. The "Healthy China 2030"

Action Plan prioritizes innovation and reform, with a focus on advances in information and technology. A thorough health governance system is significantly strengthened by quick technical breakthroughs. However, there is always room for development. These include creating Big Data platforms, improving zoonotic disease surveillance and early warning systems, and enabling the interchange of biological samples and environmental health resources. The "One Health Joint Plan of Action (2022–2026)" was created in response to global demands for action. The World Health Organization (WHO), the World Organization for Animal Health (WOAH, formerly OIE), the Food and Agriculture Organization of the United Nations (FAO), and the United Nations Environment Programme (UNEP) have committed to the goals outlined in this plan. The One Health High-Level Expert Panel (OHHLEP), which is made up of 26 specialists who offer evidence-based counsel on newly developing diseases and promote a thorough One Health strategy, is supported by these organizations.

Support system—In social governance, support systems play a critical role in the establishment and upkeep of scientifically sound and efficient systems, particularly when addressing One Health-related issues. The four essential pillars of this support system are financial assistance, talent nurturing, government coordination, and monitoring and oversight. Due to its great degree of impact, government involvement in complex health issues is frequently required through policy. The health governance system may be made to function adequately by integrating human resources when building a multidisciplinary team to handle difficulties. It is also necessary for One Health governance to have nontechnical competence in areas like capacity building, advocacy, leadership, and knowledge translation and dissemination. To promote mutually beneficial cooperation, efficient monitoring and supervision mechanisms are necessary for One Health research and governance. These systems call for mutual oversight and prompt communication among diverse stakeholders. Considering the broad reach of One Health initiatives, thorough oversight and monitoring are essential. Within the healthcare industry, social governance monitoring is extended through the One Health governance supervision system. Financial assistance and a strong support network are essential for the One Health governance structure to remain viable over time. Taking into consideration the many expenses related with technical components, medicines, management, organizational and operational variables, and One Health governance, effective resource allocation is essential when creating a budget. The need for deliberate investment is heightened by the varied character of policy implementation, particularly considering that the One Health governance structure is still in its early phases of development.

18.4.3   Challenges in implementing One Health

There are several obstacles that must be overcome before the One Health strategy can be implemented under the present healthcare system in response to potential zoonotic pandemics.

Concept—In worldwide policies and frameworks pertaining to preparation and zoonotic control, the needs particular to individual diseases are often given priority. However, it is challenging to include all of these intricate components and interactions into a single strategy due to the intricacy of the relationships between animal spillover and human transmission, as well as population demographics, public health measures, and environmental factors. Because of this, current approaches—which focus on the epidemic aspects—are inadequate for managing risks and evaluating results, especially when it comes to current health policies and frameworks. However, a number of recent analyses show that to bridge various sectoral mandates and coordinate national activities holistically in the fight against epidemics, international communities need a complete, evidence-based One Health strategy. To prevent potential dangers that might be exacerbated by this dynamic complexity, it is imperative to have a comprehensive understanding of the evolving risks involved in pandemic prevention, detection, response, and recovery.

Communication—The majority of nations lack systems for organizing programs and communicating administratively and technically in the event of a pandemic. The implementation of effective interventions may be hampered by a lack of established systems for sharing data, cooperative planning, and coordinated activities. This might lead to task overlapping or data mismatch, which would delay the successful containment of epidemics at their early stages. Opinions from the veterinary health and environmental sectors may not reach those responsible for human public health while choices are being made, even if they are pertinent to transmission cycle and enforcement actions. Establishing credibility among multiple stakeholders and creating an efficient mechanism for cooperation at the local, national, and international levels requires transparent, efficient, sufficient, and ongoing communication between the authorities and all pertinent stakeholders, including those spearheading civic engagement. Enhancing the transmission of science-based information and taking into account any potential negative effects are crucial when providing stakeholders with pertinent information about epidemics.

Coherence—The management of current and future pandemic risks requires a cohesive emergency response system, especially at the intersection of the human, animal, and environmental sectors, given the possibility of local and worldwide multisectoral infiltration. Public institutions are divided according to their particular fields, which causes them to function independently. This is an issue that will inevitably persist and cause gaps or overlaps. Basic institutional linkages should be arranged correctly in public health systems to prevent such problems. The integrated and synchronized public health systems ought to formulate policies through gradual stages, furnishing managers with a robust framework for educating policymakers.

Continuity—To counter the threat of zoonotic epidemics, certain programs have adopted the multisectoral One Health approach; however, as the emergency subsided, the method is dropped. To guarantee long-term viability, these methods need to be regarded as regular practices and therefore be made sustainable. A viable strategy for maintaining the efficacy of a One Health approach may involve adhering to existing global and local laws, such as the Sustainable Development Goals and International Health Regulations. It will be simpler to execute cooperative activities at the regional and worldwide levels and to encourage teamwork if these international references, norms, and laws are followed. This would, in fact, aid in epidemic management and maintain the sustainability of the One Health strategy.

Governance—To guarantee the collaborative involvement of many sectors and disciplines, a more comprehensive, well-coordinated, and efficient governance framework should be established, complete with top-tier leadership, function accountability, and policies. To establish a global and cohesive vision of One Health, initiatives pertaining to political will, intersectoral governance, and a regulatory framework based on global healthcare regulation may take precedence. Increasing the amount of money invested, developing and defining pathways for scientific input, exchanging data, and extending reporting requirements are a few particular places to start.

Surveillance—To effectively prevent a potential pandemic, it is crucial to establish a monitoring framework and enhance intelligence systems that can closely track the transmission of zoonotic diseases. An efficient monitoring system designed to anticipate the occurrence and magnitude of a zoonotic pandemic should depend on three key factors: (1) identifying the mechanism by which viruses infect humans by establishing a connection between the animal environment and human hosts; (2) early identification of infection in individuals exposed to the virus; and (3) investigating the ability of the virus to spread from animal to animal, animal to human, and human to human, potentially posing a hazard.

Research—The proliferation of COVID-19 not only revealed the vulnerability of the worldwide readiness and reaction to the pandemic but also highlighted deficiencies in specialized scientific investigation. Acquiring information in One Health research is crucial for preventing and controlling pandemics. This should be aligned with interdisciplinary and cross-sectoral international programs, which would expedite the collection of information in a synchronized way. Conducting a comprehensive study is essential to understand the transmission process and investigate evidence-based methods.

18.4.4   One Health and Global Health

The emergence and resurgence of infectious illnesses are indirectly impacted by globalization. In the last 10 years, the WHO has issued six declarations of public health emergencies of international concern. The necessity for enhancements in the handling and control of infectious diseases at the international level is highlighted by past pandemics and significant outbreaks including influenza A (H1N1), Zika, Ebola, and COVID-19 [25]. Less than 50% of people on the entire planet do not have access to all necessary medical care, according to the WHO. For a nation's economy to prosper and its society to advance, health is essential. Still, the process of globalization has led to significant social and health issues as well as international inequities, particularly in the less developed nations that do not dominate the world economy. These nations often have greater health-related problems because of governance shortcomings and limits on their ability to create and carry out social and health programs. Table 18.11 shows the comparison of health spending per capita and total health spending per GDP of South Asia, East Asia, and Southeast Asia countries.

To evaluate and comprehend the effects of global interconnectivity on health, two ideas are being utilized more and more. Since social, political, and even ideological circumstances have an impact on public health, the notion of Global

Table 18.11   Health spending of countries in South Asia, East Asia, and Southeast Asia in the year 2017 [14]

Region	Countries	Health spending per capita (US $)	Total health spending per GDP (%)
South Asia	Bangladesh	40	2.5
	Bhutan	108	3.2
	India	69	3.5
	Nepal	50	5.5
	Pakistan	37	2.8
East Asia	China	455	5.1
	North Korea	77	5.6
	South Korea	2118	7.2
	Taiwan	1477	6.2
Southeast Asia	Cambodia	83	5.7
	Indonesia	120	3.2
	Laos	58	2.4
	Malaysia	409	3.9
	Maldives	988	8.1
	Mauritius	606	5.7
	Myanmar	52	4.4
	Philippines	133	4.4
	Sri Lanka	152	3.9
	Seychelles	754	4.6
	Thailand	271	3.8
	Timor-Leste	86	3.6
	Vietnam	135	5.5

Health (GH) is predicated on the idea of supraterritoriality, which makes linkages between the global and local levels of knowledge. It emphasizes the values of social justice, fairness, and appreciation for the diversity of human social and cultural expression as well as the growth of individual and collective autonomy. The second idea, One Health (OH), unifies environmental factors and the health of people and animals on a local, national, and international scale. With the need for multi-disciplinary, interprofessional, and transdisciplinary viewpoints, this idea has become more and more important in understanding the intricate relationships between the various health-related aspects and domains of knowledge. Global population vulnerability is heightened by environmental and sociodemographic shifts, which pose public health issues for all nations. Given that both OH and GH views are concerned with the variables that impact the emergence, spread, and management of illnesses, a more robust link between them is therefore desirable. Forging stronger ties between OH and GH would be beneficial to achieving the SDGs.

18.5   Verdict

Research was conducted on the implementation of the One Health concept in Sweden and Italy. The study included studying the thoughts and perspectives of experts who work at public health, veterinary, environment, and food agencies in Sweden and Italy. The research also looked at the knowledge of the various institutional and political contexts in which One Health activities are executed, as well as the contrasts and similarities that exist with respect to the settings of institutions. They arrived at the conclusion that the existing issues in implementing One Health can be mapped under two categories: coordination and governance. To guarantee successful cooperation across sectors, it is necessary to define shared objectives. Institutes will profit from teaching or hiring experts who can facilitate the link between diverse industries and encourage knowledge translation.

The One Health governance framework seeks to improve the general health and health equity of the world's population rather than catering to the unique health requirements of each person. Given the complexity of this idea, a clear chain of command and centralized, effective leadership are essential components of a well-defined governance organization.

Most serious and persistent issues pertaining to the health of humans and animals are brought on by agents whose epidemiology and control and preventative techniques are well established. Due to their complex nature and the fact that leadership and communication deficiencies are major obstacles to the adoption of effective treatments, these diseases continue to be a burden on our society. Our capacity to address such issues successfully is greatly enhanced by the One Health concept.

The One Health idea has not yet gained widespread traction in the commercial sector. Once accepted, there will be a significant increase in the ability to manage public health issues and facilitate efficient planning, prevention, and quick reaction and management during epidemics or pandemics. This in turn will directly and indirectly lead to attaining the goals of ESG and SDG.

References

[1] Y. Tsang, F. Youqing and Z. Feng, "Bridging the gap: Building environmental, social and governance capabilities in small and medium logistics companies," *Journal of Environmental Management*, vol. 338, p. 117758, 2023.

[2] J. J. Liou, P. Y. Liu and S.-W. Huang, "Exploring the key barriers to ESG adoption in enterprises," *Systems and Soft Computing*, vol. 5, p. 200066, 2023.

[3] A. A. Egorova, S. V. Grishunin and A. M. Karminsky, "The Impact of ESG factors on the performance of Information Technology Companies," *Procedia Computer Science*, vol. 199, pp. 339–345, 2022.

[4] S. Ozkan, S. Romagnoli and P. Rossi, "A novel approach to rating SMEs' environmental performance: Bridging the ESG gap," *Ecological Indicators*, vol. 157, p. 111151, 2023.

[5] W. M. Wan Mohammad and S. Wasiuzzaman, "Environmental, social and governance (ESG) disclosure, competitive advantage and performance of firms in Malaysia," *Cleaner Environmental Systems*, vol. 2, p. 100015, 2021.

[6] C. J. Bruno de Oliveira and W. A. Gebreyes, "One health: connecting environmental, social and corporate governance (ESG) practices for a better world," *One Health*, vol. 15, p. 100435, 2022.

[7] N. Martin and S. Mulligan, "Environmental sustainability through good-quality oral healthcare," *International Dental Journal*, vol. 72, no. 1, pp. 26–30, 2022.

[8] D. Chaudhry, "Climate change and health of the urban poor: The role of environmental justice," *The Journal of Climate Change and Health*, vol. 15, p. 100277, 2024.

[9] K. Wan, M. Lane and Z. Feng, "Heat-health governance in a cool nation: A case study of Scotland," *Environmental Science & Policy*, vol. 147, pp. 57–66, 2023.

[10] A. F. Aysan, Y. Bakkar, S. Ul-Durar and U. N. Kayani, "Natural resources governance and conflicts: Retrospective analysis," *Resources Policy*, vol. 85, p. 103942, 2023.

[11] J. Koczar, D. Zakhmatov and V. Vagizova, "Tools for considering ESG factors in business valuation," *Procedia Computer Science*, vol. 225, pp. 4245–4253, 2023.

[12] J. Addo-Atuah, B. Senhaji-Tomza, D. Ray, P. Basu, F.-H. Loh and F. Owusu-Daaku, "Global health research partnerships in the context of the sustainable development goals (SDGs)," *Research in Social and Administrative Pharmacy*, vol. 16, no. 11, pp. 1614–1618, 2020.

[13] Y. M. Asi and C. Williams, "The role of digital health in making progress toward Sustainable Development Goal (SDG) 3 in conflict-affected populations," *International Journal of Medical Informatics*, vol. 114, pp. 114–120, 2018.

[14] G. B. o. D. H. F. C. Network, "Health sector spending and spending on HIV/AIDS, tuberculosis, and malaria, and development assistance for health: progress towards Sustainable Development Goal 3," *The Lancet*, vol. 396, no. 10252, pp. 693–724, 2020.

[15] S. A. Raji and M. O. Demehin, "'Long walk to 2030': A bibliometric and systematic review of research trends on the UN sustainable development goal 3," *Dialogues in Health*, vol. 2, p. 100132, 2023.

[16] W. L. Filho, T. Lisovska, M. Fedoruk and D. Taser, "Medical waste management and the UN Sustainable Development Goals in Ukraine: An assessment of solutions to support post-war recovery efforts," *Environmental Challenges*, vol. 13, p. 100763, 2023.

[17] A. K. Yadav, H. K. Yadav, A. Naz, M. Koul, A. Chowdhury and S. Shekhar, "Arsenic removal technologies for middle- and low-income countries to achieve the SDG-3 and SDG-6 targets: A review," *Environmental Advances*, vol. 9, p. 100262, 2022.

[18] K. Obaideen, N. Shehata, E. T. Sayed, M. A. Abdelkareem, M. S. Mahmoud and A. Olabi, "The role of wastewater treatment in achieving sustainable development goals (SDGs) and sustainability guideline," *Energy Nexus*, vol. 7, p. 100112, 2022.

[19] M. Arfanuzzaman, "Harnessing artificial intelligence and big data for SDGs and prosperous urban future in South Asia," *Environmental and Sustainability Indicators*, vol. 11, p. 100127, 2021.

[20] P. Mohankumar, L. Wai Yie and B. Y. Suprapto, "Virtual and augmented reality in Industry 4.0," in *The Nine Pillars of Technologies for Industry 4.0*, IET Digital Library, 2020, pp. 61–78.

[21] A. Sharifi, Z. Allam, S. E. Bibri and A. R. Khavarian-Garmsir, "Smart cities and sustainable development goals (SDGs): A systematic literature review of co-benefits and trade-offs," *Cities*, vol. 146, p. 104659, 2024.

[22] G. Halkos and E.-C. Gkampoura, "Where do we stand on the 17 Sustainable Development Goals? An overview on progress," *Economic Analysis and Policy*, vol. 70, pp. 94–122, 2021.

[23] L. Huang, J. He, C. Zhang, *et al.*, "China's One Health governance system: The framework and its application," *Science in One Health*, vol. 2, p. 100039, 2023.

[24] J. Chen, J. He and R. Bergguist, "Challenges and response to pandemics as seen in a One Health perspective," *Science in One Health*, vol. 1, p. 100010, 2022.

[25] P. C. Pungartnik, A. Abreu, C. V. Brito dos Santos, J. R. Cavalcante, E. Faerstein and G. L. Werneck, "The interfaces between One Health and Global Health: A scoping review," *One Health*, vol. 16, p. 100573, 2023.

Chapter 19

Building the future: innovations and challenges of ESG in transforming the construction industry at the nexus of sustainability

Lee Sun Heng[1]

Environmental, social, and governance (ESG) is becoming a hot topic and increasingly important for companies in the construction industry. As a business strive to meet long-term sustainability for ESG standards, they are facing simultaneous challenges to integrate sustainable practices and innovative technology to create buildings that are eco-friendly, socially conscious, and economically viable. The construction sector plays a critical role in driving the economic growth of a country, and as the industry continues to grow, it is important to incorporate ESG principles to achieve long-term sustainability and success. To contribute to a better future and fulfil the regulatory requirements, construction companies can always practice wisely in resource optimisation by utilising the current technology and innovative practices. Construction companies demonstrating their commitment to environmental stewardship, social responsibility, and good governance can lead to long-term success in a rapidly changing industry.

This chapter will explore an overview of how ESG principles are being integrated and driving innovation in the construction industry, a sector traditionally associated with significant environmental impacts and social implications. There is no universal solution for managing ESG matters, and this is completely understandable. Every industry must navigate innovations to overcome the challenges. The narrative will begin by outlining the historical context of environmental concerns in construction, including resource consumption, waste production, and carbon emissions. It will then transition into the modern demands and regulatory pressures that are reshaping industry standards towards more sustainable practices.

19.1 Overview of construction industry

The UN Environment Programme (UNEP) published the Global Status Report for Buildings and Construction (Buildings-GSR) in March 2024, highlighting the

[1]HLS Pro Construction Sdn. Bhd., Penang, Malaysia

sector's significant influence on global climate change. The construction industry accounts for 37% of carbon emissions. Buildings are also responsible for approximately 21% of global greenhouse gas emissions (GHGs) and 34% of global energy demand in 2022 [1]. The heavy footprint of industry is under increasing pressure to adopt more sustainable practices and reduce its environmental impact, to mitigate climate change and protect the planet for future generations. The importance of sustainability in the construction industry goes beyond just construction itself. It encompasses every step of the process, including planning, design, execution, operation, maintenance, demolition, and disposal. To effectively address carbon emissions and climate change, sustainable practices need to be adopted at each stage. However, there is a substantial disparity between the current state and the goal of achieving decarbonisation by 2030 for the target.

Malaysia's construction sector has been a major contributor to the country's economy, attracting international investment and contributing significantly to GDP. As stated in 2023, construction growth is gradually expanding, accounting for around RM132.2 billion and contributing 8.4% of GDP, as shown in Figure 19.1. According to the most recent data published by the Department of Statistics Malaysia (DOSM) in February 2024, the construction sector increased by 6.8% in the fourth quarter of 2023 and was contributed significantly by the civil engineering works done [2]. The private sector remains the primary driver of growth for the construction industry, accounting for 60% of the value of work completed. This increase in the private sector reflects strong industrial trends and demonstrates the confidence of foreign and local investors. This is mainly driven by a few sectors including civil engineering, particularly in utilities, road and railway projects, residential buildings, non-residential buildings, and other special trade activities. Among the sub-sectors, the civil engineering and non-residential buildings sub-sectors are the major contributors to the total value of work done in Q4 2023, which gave an overall value of RM22.7 billion in that year (DOSM, 2023).

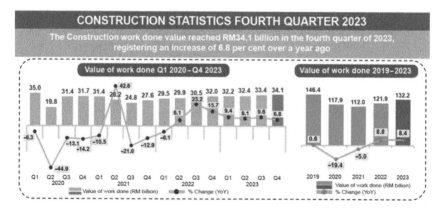

Figure 19.1 Construction Statistics, Fourth Quarter 2023. Source: Department of Statistics Malaysia (DOSM).

19.2   ESG practices in the construction industry

The legendary of the building's construction is always related to what, why, and how the building is being constructed. Therefore, ESG practices go beyond a typical project life cycle, where the type of design, materials, and process being selected will directly or indirectly contribute to the ESG impacts. This is a critical consideration for sustainable development and reducing the carbon footprint of the industry. The primary focus areas of ESG are the characteristics of sustainability, durability, and resilience, as well as energy efficiency, waste reduction, and water conservation. As a result, there is a growing emphasis on incorporating ESG, green building practices, and sustainable technologies into construction projects [3,4]. Furthermore, this shift towards environmentally friendly practices is reshaping the construction industry and leading to the development of innovative green building solutions. Industry players are to strive harder in optimising profit and balancing the societal and environmental impacts with the integration of these practices.

Environmental overview in construction, draw the biggest picture among the three pillars of sustainable development. These principles across the life cycle of construction include preservation and conservation of resources, greenhouse carbon emissions, energy efficiency, waste management, and water management, which are also the parameters that are most concerning in the balancing of natural resources and development.

Social sustainability in the construction industry is concerned with the workers' safety, health, lives, and human rights to ensure a conducive working environment. It covers aspects such as adhering to fair labour practices, promoting community engagement, and encouraging diversity and inclusion. The construction industry can contribute to a more balanced and socially responsible industry by prioritising the well-being of workers, strictly participating in fair labour wages and reasonable working hours and fostering positive relationships with communities by engaging in community development activities.

Governance in construction practices refers to the strategies and policy, business ethics, and transparency measurement to guide the construction company to adhere to the highest ethical conduct within their organisation. The construction industry is prone to a high risk of bribery cases compared to other industries. According to research, Cicchiello *et al.* [5] revealed that out of the 427 cases examined, 15% were related to the construction industry. This highlighted the importance of implementing strong governance practices in helping to mitigate these risks and ensure that construction companies operate with integrity. By promoting transparency and accountability, construction companies can build trust with stakeholders and uphold ethical standards in their operations.

As shown in Figure 19.2, ESG considerations in the context of the construction industry take into account the following aspects:

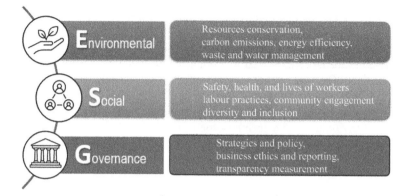

Figure 19.2   ESG in the context of the construction industry

19.3   Smart technological innovations promoting ESG principles

Integrating smart technology into design engineering can enhance efficiency, improve design visualisation and collaboration, streamline project planning and coordination, enhance construction progress and inspections, and ensure the sustainability and safety of construction projects. This integration can also lead to cost savings and reduced waste by optimising resource usage and minimising errors during the construction process. In addition, intelligent technology can offer real-time data and analytics to assist construction firms in making well-informed choices for improved project results. Some typically utilised forms of smart technology in the construction industry are building information modelling (BIM), prefabrication, modular construction apart from the commonly used 3D software, drones for site monitoring, energy-efficient management, and many more. These technologies have the potential to boost collaboration among individuals involved in a project, improve communication, and make project management processes more efficient. By integrating intelligent technologies into building projects, firms may enhance efficiency and productivity while delivering superior outcomes. A holistic approach to BIM integration in the project life cycle is shown in Figure 19.3. Numerous studies on the investment of BIM [6,7] have shown that smart technology integration has given a better return on investment in a project life cycle.

19.3.1   Building information modelling

The construction industry is making significant efforts to reduce the amount of time it takes to execute a project, minimise risks, and simultaneously improve efficiency through the integration of professional and functional components [8]. One of these efforts is the adoption of BIM into design engineering. This process helps to reduce the number of errors and conflicts that occur during construction, which ultimately results in cost savings and an improvement in the quality of the

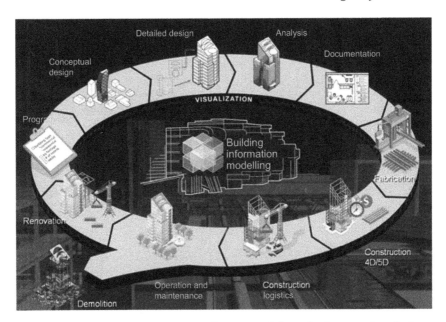

Figure 19.3   BIM integration in the project life cycle. Source: Autodesk.

project. Optimising the workflow is a challenging task as it involves multiple disciplines in engineering and other professions. BIM facilitates enhanced collaboration among stakeholders throughout the duration of a project by providing visualisation capabilities and recommended solutions that support well-informed decision-making. This indirectly provides opportunities for value engineering and ultimately leads to more efficient project cost delivery and improved overall project outcomes [9].

BIM gives a visualisation of a constructed building model, in which this virtual model can be used throughout the lifecycle of projects starting from planning, design, construction, operations stage, and end of life for demolition works [10]. It can be viewed as a simulated, constructed building in a virtual environment. It assists architects, engineers, and contractors (AEC) to visualise and identify the potential issues raised during the design stage and possible risks in the construction or operation stage. Thus, enhance the design or suggest remedies before it is built. It also assists in optimising resource allocation and increasing project efficiency. BIM signifies a paradigm shift in the AEC industry by bridging the gap between prediction and actual execution and enhancing stakeholders' collaboration [11]. Figure 19.4 depicts the typical detailed applications of BIM integration at different components of the project life cycle.

The BIM integration not merely increases the productivity and quality of the project, reduces the project timeline, and preserves better profitability of the project cost, but also aligns with the sustainable development goals by optimising resources and minimising environmental impact.

19.3.2 Prefabrication and modular construction

The construction industry is continuing evolving especially on the path to sustainable development and the environment. Over time, the construction industry has streamlined and become more efficient with the process of prefabrication and modular construction. It is undeniable that modular construction is getting more attention and demand from the market as it emerges as an important alternative approach for hotel and high-rise development projects [12]. Prefabrication always refers to off-site fabrication, where the elements or components are manufactured in a factory before assembly at the construction site, as shown in Figure 19.5. Normally, the building elements are used to accelerate the process and reduce the construction time, and these include beams, columns, slabs, wall panels, staircases, and roof trusses.

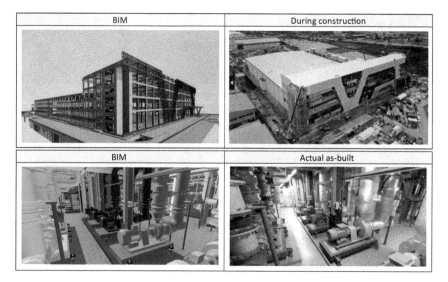

Figure 19.4   Different components of a BIM – integration of architectural, structural, and MEP

Figure 19.5   Prefabrication of steel structures off-site and pre-cast concrete installation at site

Modular construction refers to construction in which parts of prefabricated elements are integrated into a volumetric space or a total system in a larger piece to complete a finished structure [13]. The study by Bertram *et al.* [14] revealed the advantages brought by modular construction; it has the potential to reduce project costs by 20% in the ideal setting and speed up the construction process by 50% compared to traditional construction. Despite the many advantages that have been laid out, a statistic by Statisca [15] reported that the global modular construction market is predicted to expand as developing countries experience rapid urbanisation. Modular construction is widely employed in these areas, and the global modular construction market is anticipated to be worth approximately $175 billion by 2025. Especially in the Asia Pacific market for modular construction, it is projected to grow rapidly due to growing urbanisation. As more and more people move to cities, the demand for more comfortable housing and quality of life is increasing, thus creating the popularity of modular construction across the region.

19.3.3   Sustainable construction materials and practices

Sustainable building materials include every selection of products from production to integration as construction materials are used efficiently to minimise the environmental impact. There is an increasing number of research on sustainable construction materials (SCMs), and this implies the market acceptance level of SCMs has gained popularity in the industry recently [16]. Innovations in materials promote energy efficiency and durability that are essential for mitigating the negative impact of buildings on the environment, especially in reducing GHGs, conserving natural resources, and promoting social well-being. It plays a crucial role in ensuring buildings' durability and environmental compatibility. This SCM includes but is not limited to eco-friendly materials, recycled materials, and also renewable materials that produce a lower carbon footprint compared to conventional materials. Using SCMs in ESG construction can help mitigate the environmental impact of buildings, conserve natural resources, reduce GHGs, and improve energy efficiency. In general, sustainable materials could help to prolong the life cycle of the materials and thus improve the circular economy [17].

19.3.4   Sustainable energy in construction industry

Sustainable energy presents an opportunity to improve people's lives and economies while simultaneously protecting the environment. Sustainable Development Goal (SDG), focuses on the sustainable, affordable, ensuring reliable and modern energy for all by 2030. Under the environmental dimension of the SDGs, SDG 7 underlines three key targets: SDG 7.1 emphasises affordable and clean energy; SDG 7.2 emphasises to increase substantially the share of renewable energy; SDG 7.3 aims to achieve significant global progress in energy efficiency by doubling the rate of improvement worldwide by 2030.

In the construction industry, energy efficiency principles refer to the design strategies and techniques implemented throughout the architectural, construction, and maintenance phases of a structure or building to reduce energy consumption. The integration of renewable energy sources into building designs, such as solar panels, wind energy solutions, biomass, and geothermal energy, plays a significant role in the reduction of emissions and the improvement of energy efficiency in buildings. Through the utilisation of pre-existing natural resources and technological-economic measures, it enriches and encourages the sustainable development of buildings in the energy sector and, eventually, on a global scale. By incorporating renewable energy sources into building designs, not only can energy consumption be reduced, but GHGs can also be minimised. This shift towards sustainable building practices is essential to addressing climate change and creating a more environmentally friendly future.

19.3.5   *Water conservation in construction industry*

The construction sector is well known for its significant energy usage and water consumption amount depending on the type of project in nature, magnitude, and phases. With the aim of promoting environmental sustainability, professionals in the construction industry have developed more innovative methods to optimise water usage during the building process. Although each company faces unique challenges when it comes to water usage, they can and should take different steps to reduce water usage during the construction and material manufacturing processes. Some of the practical methods followed in the industry include rain harvesting, grey-water harvesting, and water-efficient practices during construction, including responsible water use, erosion control measures, and the use of water-saving technologies.

In a water management system, a rain-harvesting system is designed to efficiently collect and store runoff from various surfaces such as roofs and paved areas. It typically includes a capacity of storage tank, supply and distribution networks, and an overflow unit to ensure optimal functionality during the design phases. On the other hand, greywater is a type of wastewater that is produced by various activities within buildings, including handwashing, showering, and doing laundry. This wastewater can be recycled on-site for various purposes such as toilet systems and garden irrigations. In the study reported by Malinowski *et al.* [18], there is a direct correlation between energy conservation and water conservation. They proposed an integrated water management system for households that incorporates rainwater harvesting and greywater reuse, which has the potential to achieve substantial water conservation.

19.4   Challenges in implementing ESG in construction industry

19.4.1   *Economic barriers*

19.4.1.1   High initial cost

Costs associated with adopting new technologies and practices are one of the main considerations for most companies when adopting ESG principles. Innovations like smart construction materials, renewable energy systems, and advanced manu-facturing techniques usually require a large initial investment. For example, pre-fabrication requires an initial investment in setting up manufacturing facilities, transportation logistics, and buying specialised equipment, which can be costly. Additionally, smart materials may initially be more expensive than traditional choices, even if they have long-term cost savings. This can be especially challen-ging for small and medium enterprises (SME) companies that may not have the financial resources to support such costs, as they often operate with a limited budget and resources compared to larger companies. Given the tight financial situation, the SME may also exhibit low risk-taking in implementing new tech-nology as they prefer conventional technology that they are more familiar with and cost-effective, thus providing a predictable return. However, it is important for SMEs to consider the potential long-term benefits and competitive advantages that smart materials can offer in terms of efficiency, sustainability, and innovation. Investing in smart materials could ultimately lead to cost savings, improved pro-duct quality, and increased market competitiveness in the long run. It may be beneficial for SMEs to explore funding options or partnerships that can help offset the initial costs of adopting smart material technology.

19.4.1.2 Uncertainty of return of investment

The uncertainty of return of investment (ROI) is a significant barrier to the adoption of ESG initiatives within the construction industry. This uncertainty primarily arises from the difficulty in quantifying the economic benefits of ESG practices, the time lag in realising these benefits, and the inherent variability of financial returns associated with such investments. ESG practices often involve benefits that may require monitoring or long-term results. Investments such as energy-efficiency systems, sustainability materials, and green building practices require huge upfront costs. In return, the savings of the practices from lesser maintenance costs may occur over time and need to be monitored, making the impact on the financial benefits insignificant. The timeframe in most construction projects normally takes several years to complete, and the payback from the ESG investment may be lengthy. This hinders investors who aim for short-term investment and face pres-sure from stakeholders for a quick return. However, for investors with a long-term perspective, the benefits of ESG investments can be substantial in terms of cost savings and environmental impact. It is important for investors to carefully con-sider their investment goals and timelines before committing to ESG initiatives in the construction industry.

19.4.2 Social barriers

19.4.2.1 Market acceptance and resistance to change

This reflects the acceptance and readiness level of investors, stakeholders, and consumers to embrace and support new technologies and innovations. There can be hesitance from stakeholders to adopt new technologies and ESG practices due to concerns about cost, durability, effectiveness, or potential risks. Human behaviour with resistance to change is one of the biggest challenges in social barriers, where it directly impacts the organisational culture and becomes an obstacle within the environment. This resistance can arise due to a lack of awareness and understanding of the ESG benefits, unfamiliarity, and the preservation of the status quo. Many of them may not fully grasp the benefits, as they are long term and intangible, such as carbon footprint reduction and enhanced community well-being. Furthermore, fear of the unknown is associated with new technology, which may bring new challenges and risks that expose uncertainty and delay projects. ESG and sustainability issues are seen as multifaceted challenges that can impact consumer demand and trust in the industry. Therefore, it is crucial for businesses to effectively communicate the tangible benefits of ESG practices, such as cost savings and improved brand reputation, to overcome this resistance. Additionally, implementing transparent reporting mechanisms can help build trust with stakeholders and demonstrate a commitment to long-term sustainability goals.

19.4.2.2   Skilled gaps and training

As the industry begins to adopt new technology and innovations, it is becoming increasingly obvious that there are skill and knowledge gaps within the labour force, which can impact the quality and speed of construction. The existing workforce may not have adequate experiences, capabilities, and know-how to support the new methods and adopted technologies. One of the main reasons is the lack of investment in human capital training and the development of emerging technologies, which leads to gaps. Therefore, technical training and awareness of cultural shifts in preparing a workforce for future technologies and sustainability are essential, especially in the way they manage and handle the project and embrace the change for future sustainability. Additionally, companies should prioritise ongoing education and upskilling programs to ensure their workforce remains competitive in the industry. By investing in human capital development, organisations can better adapt to new technologies and improve overall project efficiency.

19.4.3   *Governance and compliance challenge*
19.4.3.1   Regulatory and policy

Regulatory and policy challenges are significant in integrating ESG principles in the construction industry, as they vary across regions and countries. Navigating varying regulations across regions could impact the pace of adopting new technology and innovations, especially in project planning, execution, and compliance. Among the basic regulations in the construction industry are building codes, safety and health standards, environmental regulations, and labour law, which require a company to have high flexibility and adaptability in project implementation. In ESG practices, to provide measurable achievement and independent ratings for the design and construction of buildings, the most common achievement is obtaining certification from Leadership in Energy and Environmental Design (LEED), Green Building Index (GBI), and Building Research Establishment Environmental Assessment Method (BREEAM) certification. These certifications ensure that construction projects are built with sustainability in mind, reducing their environmental impact and promoting a healthier work environment for employees. Additionally, companies that adhere to ESG practices often see improved reputations and increased investor interest due to their commitment to responsible business practices.

19.4.3.2 Transparency in sustainability reporting

A sustainability report is a report that a company presents, tracks, and publishes its ESG performance and its impact to enable stakeholders to understand further the opportunities and risks faced by the company. Transparency in sustainability reporting is a crucial component in the ESG management process, to be part of the responsible community in preserving the environment. Additionally, it is also a tool to convince the investors on the company's direction and action on how sincere they are. It is about the disclosure on the sustainability strategies on their ESG move. As the importance of sustainability reporting is becoming more and more crucial, the construction industry is facing equivalent pressures from many directions and facing several challenges, such as cost pressures, technical pressures, data quality issues, and uncertainty in the regulatory requirements.

Various standards, guidelines, and procedures are used in external sustainability reporting to enhance transparency in the sustainable growth of organisations and corporations. The most prominent and widely utilised standards include the Global Reporting Initiative (GRI), the Sustainability Accounting Standards Board (SASB), and the United Nations Sustainable Development Goals (SDGs). According to Glass [19], the standard of disclosure world's most widely used for sustainability reporting is the GRI standards for a company reporting approach and is governed by the Global Sustainability Standards Board (GSSB). The GRI standards are comprehensive and extensive guidelines for reporting sustainability, with the goal of providing information to all parties involved. In contrast, the SASB standards are more focused on certain sectors and primarily designed for investors and capital providers. The UN SDGs are broad in scope and are not limited to certain sectors. They are utilised by both companies and countries [20].

In summary, these challenges are push factors for the construction industry to grow at an exponential rate as they represent the pillars of a country's development. It must be growth in parallel, with advancements in technology and a shift towards more sustainable practices to ensure long-term success and a positive impact on both the environment and society. Additionally, collaboration among stakeholders, including government, industry leaders, and consumers, is essential to drive widespread adoption of these sustainable building practices and create a more sustainable future for all.

19.5 Future outlook

Integrating ESG practices into the construction industry is a continuous journey filled with obstacles. The impact of ESG on the construction industry cannot be underestimated. Emerging trends in technology and sustainability have the potential to greatly accelerate and reshape the industry. It will require the collaboration of various stakeholders, such as the government, policymakers, industry veterans, and the community. The continuous development of awareness and regulatory bodies around ESG is expected to have a greater impact on construction practices. By integrating ESG principles, the construction industry is pushing innovation and working towards a more sustainable future, making a beneficial impact on the building sector. There are a few key benefits that could be gained from the prioritisation in ESG principles: 1. Improve company reputation and brand loyalty through demonstrating sustainability practices and enhancement in the industry. 2. Attracting investors via a commitment to ESG standards and alignment of their investment goals. 3. Increase profitability through resource optimisation and energy-efficiency management. 4. Achieve stakeholders' trust with transparency and accountability in good governance and reporting.

ESG principles have the potential to fundamentally transform the construction industry by embedding sustainability, ethical practices, and social responsibility into the core of industry operations. This transformation can lead to a more resilient industry that can deliver high-quality, sustainable, and ethically constructed infrastructure that meets the needs of present and future generations. Businesses that adopt ESG principles now will have a competitive advantage in this ever-changing environment. By aligning closely with ESG principles, we not only push the industry towards more sustainable practices but also offer a blueprint for how construction might evolve to meet the demands of the future – balancing efficiency, waste reduction, and social responsibility. Smart construction materials are pivotal in driving the construction industry towards more sustainable practices. By enhancing energy efficiency, utilising sustainable resources, and improving the durability and adaptive capabilities of building materials, these innovations not only support environmental objectives but also contribute to the social and governance pillars of ESG by promoting healthier, more resilient communities and adhering to stricter environmental standards.

References

[1] United Nation Environment Programme (UNEP). (2024). *Global Status Report for Buildings and Construction.* https://www.unep.org/resources/report/global-status-report-buildings-and-construction.
[2] Department of Statistics Malaysia. (2024). *Construction Statistics, Fourth Quarter 2023.* https://www.dosm.gov.my/portal-main/release-content/construction-statistics-fourth-quarter-2023.
[3] Sulaiman, N. N., Jalil Omar, A., Pakir, F., and Mohamad, M. (2024). Evaluating environmental, social and governance (ESG) practice among Malaysian public listed construction companies using FTSE Russell rating model. *International Journal of Sustainable Construction Engineering and Tech,* 15(1), 25–40. https://doi.org/10.30880/ijscet.2024.15.01.003.

[4] Das, A. (2023). Predictive value of supply chain sustainability initiatives for ESG performance: a study of large multinationals. *Multinational Business Review*, 32(1), 20–40, https://doi.org/10.1108/MBR-09-2022-0149.

[5] Cicchiello, A. F., Kazemikhasragh, A., Perdichizzi, S., and Rey, A. (2023). The impact of corruption on companies' engagement in sustainability reporting practices: an empirical examination. *International Journal of Emerging Markets*. https://doi.org/10.1108/IJOEM-03-2022-0418.

[6] Sompolgrunk, A., Banihashemi, S., and Mohandes, S. R. (2021). Building information modelling (BIM) and the return on investment: a systematic analysis. *Construction Innovation*, 23(1), 129–154. https://doi.org/10.1108/ci-06-2021-0119.

[7] Stowe, K., Zhang, S., Teizer, J., and Jaselskis, E. J. (2015). Capturing the return on investment of all-in building information modeling: structured approach. *Practice Periodical on Structural Design and Construction*, 20(1). https://doi.org/10.1061/(asce)sc.1943-5576.0000221.

[8] Lu, Y., Gong, P., Tang, Y., Sun, S., and Li, Q. (2021). BIM-integrated construction safety risk assessment at the design stage of building projects. *Automation in Construction*, 124, 103553. https://doi.org/10.1016/J.AUTCON.2021.103553.

[9] Li, X., Wang, C., and Alashwal, A. (2021). Case study on BIM and value engineering integration for construction cost control. *Advances in Civil Engineering*, 2021, 1–13. https://doi.org/10.1155/2021/8849303.

[10] Honcharenko, T., Terentyev, O., Malykhina, O., Druzhynina, I., and Gorbatyuk, I. (2021). BIM-concept for design of engineering networks at the stage of urban planning. *International Journal on Advanced Science, Engineering and Information Technology*, 11(5), 1728–1735. https://doi.org/10.18517/ijaseit.11.5.13687.

[11] Azhar, S., and Asce, A. M. (2011). Building information modeling (BIM): trends, benefits, risks, and challenges for the AEC industry. In *Leadership Management and Engineering*, 11(3), 241–252.

[12] Noordzy, G., Whitfield, R., Saliot, G., and Ricaurte, E. (2021). Modular construction an important alternative approach for new hotel development project. *Journal of Modern Project Management*, 10(1), 217–235. https://doi.org/10.19255/JMPM02715.

[13] Choi, J. O., Chen, X. B., and Kim, T. W. (2017). Opportunities and challenges of modular methods in dense urban environment. *International Journal of Construction Management/the International Journal of Construction Management*, 19(2), 93–105. https://doi.org/10.1080/15623599.2017.1382093.

[14] Bertram, N., Fuchs, S., Mishke, J., Palter, R., Strube, G., and Woetzel, L. (2019). *Modular Construction: From Projects to Products*. McKinsey & Company. Available: https://www.mckinsey.com/capabilities/operations/our-insights/modular-construction-from-projects-to-products.

[15]   Statista. (2022) Global modular construction market size 2018, with a fore-cast for 2025. Available: https://www.statista.com/statistics/1059903/global-modular-construction-market-value/.

[16]   Yap, C. K., Leow, C. S., and Goh, B. (2024). Sustainable construction materials under ESG: a literature review and synthesis. *MOJ Biology and Medicine*, *9*(1), 1–6. https://doi.org/10.15406/mojbm.2024.09.00208.

[17]   Ghaffar, S. H., Burman, M., and Braimah, N. (2020). Pathways to circular construction: an integrated management of construction and demolition waste for resource recovery. *Journal of Cleaner Production*, *244*, 118710. https://doi.org/10.1016/j.jclepro.2019.118710.

[18]   Malinowski, P. A., Stillwell, A. S., Wu, J. S., and Schwarz, P. (2015). Energy-water nexus: potential energy savings and implications for sustain-able integrated water management in urban areas from rainwater harvesting and gray-water reuse. *Journal of Water Resources Planning and Manage-ment*, *141*(12). https://doi.org/10.1061/(asce)wr.1943-5452.0000528.

[19]   Glass, J. (2012). The state of sustainability reporting in the construction sector. *Smart and Sustainable Built Environment*, *1*(1), 87–104. https://doi.org/10.1108/20466091211227070.

[20]   Nordea. (2021). *GRI, SASB, CDP – Making Sense of the Overlapping Sustainability and Climate Disclosures*. Retrieved April 20, 2024, from https://www.nordea.com/en/news/gri-sasb-cdp-making-sense-of-overlapping-sustainability-and-climate-disclosures.

Index

Printed in the USA
CPSIA information can be obtained
at www.ICGtesting.com
JSHW011917271124
74431JS00002B/8